Lecture Notes in Business Information Processing 397

More information about this series at http://www.springer.com/series/7911

Adela Del Río Ortega · Henrik Leopold ·
Flávia Maria Santoro (Eds.)

Business Process Management Workshops

BPM 2020 International Workshops
Seville, Spain, September 13–18, 2020
Revised Selected Papers

Springer

Editors
Adela Del Río Ortega ⓘ
Universidad de Sevilla
Seville, Spain

Henrik Leopold ⓘ
Kühne Logistics University
Hamburg, Germany

Flávia Maria Santoro
Universidade do Estado do Rio
de Janeiro – UERJ
Rio de Janeiro, Brazil

ISSN 1865-1348 ISSN 1865-1356 (electronic)
Lecture Notes in Business Information Processing
ISBN 978-3-030-66497-8 ISBN 978-3-030-66498-5 (eBook)
https://doi.org/10.1007/978-3-030-66498-5

This Springer imprint is published by the registered company Springer Nature Switzerland AG
The registered company address is: Gewerbestrasse 11, 6330 Cham, Switzerland

Preface

The International Conference on Business Process Management (BPM) was established about 18 years ago as the conference where people from academia and industry meet and discuss the latest developments in the area of business process management. The conference has a high record of attracting innovative research of the highest quality from scholars and industrial researchers from the computer science, information systems, and management fields.

While the conference was planned to take place in Sevilla, Spain, this year, things turned out to be quite different. Just like many other academic conferences, BPM also had to move online due to the COVID-19 pandemic. While this was unfortunate from many perspectives, it did not affect the program offered by BPM. This year's BPM also featured seven different workshops that provided a forum for novel research ideas. Each of the seven workshops focused on particular aspects of business process management, either from a technical or from a domain perspective. These proceedings present and summarize the work that was discussed in the context of the workshops.

In total, the BPM 2020 workshops were able to attract 53 submissions. 28 full papers and 1 short paper were accepted for publication, leading to a total acceptance rate of 52.8%. This is slightly higher than in previous years for two main reasons. First, the workshops attracted fewer submissions than usual because of the hectic times caused by the COVID-19 pandemic. Second, there was a very large number of highly rated submissions. Given the exceptional circumstances of BPM 2020, we therefore decided to provide the opportunity to discuss and present their research to as many researchers as possible. The resulting workshop programs were complemented by keynotes from various experts, which are included in these proceedings. BPM 2020 featured the following workshops:

- Workshop on Security and Privacy-enhanced Business Process Management (SPBP)
- Workshop on Social and Human Aspects of Business Process Management (BPMS2)
- Workshop on Business Processes Meet the Internet-of-Things (BP-Meet-IoT)
- Workshop on Artificial Intelligence for Business Process Management (AI4BPM)
- Workshop BPM in the era of Digital Innovation and Transformation (BPMinDIT)
- Workshop on Business Process Intelligence (BPI)
- Workshop on DEClarative, DECision and Hybrid approaches to processes (DEC2H)

As in previous years, the BPM workshop proceedings are post-proceedings. This means that the authors were given the opportunity to revise their papers based on the feedback they received at the workshops. We are confident that this process of selection, presentation, and revision has led to a collection of high-quality papers. We hope that the reader will enjoy this volume and find some inspiration for future work.

We would like to thank all the people from the BPM community that helped make the BPM 2020 workshops a success. We particularly thank the general chairs, Manuel Resinas and Antonio Ruiz-Cortés, for organizing such an outstanding conference despite all the challenges that came with the COVID-19 pandemic. We also thank the workshop organizers, the numerous reviewers, and, of course, the authors for making the BPM 2020 workshops such a success even in this difficult time.

November 2020   Adela Del Río Ortega
                                                                      Henrik Leopold
                                                               Flávia Maria Santoro

Organization

Program Committee

Lars Ackermann	Bayreuth University, Germany
Marco Aiello	University of Stuttgart, Germany
Pedro Alvarez	University of Zaragoza, Spain
Daniel Amyot	University of Ottawa, Canada
Sorin Anagnoste	Bucharest University of Economic Studies, Romania
Robert Andrews	Queensland University of Technology, Australia
Annalisa Appice	University of Bari Aldo Moro, Italy
Abel Armas Cervantes	The University of Melbourne, Australia
Mayutan Arumaithurai	University of Göttingen, Germany
Aleksandre Asatiani	Gothenburg University, Sweden
Ahmed Awad	University of Tartu, Estonia
Banu Aysolmaz	TU Eindhoven, The Netherlands
Hyerim Bae	Pusan National University, South Korea
Bart Baesens	Katholieke Universiteit Leuven, Belgium
Saimir Bala	WU Vienna, Austria
Matteo Baldoni	University of Turin, Italy
Wasana Bandara	Queensland University of Technology, Australia
Irene Barba	University of Seville, Spain
Jörg Becker	University of Münster, Germany
Boualem Benatallah	University of New South Wales, Australia
Nick Berente	University of Notre Dame, USA
Ralph Bergmann	University of Trier, Germany
Daniel Beverungen	Universität Paderborn, Germany
Jan Bosch	Chalmers University of Technology, Sweden
Vesna Bosilj Vuksic	University of Zagreb, Croatia
Alessio Maria Braccini	Tuscia University, Italy
Marco Brambilla	Politecnico di Milano, Italy
Achim D. Brucker	University of Exeter, UK
Andrea Burattin	Technical University of Denmark, Denmark
Cristina Cabanillas	University of Seville, Spain
Josep Carmona	Universitat Politècnica de Catalunya, Spain
Fabio Casati	ServiceNow, USA
Friedrich Chasin	University of Münster, Germany
Sylvain Cherrier	Université Paris-Est Marne-La-Vallée, France
Federico Chesani	University of Bologna, Italy
Marco Comuzzi	Ulsan National Institute of Science and Technology, South Korea
Francesco Corcoglioniti	Free University of Bozen-Bolzano, Italy

Ann-Kristin Cordes	University of Münster, Germany
João Costa Seco	Universidade NOVA de Lisboa, Portugal
Christian Czarnecki	Hochschule Hamm-Lippstadt, Germany
Florian Daniel	Politecnico di Milano, Italy
Massimiliano de Leoni	University of Padua, Italy
Johannes De Smedt	The University of Edinburgh, UK
Jochen De Weerdt	Katholieke Universiteit Leuven, Belgium
Søren Debois	IT University of Copenhagen, Denmark
Gero Decker	Signavio, Germany
Carmelo Del Valle	University of Seville, Spain
Benoît Depaire	Hasselt University, Belgium
Jörg Desel	FernUniversität in Hagen, Germany
Claudio Di Ciccio	Sapienza University of Rome, Italy
Chiara Di Francescomarino	Fondazione Bruno Kessler, Italy
Remco Dijkman	Eindhoven University of Technology, The Netherlands
Rod Dilnutt	The University of Melbourne, Australia
Francisco Jose Dominguez Mayo	University of Seville, Spain
Alevtina Dubovitskaya	Lucerne University of Applied Sciences and Arts (HSLU), Switzerland
Marlon Dumas	University of Tartu, Estonia
Amador Durán Toro	University of Seville, Spain
Mahendrawathi Er	Sepuluh Nopember Institute of Technology, Indonesia
María J. Escalona Cuaresma	University of Seville, Spain
Rik Eshuis	Eindhoven University of Technology, The Netherlands
Joerg Evermann	Memorial University of Newfoundland, Canada
Dirk Fahland	Eindhoven University of Technology, The Netherlands
Stephan Fahrenkrog-Petersen	Humboldt University of Berlin, Germany
Michael Fellmann	University of Rostock, Germany
Pablo Fernandez	University of Seville, Spain
Diogo R. Ferreira	Universidade de Lisboa, Portugal
Peter Fettke	German Research Center for Artificial Intelligence (DFKI) and Saarland University, Germany
Hans-Georg Fill	University of Fribourg, Switzerland
Elgar Fleisch	ETH Zurich, Switzerland
Francesco Folino	ICAR-CNR, Italy
Vladislav Fomin	Vilnius University, Lithuania
Virginia Franqueira	University of Kent, UK
Walid Gaaloul	Télécom SudParis, France
Renata Gabryelczyk	University of Warsaw, Poland
Frederik Gailly	Ghent University, Belgium
Avigdor Gal	Technion, Israel
José María García	University of Seville, Spain
Félix Oscar García Rubio	University of Castilla-La Mancha, Spain
Luciano García-Bañuelos	Tecnológico de Monterrey, Mexico

Agnes Koschmider	University of Kiel, Germany
John Krogstie	Norwegian University of Science and Technology, Norway
Akhil Kumar	Penn State University, USA
Rob Kusters	Open University of the Netherlands, The Netherlands
Julius Köpke	Alpen-Adria-Universität Klagenfurt, Austria
Andrea Kö	Corvinus University of Budapest, Hungary
Marcello La Rosa	The University of Melbourne, Australia
Manuel Lama Penín	University of Santiago de Compostela, Spain
Sander J. J. Leemans	Queensland University of Technology, Australia
Susanne Leist	University of Regensburg, Germany
Henrik Leopold	Kühne Logistics University, Germany
Francesco Leotta	Sapienza University of Rome, Italy
Michael Leyer	University of Rostock, Germany
Irina Lomazova	Higher School of Economics, Russia
Peter Loos	IWi at DFKI, Saarland University, Germany
Qinghua Lu	CSIRO, Australia
Xixi Lu	Utrecht University, The Netherlands
Zakaria Maamar	Zayed University, UAE
Fabrizio Maria Maggi	Free University of Bozen-Bolzano, Italy
Monika Malinova	Vienna University of Economics and Business, Austria
Felix Mannhardt	Norwegian University of Science and Technology, Norway
Elisa Marengo	Free University of Bozen-Bolzano, Italy
Andrea Marrella	Sapienza University of Rome, Italy
Niels Martin	Hasselt University, Belgium
Raimundas Matulevicius	University of Tartu, Estonia
Martin Matzner	Friedrich-Alexander-Universität Erlangen-Nürnberg, Germany
Christoph Mayr-Dorn	Johannes Kepler University Linz, Austria
Massimo Mecella	Sapienza University of Rome, Italy
Paola Mello	University of Bologna, Italy
Jan Mendling	Vienna University of Economics and Business, Austria
Giovanni Meroni	Politecnico di Milano, Italy
Roberto Micalizio	Università degli Studi di Torino, Italy
Tommi Mikkonen	University of Helsinki, Finland
Fredrik Milani	University of Tartu, Estonia
Rabeb Mizouni	Khalifa University, UAE
Marco Montali	Free University of Bozen-Bolzano, Italy
Juergen Moormann	Frankfurt School of Finance & Management, Germany
Hamid Motahari	Ernst & Young, USA
Jorge Munoz-Gama	Pontificia Universidad Católica de Chile, Chile
Juan Manuel Murillo Rodríguez	University of Extremadura, Spain
John Mylopoulos	University of Toronto, Canada

Alexander Mädche	Karlsruhe Institute of Technology, Germany
Michael Möhring	Munich University of Applied Sciences, Germany
Charles Møller	Aalborg University, Denmark
Alex Norta	Tallinn University of Technology, Estonia
Selmin Nurcan	Université Paris 1 Panthéon-Sorbonne, France
Werner Nutt	Free University of Bozen-Bolzano, Italy
Markus Nüttgens	University of Hamburg, Germany
Andreas Oberweis	Karlsruhe Institute of Technology, Germany
Sven Overhage	University of Bamberg, Germany
Baris Ozkan	Eindhoven University of Technology, The Netherlands
Hye-Young Paik	University of New South Wales, Australia
Sooyong Park	Sogang University, South Korea
Oscar Pastor Lopez	Universitat Politècnica de València, Spain
Cesare Pautasso	University of Lugano, Switzerland
Juan Pavón	Universidad Complutense de Madrid, Spain
Vicente Pelechano	Universitat Politècnica de València, Spain
Brian Pentland	Michigan State University, USA
Esko Penttinen	Aalto University School of Business, Finland
Günther Pernul	University of Regensburg, Germany
Giulio Petrucci	Google, Switzerland
Ralf Plattfaut	Fachhochschule Südwestfalen, Germany
Pierluigi Plebani	Politecnico di Milano, Italy
Geert Poels	Ghent University, Belgium
Pascal Poizat	Paris Nanterre University, LIP6, France
Artem Polyvyanyy	The University of Melbourne, Australia
Luigi Pontieri	ICAR-CNR, Italy
Rüdiger Pryss	Ulm University, Germany
Luise Pufahl	TU Berlin, Germany
Jens Pöppelbuß	Ruhr-Universität Bochum, Germany
Mohammad Ehson Rangiha	City University of London, UK
Milla Ratia	Tampere University of Technology, Finland
Hans-Peter Rauer	Bielefeld University of Applied Sciences, Germany
Barbara Re	University of Camerino, Italy
Jan Recker	University of Cologne, Germany
Jana-Rebecca Rehse	University of Mannheim, Germany
Manfred Reichert	Ulm University, Germany
Hajo Reijers	Utrecht University, The Netherlands
Wolfgang Reisig	Humboldt University of Berlin, Germany
Manuel Resinas	University of Seville, Spain
Kate Revoredo	Universidade Federal do Rio de Janeiro, Brazil
Dennis M. Riehle	University of Münster, Germany
Stefanie Rinderle-Ma	University of Vienna, Austria
António Rito Silva	Universidade de Lisboa, Portugal
Daniel Ritter	University of Vienna, Austria
Andrey Rivkin	Free University of Bozen-Bolzano, Italy

Williams Rizzi	Free University of Bozen-Bolzano and Fondazione Bruno Kessler, Italy
Pedro Robledo	International University of La Rioja and ABPMP, Spain
Maximilian Roeglinger	Research Center Finance & Information Management, Germany
Michael Rosemann	Queensland University of Technology, Australia
Gustavo Rossi	National University of La Plata, Argentina
Matti Rossi	Aalto University, Finland
Uwe Roth	Luxembourg Institute of Science and Technology, Luxembourg
Antonio Ruiz-Cortés	University of Seville, Spain
Stefan Sackmann	Martin-Luther-University Halle-Wittenberg, Germany
Shazia Sadiq	The University of Queensland, Australia
Mattia Salnitri	Politecnico di Milano, Italy
Flávia Maria Santoro	Rio de Janeiro State University, Brazil
Rainer Schmidt	Munich University of Applied Sciences, Germany
Werner Schmidt	Technische Hochschule Ingolstadt, Germany
Theresa Schmiedel	University of Applied Sciences and Arts Northwestern Switzerland, Switzerland
Stefan Schulte	Vienna University of Technology, Austria
Dennis Schunselaar	Bvolve, The Netherlands
Stefan Schönig	University of Regensburg, Germany
Arik Senderovich	Technion, Israel
Arik Senderovich	University of Toronto, Canada
Marcos Sepúlveda	Pontificia Universidad Católica de Chile, Chile
Estefanía Serral	Katholieke Universiteit Leuven, Belgium
Miguel-Angel Sicilia	Alcalá University, Spain
Natalia Sidorova	Technische Universiteit Eindhoven, The Netherlands
Volker Skwarek	Hamburg University of Applied Sciences, Germany
Tijs Slaats	Copenhagen University, Denmark
Pnina Soffer	University of Haifa, Israel
Minseok Song	Pohang University of Science and Technology, South Korea
Biplav Srivastava	University of South Carolina, USA
Mark Staples	CSIRO, Australia
Mark Strembeck	Vienna University of Economics and Business, Austria
Heiner Stuckenschmidt	University of Mannheim, Germany
Jianwen Su	University of California, Santa Barbara, USA
Emilio Sulis	Università di Torino, Italy
Ali Sunyaev	University of Cologne, Germany
Rehan Syed	Queensland University of Technology, Australia
Niek Tax	Booking.com, The Netherlands
Irene Teinemaa	Booking.com, The Netherlands
Ernest Teniente	Universitat Politècnica de Catalunya, Spain
Arthur Ter Hofstede	Queensland University of Technology, Australia

Additional Reviewers

Acheli, Mehdi
Bellan, Patrizio
Benke, Ivo
Bogodistov, Yevgen
Cabanillas, Cristina
Colombo Tosatto, Silvano
Danish, Syed Muhammad
Di Francescomarino, Chiara
Feine, Jasper
Gilsing, Rick
Halsbenning, Sebastian
Heindel, Tobias
Herm, Lukas-Valentin
Heuchert, Markus
Hoffmeister, Benedikt
Härer, Felix
Iqbal, Mubashar
Johannsen, Florian
Kalenkova, Anna
Kammerer, Klaus
Koch, Julian
Korkut, Safak
Kregel, Ingo
Käppel, Martin
Lambusch, Fabienne
Laue, Ralf
Leist, Susanne

Lutz, Jonas
López, Hugo-Andrés
Mertens, Steven
Meza, Miguel
Nesterov, Roman
Oriol, Xavier
Ostern, Nadine
Perscheid, Guido
Poeppelbuss, Jens
Poppe, Michael
Putz, Benedikt
Rangiha, Mohammad Ehson
Rauer, Hans Peter
Rietz, Tim
Roeglinger, Maximilian
Rossi, Lorenzo
Schiller, Alexander
Schlette, Daniel
Schönig, Stefan
Shafiee, Fatemeh
Steinau, Sebastian
Toreini, Peyman
Trampler, Michael
Turkmen, Fatih
Wanner, Jonas
Wilbik, Anna
Yalman, Sakine

Contents

The Third International Workshop on Security and Privacy-Enhanced Business Process Management (SPBP'20)

Decentralized Data Access with IPFS and Smart Contract Permission Management for Electronic Health Records 5
Michaël Verdonck and Geert Poels

Data Minimisation as Privacy and Trust Instrument in Business Processes ... 17
Rashid Zaman, Marwan Hassani, and Boudewijn F. van Dongen

Verifiable Multi-Party Business Process Automation 30
Joosep Simm, Jamie Steiner, and Ahto Truu

The Thirteenth Workshop on Social and Human Aspects of Business Process Management (BPMS2'20)

A Classification of Digital-Oriented Work Practices................. 49
Pooria Jafari and Amy Van Looy

Competency Cataloging and Localization to Support Organizational Agility in BPM 60
Olivier Hotel, Lilia Gzara, Hervé Verjus, and Wafa Triaa

Enterprise System Capabilities for Organizational Change in the BPM Life Cycle 70
Johannes Tenschert, Willi Tang, and Martin Matzner

SentiProMo: A Sentiment Analysis-Enabled Social Business Process Modeling Tool 83
Egon Lüftenegger and Selver Softic

4th International Workshop on Business Processes Meet Internet-of-Things (BP-Meet-IoT 2020)

Using Physical Factory Simulation Models for Business Process Management Research.................................... 95
Lukas Malburg, Ronny Seiger, Ralph Bergmann, and Barbara Weber

Modelling Notations for IoT-Aware Business Processes: A Systematic Literature Review 108
Ivan Compagnucci, Flavio Corradini, Fabrizio Fornari, Andrea Polini, Barbara Re, and Francesco Tiezzi

Workshop on Artificial Intelligence for Business Process Management (AI4BPM)

XNAP: Making LSTM-Based Next Activity Predictions Explainable
by Using LRP   129
   Sven Weinzierl, Sandra Zilker, Jens Brunk, Kate Revoredo,
   Martin Matzner, and Jörg Becker

Unsupervised Contextual State Representation for Improved Business
Process Models.   142
   Prerna Agarwal, Daivik Swarup, Sushruth Prasannakumar,
   Sampath Dechu, and Monika Gupta

Root Cause Analysis in Process Mining Using Structural
Equation Models.   155
   Mahnaz Sadat Qafari and Wil van der Aalst

On the Complexity of Resource Controllability in Business
Process Management   168
   Matteo Zavatteri, Romeo Rizzi, and Tiziano Villa

D3BA: A Tool for Optimizing Business Processes Using
Non-deterministic Planning.   181
   Tathagata Chakraborti, Shubham Agarwal, Yasaman Khazaeni,
   Yara Rizk, and Vatche Isahagian

Conceptualizing a Capability-Based View of Artificial Intelligence
Adoption in a BPM Context.   194
   Aleš Zebec and Mojca Indihar Štemberger

A General Framework for Action-Oriented Process Mining   206
   Gyunam Park and Wil M. P. van der Aalst

Analyzing Comments in Ticket Resolution to Capture Underlying
Process Interactions.   219
   Monika Gupta, Prerna Agarwal, Tarun Tater, Sampath Dechu,
   and Alexander Serebrenik

Automated Business Process Discovery from Unstructured Natural-
Language Documents   232
   Alexander J. Chambers, Amy M. Stringfellow, Ben B. Luo,
   Sophie J. Underwood, Tony G. Allard, Ian A. Johnston,
   Sarah Brockman, Leslie Shing, Allan Wollaber, and Courtland VanDam

Workshop on Business Process Management in the Era of Digital Innovation and Transformation (BPMinDIT)

Using Blockchain Technology to Redesign Know-Your-Customer
Processes Within the Banking Industry   251
  Kristin Kamilla Kirss and Fredrik Milani

Increasing Control in Construction Processes: The Role of Digitalization   263
  Arif Ur Rahman, Syed Mehtab Alam, Patrick Dallasega, Elisa Marengo,
  and Werner Nutt

16th International Workshop on Business Process Intelligence (BPI)

Prototype Selection Using Clustering and Conformance Metrics
for Process Discovery   281
  Mohammadreza Fani Sani, Mathilde Boltenhagen, and Wil van der Aalst

Enhancing Discovered Process Models Using Bayesian Inference and
MCMC   295
  Gert Janssenswillen, Benoît Depaire, and Christel Faes

A Generic Framework for Attribute-Driven Hierarchical Trace Clustering . . .   308
  Sebastiaan J. van Zelst and Yukun Cao

Process Outcome Prediction: CNN vs. LSTM (with Attention)   321
  Hans Weytjens and Jochen De Weerdt

Improving the State-Space Traversal of the eST-Miner by Exploiting
Underlying Log Structures   334
  Lisa L. Mannel, Yannick Epstein, and Wil M. P. van der Aalst

8th International Workshop on Declarative, Decision and Hybrid Approaches to processes (DEC2H 2020)

Evaluation of Heuristics for Product Data Models   355
  Konstantinos Varvoutas and Anastasios Gounaris

Text2Dec: Extracting Decision Dependencies from Natural Language Text
for Automated DMN Decision Modelling.   367
  Vedavyas Etikala, Ziboud Van Veldhoven, and Jan Vanthienen

Data Object Cardinalities in Flexible Business Processes   380
  Stephan Haarmann and Mathias Weske

Author Index   393

The Third International Workshop on Security and Privacy-Enhanced Business Process Management (SPBP'20)

The Third International Workshop on Security and Privacy-Enhanced Business Process Management (SPBP'20)

Despite the growing demand for business processes that comply with security and privacy policies, security and privacy incidents caused by erroneous workflow specifications are regrettably common. This is because business process management and security and privacy are seldom addressed together, thereby hindering the development of trustworthy, privacy and security-compliant business processes. The goal of the third edition of the International Workshop on Security and Privacy-Enhanced Business Process Management (SPBP'20) is to establish a venue to discuss business process privacy and integrity management. The workshop has attracted eight submissions and seven submissions were passed to the review process. Each paper was reviewed by three members of the program committee.

The top three submissions are accepted for presentation at the workshop and for the inclusion into the pre-proceedings. The paper by Zaman *et al.* illustrate how data minimisation should be performed to respect principles General Data Protection Regulation. In the next paper, Verdonck and Poels present a decentralised solution to enhance the privacy-control of patient records using smart contracts and blockchain technology. Research of the blockchain technology and its interplay to the business processes is continued in the next paper, where Simm *et al.* discuss how the authenticated data structures could improve trust among the business participants.

We wish to thank all those who contribute to making SPBP'20 a success: the authors who submitted papers, the members of the Program Committee who carefully reviewed the submissions, and the speakers who presented their work at the workshop. We also express our gratitude to the BPM 2020 Workshop Chairs for their support in preparing the workshop.

October 2020
Raimundas Matulevičius
Nicolas Mayer

Organization

Workshop Chairs

Raimundas Matulevičius   University of Tartu, Estonia
Nicolas Mayer   Luxembourg Institute of Science
    and Technology, Luxembourg

Publicity Chair

Mubashar Iqbal   University of Tartu, Estonia

Program Committee

Achim D. Brucker   University of Exeter, UK
Josep Carmona   Universitat Politècnica de Catalunya, Spain
Claudio Di Ciccio   Sapienza University of Rome, Italy
Vladislav Fomin   Vilnius University, Lithuania
Virginia Franqueira   University of Kent, UK
Michael Henke   Fraunhofer, Germany
Jan Jürjens   University of Koblenz-Landau, Germany
Christos Kalloniatis   University of the Aegean, Greece
Dimka Karastoyanova   University of Groningen, the Netherlands
Fredrik Milani   University of Tartu, Estonia
Alex Norta   Tallinn University of Technology, Estonia
Oscar Pastor   Universitat Politècnica de València, Spain
Constantinos Patsakis   University of Piraeus, Greece
Jens Myrup Pedersen   Aalborg University, Denmark
Günther Pernul   Universität Regensburg, Germany
Uwe Roth   Luxembourg Institute of Science
    and Technology, Luxembourg
Mark Strembeck   Vienna University of Economics and Business,
    Austria
Xiwei Xu   CSIRO, Australia

Additional Reviewers

Benedikt Putz
Sakine Yalman
Daniel Schlette
Fatih Turkmen

Decentralized Data Access with IPFS and Smart Contract Permission Management for Electronic Health Records

Michaël Verdonck(⊠) and Geert Poels(⊠)

Faculty of Economics and Business Administration, Ghent University, Gent, Belgium
{michael.verdonck,Geert.Poels}@ugent.be

Abstract. Through the adoption of smart contracts as a permission management database for electronic health records (EHRs), blockchain technology has introduced several advantages such as traceability, transparency and enhanced security. Despite that the permissions to a patient's record are being managed in a more secure and decentralized manner, these records remain stored within the servers of local hospitals and health providers. As such, it remains perfectly possible to query an EHR from within the hospital infrastructure, without the query being stored or even be detected by the smart contract. Moreover, the introduction of smart contracts to facilitate the management of permissions results in a new challenge: how to verify an actor's identity with public key cryptography and the handling of stolen or lost private keys. In this article, we propose a different design where we integrate the research performed by previous efforts to adopt blockchain technology for the management of permissions of EHRs together with the InterPlanetary File system (IPFS) as a peer-to-peer distributed network to provide decentralized storage for EHRs, and a governing body that performs the task of public key governance and identity verification.

Keywords: Blockchain technology · Smart contracts · Electronic health records · IPFS · Public key governance

1 Introduction

Blockchain technology has given rise to transform current health care systems and processes by placing the patient at the center of the health system and increasing the security, privacy, and interoperability of health data. This technology has the potential to provide a new model for health information exchange by making electronic health records (EHRs) more efficient and secure [1]. Adopting healthcare information systems and EHRs result in various benefits for the healthcare sector such as real-time decision support for clinicians or making critically clinical information available to health providers [2]. Besides healthcare advantages, health information exchange in the form of EHRs are estimated to have substantial financial benefits [3]. Despite the many benefits that are associated with adopting EHRs, the transition to digitally stored and shared records holds various

© Springer Nature Switzerland AG 2020
A. Del Río Ortega et al. (Eds.): BPM 2020 Workshops, LNBIP 397, pp. 5–16, 2020.
https://doi.org/10.1007/978-3-030-66498-5_1

challenges regarding the privacy and security of medical data [4, 5]. An EHR is typically stored by and spread among multiple hospitals, clinics, and health providers [6]. As such, these providers maintain primary access to the records, preventing easy access to past data by patients or other healthcare providers. This results in situations where patients and healthcare providers alike interact with a fragmented and incomplete health record. Moreover, data that is stored electronically is prone to be copied, distributed, and mined for confidential information. Securing sensitive data is a tedious and costly endeavour and is not considered as the core activity of a typical healthcare provider, whom often lack the proper resources to do so. Data breaches and the consequent loss or misappropriation of data has already exposed patients' confidential information and have led to hefty fines for hospitals[1].

In order to tackle these problems, recent research efforts have been investigating the application of distributed ledger technology, more in particular *blockchain* technology. While originally introduced as a technology to support new forms of digital currency [7], blockchain has evolved as a promising foundation to support any type of transactions in society. In its essence, a blockchain is a data structure that is composed of an ordered, back-linked list of blocks of transactions [8]. Through the years, several new blockchain technologies have emerged, that act both as a database that records data transactions between parties, while also providing a computational platform for executable programs, i.e., *smart contracts*. Smart contracts can carry and conditionally transfer digital assets or tokens between parties [9]. Since they are stored and executed on the blockchain platform, they can be publicly viewed (assuming a public blockchain) by parties having access to this platform. This feature also guarantees that the execution of a smart contract runs in a deterministic and decentralized manner. Consequently, these unique features give blockchains and smart contracts certain advantages such as traceability, transparency and enhanced security. While certain research efforts [10–12] have already managed to leverage blockchain technology to enhance the security of an EHR system – mainly through the adoption of a smart contract as a permission management database – the fundamental problems associated with EHR systems persist. Despite that these permissions to a patient's record are being managed in a more secure and decentralized manner, these records remain stored within the servers of local hospitals and health providers. As such, it remains perfectly possible to query an EHR from within the hospital infrastructure, without the query being stored or even be detected by the smart contract.

Moreover, introducing additional decentralization and adopting the public key cryptography mechanism for authentication implies however several important design decisions concerning the governance of public and private keys. Since any entity in a decentralized system is identified by their corresponding public key (both requestor and patient), the consequential loss or theft of the private key results in a complete loss of control of a patient's EHR. As there is no central authority, there is no way for a patient to regain access to the electronic health record. Moreover, the loss or theft of the private key corresponding to a healthcare provider can result in a malicious requestor obtaining access to numerous EHRs due to patients unknowingly accepting access to

---

[1] https://eurocloud.org/news/article/fine-of-eur-460000-imposed-on-dutch-haga-hospital-by-dutch-data-protection-officer-the-first-dutch/.

their healthcare records. Furthermore, an additional issue that arises through public key cryptography is the verification associated to one's identity that belongs to a particular public key. Notwithstanding the theft of keys by malicious actors, how can we verify and safely presume that a licensed healthcare provider is associated with a certain public key? Therefore, it seems that moving to a fully decentralized system imposes the problematic issue of verifying a requestor's or patient's identity associated with a particular public key, and the governance of private keys including their theft or loss. Hence, the existing fundamental issues of EHR systems are maintained in such a way that: (1) in essence, a patient has no absolute privacy-control over his or her own EHR and (2) EHR data remains stored with different providers, each remaining responsible for the security and privacy of a patient's sensitive data. Moreover, the introduction of smart contracts to facilitate the management of permissions results in a new challenge: (3) how to verify an actor's identity with public key cryptography and the handling of stolen or lost private keys.

Based upon these issues, we formulate the following research question as a basis for this article: *how can we design a blockchain-based system that provides a decentralized and secure way of managing permissions while enabling patients to have full privacy-control and storage of their own electronic health record?* Therefore, as the main contribution of this article, we seek to answer this research question by proposing a novel and more decentralized design to enhance the privacy-control of a patient over his or her own record, while also maintaining the additional security that arrives through implementing smart contracts as a permission management database. Adopting a more decentralized design gives rise to challenges as how to deal with the immutability of data on public blockchains, and regulatory frameworks such as the General Data Protection Regulation (GDPR, EU), which asserts the right of the user to have control over, and access to their data upon request, as well as the 'right to be forgotten'. Similar legislation is also formulated by the HIPAA Privacy Rule (USA). Hence, if personal data is stored directly on the chain, this means that the EHR system would be unable to fully comply with the required legislation. In this article, we propose a design based upon the InterPlanetary File System (IPFS) as a decentralized storage solution that also complies with the existing legislation.

An additional contribution of this article is that we will provide a process-aware perspective of how EHR request are handled through smart contracts and blockchain technology. While most research efforts demonstrate their architecture and design through a more data-driven representation, this article seeks to highlight the process that takes place while handling EHR requests with blockchain technology.

In Sect. 2, we will discuss several research efforts that each propose a rather different design to manage EHRs through blockchain technology. Next, in Sect. 3, we will provide a more detailed description in the form of a process description of the issues that can still be identified with adopting smart contracts for decentralized permission management of EHRs. In Sect. 4 we describe how a distributed peer-to-peer network, i.e. IPFS, can be adopted to provide decentralized storage for EHRs. Moreover, due to the additional layer of decentralization, we propose the role of a governing body to aid with public key governance and identity verification. Section 5 then compares this new design with the existing research efforts that we have discussed in Sect. 2, while also providing a process model to illustrate their differences. Finally, in Sect. 6 we conclude the research of this article and discuss in more detail the implications of the proposed design decision and the associated limitations.

2 Related Research

In this section we aim to discuss several relevant research efforts that have been published to provide a solution to the security issues that occur with EHR management systems through the adoption of blockchain technology. We haven chosen these specific research efforts since they each propose a rather different architecture and design to implement blockchain technology for the management of EHRs.

One solution that implements a public blockchain to EHR management is MedRec [10]. MedRec is designed as a decentralized record management system to handle EHRs, using the Ethereum blockchain. More specifically, the system assembles references to disparate medical data and encodes these onto the Ethereum ledger. These references are organized to create an accessible 'bread crumb trail' for medical history. Moreover, the MedRec system supplements these pointers with on-chain permissioning and data integrity logic, as such empowering individuals with record authenticity, auditability and data sharing. Additionally, the modular design of MedRec enables APIs to integrate with existing provider databases for interoperability. An important emphasis of MedRec is to create a data-mining scheme in order to bring open, big data to medical researchers.

Another solution proposed by Fan et al. [11] and given the name Medblock applies blockchain technology as a type of index database, that stores encrypted summary data of a patient's data from a specific healthcare provider. Patients can then query the blockchain for finding certain data at the server of a specific healthcare provider, where their data is encrypted with an encryption key. Key in their design is a central, governing role called the Certificate Authority that can be seen as both a system administrator and an authority management agency. It safeguards the blockchain network from malicious nodes and holds the private keys of patients.

Xia et al. [12] propose a different architecture, more specifically a data sharing model that relies on cloud service providers that adopt the blockchain. Their design employs the use of smart contracts and an access control mechanism to effectively trace the behavior of a patient's data as well as revoke access to violated rules and permissions on data. Their proposed system – i.e. MeDShare – relies on big data entities and their adoption of blockchain to provide data provenance, auditing, and control for shared medical data in cloud repositories.

The above-mentioned research efforts have successfully managed to improve security of EHRs through the implementation of smart contracts to manage the permissions of a patient's EHR and the immutable ledger to store these granted/revoked permissions. Nonetheless, several fundamental issues remain unsolved. While permissions are managed through smart contracts, the actual EHR is still stored in a centralized manner, either with the care provider or with a data custodian such as a big data entity. For instance, MedRec assumes that many care providers already trustfully manage their databases. Their design therefore allows caretakers to store patient's EHRs within their own databases and servers. Every healthcare provider is expected to implement an off-chain (i.e. outside the blockchain) gatekeeper to manage access to patient data of that particular provider. Hence, access to patient files are managed and guarded offside the blockchain, which defeats the purpose of utilizing blockchain technology. As such, it remains perfectly possible to query an EHR from within the hospital data infrastructure, without that this query would be recorded or even detected by the smart contract or blockchain. A similar design can be found within the MedBlock and MedShare systems. Within the architecture of MedBlock, national hospitals fulfill the major tasks in their system. National hospitals are expected to arrange the encrypted summaries of EHRs uploaded by the sub-area community hospitals and the various departments. After sorting the data, a national hospital is then required to pack this sorted data into blocks and send a request to consensus nodes to add blocks on the according blockchain. MedShare relies on big data entities and cloud providers to store the medical data on their existing database infrastructure layer. They require these data providers to truthfully register any modifications of views of a patient's data to the smart contract. The authors further explicitly assume that these database systems are only accessible by authorized personnel of the respective companies. If patient data remains stored in a centralized manner, this clearly defeats the purpose of adopting a decentralized permission management database. We cannot exclude that off-chain search queries are being performed within these centralized databases.

Moreover, another issues that arises with the adoption of smart contracts and blockchain technology to manage the permissions of EHRs, is the reliance of the system on the private/public key cryptography for authentication. Patients and requestors alike are represented in the system with a public key. While for instance MedShare specifies that requestors and patients are responsible for generating and managing their own keys, it does not describe how the system will handle the loss or theft of a private key by for example a patient. Not only would this lead to a full loss of control of a patient's EHR, in the case of a healthcare provider that loses its private keys to a malicious actor, this would result in the malicious actor obtaining access to numerous EHRs. To deal with certain situations, MedShare proposes that the Certificate Authority generates and manages the private keys of requestors and patients. This solution however results in one central authority managing all private keys of every participant, which strongly reduces the decentralized character of the design and places a strong vulnerability on the central authority. In the section below, we will describe in more detail the process that takes place when implementing a smart contract for permission management and identify the pitfall in this design. Next, we will propose a different design, one that aims to alleviate the issues associated with the current process of applying smart contracts for EHR permission management.

3 Process of Smart Contracts for Decentralized Permission Management

As to give an overall process perspective on how smart contracts are currently being applied to manage the permissions to a patient's EHR, we have created an abstract business process model with the Business Process Model and Notation (BPMN) of how requests are being handled. As demonstrated in Fig. 1, we can identify three parties: the patient, the requestor of a patient's EHR (e.g. the healthcare provider) and a smart contract (which is executed on a public blockchain). Note that for the sake of abstraction we represent only one smart contract in the process model, although it is more likely that multiple smart contracts will be created that each fulfill a distinct task (such as patient registration). In this process model, the healthcare provider is assumed to store the EHRs of its patients. In a similar process, the storage of EHRs could be fulfilled by a data custodian, as described in the MedShare design [12]. Since public blockchains operate through public/private-key cryptography (also known as asymmetric cryptography) to identify owners of an account and enable secure transaction between these accounts, each patient and healthcare provider needs to be associated with a public key. The execution of this process is similar to the design of MedRec, where a smart contract maintains a mapping of the public keys that belong to a particular patient or healthcare provider. Here patients and requestors generate their own private key, and then register their pubic key with the smart contract. Hence, in the process below, when a healthcare provider sends a request to obtain a certain patient's EHR, the smart contract will have to query this mapping in order to identify that patient's public key. Only when a public key is found, can this request be forwarded to the corresponding patient. In the case a public key has not been found, this would mean that a patient has not been registered by the smart contract and therefore is not included in the EHR system. However, as addressed in Sect. 2, the proposed designs do not suggest a mechanism to cope with stolen or lost private keys. Additionally, no identity verification process takes place in order to establish that for instance a healthcare provider is a licensed and legitimate party and not a malicious actor. How can we verify that a patient or requestor registering their public key with the smart contract are actual patients and requestors?

When the request is forwarded successfully, the patient can then respond by either accepting or rejecting the request. An accepted request will result in the smart contract adding a permission for the respective healthcare provider to access the patient's health record. These permissions are again stored within the smart contract (technically, this is stored within the corresponding blockchain) as one or multiple mappings between the public key of the healthcare provider to the public key of the patient. In either situation, the healthcare provider will be informed by the smart contract if the request was accepted or rejected by the patient. The process in Fig. 1 – and similar to the designs proposed by MedRec, Medblock and MedShare – make the crucial assumption that the centralized server of the healthcare provider (or data custodian) will always consult the permissions as stored within the smart contract. This design cannot guarantee nor prevent however that off-chain queries are being performed within the local hospital provider, even though access was denied by a particular patient. While it is clear that the implementation of blockchain technology and smart contracts result in a simple yet powerful and secure process to manage the permissions of EHRs, the pitfall of this design lies within the centralized or off-chain storage of the health records and a lack of public key governance that verifies the identities behind these public keys, and copes with any losses or thefts of the corresponding private keys.

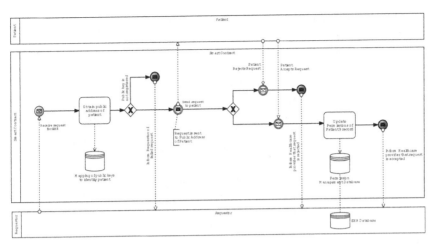

Fig. 1. BPMN process model of smart contracts as permission management database.

4   Decentralized EHR Storage and Public Key Governance

IPFS - Decentralized Storage of Electronic Health Records

While implementing central storage servers for EHRs seems to contradict the application of decentralized blockchain technology, this design decision almost seems unavoidable. Despite that a blockchain is essentially a distributed database of records [13], it is rather inefficient at storing large files. Moreover, sending and storing large files using smart contracts is expensive, while all files that are appended to the blockchain need to be executed and stored at every mining or verifying node.

There are several off-chain data storage solutions designed to overcome this trade-off between decentralization and storage capacity, such as Storj (https://storj.io/), Swarm (https://ethersphere.github.io/swarm-home/), and IPFS (https://ipfs.io/). We have decided to adopt IPFS for our design because of it general purpose applications, wide adoption and for not having any limitations on storage. More specifically, IPFS is a is a peer-to-peer distributed file system that seeks to connect all computing devices (or nodes) with the same system of files. An important characteristic of IPFS is that files are identified by their content and not their location. Therefore, every file that is managed by the IPFS protocol has a content identifier, which is a cryptographic hash that is computationally derived from the specific content of that file. IPFS then optimizes the management of files and directories between different nodes through directed acyclic graphs (DAG). Specifically, IPFS uses Merkle-DAGs, which are DAGs where each node has an identifier that is a hash of the node's contents. To build a Merkle-DAG representation of a file's content, IPFS first splits this file into blocks. This means that different parts of a file can come from different sources or nodes and can be authenticated quickly. To illustrate this with an example, when a healthcare record would be submitted to the IPFS network, it is split into different blocks, each containing at most 256 kbytes of data and/or links to other chunks. Every one of these blocks are then identified through the cryptographic hash – i.e. the content identifier – that is derived from the content of that block and enables the content-addressing feature of IPFS. The combined content identifiers from the separate blocks that compose the EHR form the Merkle-DAG, which can be thus be used to reconstruct the EHR from its distinct blocks. Once an IPFS node has divided the EHR into blocks, and the Merkle DAG has been formed, this node registers itself on the IPFS network as a provider of the EHR. In order to enforce privacy on IPFS for sensitive data such as EHRs, smart contracts can serve as access-control lists [14] with attribute-based encryption [15], allowing for the dynamic modification of sensitive data through adding, removing and updating file ownership and accesses. The role of smart contracts in this case would be very similar to the management of permissions as proposed for MedRec, Medblock and MedShare.

However, by introducing this additional level of decentralization in the form of decentralized storage, we rely entirely on the public key cryptographic system to authenticate users within our EHR design. As such, this leads us to the challenging aspect of how to deal with lost or stolen private keys, and how we can truthfully verify an actor's identity behind the public key.

Governing Body – Public Key Governance

To deal with the issue of verification of identity and possible loss/theft of private keys, an off-chain and objective actor (e.g. an oracle in blockchain community terms) would have to perform public key governance by maintaining a mapping between actual healthcare providers and patients, and their corresponding public key.

In our design, we believe an institute such as the national government of a country is the most evident choice to assign as governing body since a national government actually already performs these tasks by storing and verifying the actual and digital identities of its citizens. However, any type of institution that is capable of performing these tasks can of course be assigned as governing body. By introducing a governing body as a means to verify identities, we can overcome the issues of a decentralized storage system for EHRs. In the case of a loss or theft of the private key of a patient, the patient can then notify the governing body, whom no longer recognizes the public key as a valid digital identity of that patient. The patient can then create a new private key and share the new public key with the governing body. The governing body can then verify if the newly generated public key actually does belong to that specific patient - similar to the case where a person would lose his or her identification documents. A patient can thus, through the governing body, easily be associated with a new public key without losing forever access to their EHR in case of a loss or theft of the private key.

5   Comparison with Related Research

In our discussion of the related research efforts of MedRec, Medblock and MedShare, we argued that certain fundamental issues of EHR systems concerning privacy control, centralized storage and identity verification were still maintained. Through the introduction of IPFS and a governing body, we believe such issues can be alleviated.

Adopting IPFS would allow patients to take full control in terms of privacy and data access of their HER, while avoiding the tedious task of providing reliable and secure storage. Through the identification of a patient's private key, ownership of an EHR can be confirmed on IPFS, while files are managed by their unique cryptographic hash. Smart contracts fulfill their tasks as a highly secure permission management database, controlling the ownership and accesses to EHRs on IPFS. As such, a patient owns their EHR, which is not stored centrally within one or multiple healthcare providers or database custodians but is instead safeguarded by a decentralized network. A patient decides who can have access to their record and can revoke this access also at any given time. Since data is controlled by the patient, and not stored on a public blockchain, the implementation with IPFS does not defy regulatory requirement such as for instance GDPR. While the designs of MedRec, Medblock and MedShare cannot prevent off-chain queries being performed on a patient's EHR – due to the assumption that the centralized servers of a healthcare provider (or data custodian) will always consult the permissions as stored within the smart contract – off-chain queries are not possible without the explicit patient's consent on IPFS.

Additionally, we have included a governing body for public key governance and identity verification to cope with the loss or theft of keys, or to identify any malicious actors that may act as a healthcare provider. While the discussed research efforts have focused on creating a network without a governing body, we believe that this role is still crucial. We do not argue that a blockchain implementation without governing body cannot be accomplished, we argue however that the technology is still too immature and lacks an overall adoption in current society. We would like to emphasize however that the purpose of the governing body is to serve merely as an objective and reliable source of information concerning the identities of the different actors interacting with the blockchain-based EHR system. Unlike, for instance the Medblock design, where private keys are being stored by a central certificate authority, the governing body in our design does not store the private keys of say patients or healthcare providers. The governing body instead is responsible for maintaining a mapping between the digital identities of patients and authorized healthcare providers represented by their unique public key with their national identities (e.g. social security number). Hence, our design incorporates that a patient (and requestor) will always choose – and consequently control – their own private key and only share their public key to other parties, such as the governing body. Additionally, a second advantage of the public governance of public keys by the governing body relates to the detection of illegitimate requests to a patient record by an unauthorized healthcare provider. Since every healthcare provider has to register with the governing body in order to practice healthcare, the smart contract can easily verify that a public key corresponds to a recognized healthcare provider by consulting the governing body.

To illustrate the process of implementing IPFS for decentralized EHR storage and the incorporation of a governing body for public key governance, we have created a BPMN process model as depicted in Fig. 2. Similar to the process in Fig. 1, a requestor (e.g. a healthcare provider) initiates the process by sending a request for a particular patient's EHR to the smart contract. Unlike the initial process, where the smart contract was responsible for maintaining a mapping of the public keys of patients, the smart contract now forwards the verification of identity of both requestor and patient towards the governing body. It is the governing body that then verifies if the request was actually sent by an authorized and legitimate healthcare provider, and that the patient is also an existing citizen with a corresponding healthcare record. In the case of for instance a theft of the private key of a patient, the informed governing body no longer recognizes that particular public key as a valid digital identity of the respective patient and notifies the smart contract. When the smart contract receives a positive validation of both identities, the request is forwarded to the patient's public key. The patient can then decide to accept or reject this request. Similar to the process described in Sect. 3, and the corresponding EHR designs of MedRec, Medblock and MedShare, the smart contract serves as a per-mission management database that maintains a list of accepted requests. However, the main difference now lies with the decentralized storage of the EHR provided by IPFS. Healthcare providers (or data custodians) no longer maintain a patient's health record. It is the patient itself that controls the storage of the EHR with IPFS. Consequently, off-chain queries by unauthorized parties are no longer possible. Due to the content-addressing characteristics of IPFS, a requestor will always need to obtain the content identifier of the EHR (i.e. the cryptographic hash of the file). Since their access is man-aged by the smart contract, a requestor therefore always needs to be given permission from the patient.

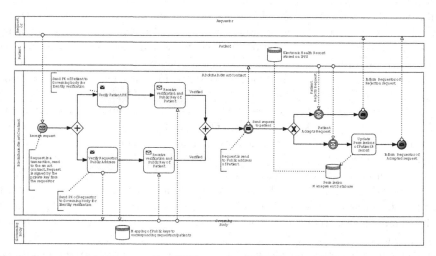

Fig. 2. Process model of Decentralized EHR storage with IPFS and public key governance through a governing body

6 Conclusion

By introducing IPFS for decentralized storage of patient's EHRs and a governing body to acts as reliable source for public key governance, we aimed to provide an answer to the research question posed at the beginning of this article on how we can design a blockchain-based system that provides a decentralized and secure way of managing permissions while enabling patients to have full privacy-control and storage of their own electronic health record. Through the adoption of IPFS for decentralized storage for

EHRs, we offer a solution to provide patients with full privacy control over their own healthcare record without reducing the security of these records. Since permissions and access to EHRs on IPFS are managed through the smart contract on the blockchain, the patient itself only has to accept or reject incoming requests from a healthcare provider, without having to care with the security or data storage of his or her health record.

Additionally, we have included a governing body for public key governance and identity verification to cope with the loss or theft of keys, or to identify any malicious actors that may act as a healthcare provider. While other research efforts have focused on creating a network without a governing body [10, 12] – we think that this role is still crucial. Therefore, our design of a blockchain-based EHR management system remains dependent on a governing body in order to be operational and to be adopted by healthcare providers and patients. Our proposed design is a decentralized system that integrates the existing tasks and responsibilities of a governing body with a more efficient, secure and automized way of sharing and managing EHRs, that benefits patients by giving them full privacy control.

Design Limitations. Several limitations can be addressed of the proposed EHR management design of this paper. First, this article has not implemented the proposed design into a functioning system or prototype. Future research efforts aim to translate the system design into a first proof-of-concept, while also evaluating its performance against similar EHR management designs. Second, we have not specified who would compose and own (in as far that his is possible) the smart contract. In a way, it can be argued that the most obvious choice would be the governing body, since the smart contract relies on the governing body to verify the identities of the public keys every time a new request is submitted. This would reduce the overall decentralized character of the EHR system however, since the governing body gains even more control of the system itself. Another concern that we have not addressed within the article is that given the full control a patient has over the EHR, how can a healthcare provider obtain the required access to an EHR if that particular patient is unable to provide access, for instance because the patient is unconscious due to an injury. One possible solution to deal with this kind of situation is to enable 'emergency rights' to healthcare providers, which can be invoked when a patient cannot accept a request. In this case, the identity of the healthcare provider can still be verified by the governing body to prevent any malicious actor gaining access to a healthcare record. The smart contract can be programmed in such a way that emergency request immediately gain permission to a certain EHR, after identity verification has been performed. Finally, the argument can be raised that the inclusion of a governing body for identity verification reduces or defeats the purpose of adopting a decentralized blockchain. While the authors concur that the degree of decentralization is reduced through a governing body, we are unaware of an existing and suitable decentralized alternative that can overcome the loss or theft of private keys or identify malicious actors.

References

1. Rabah, K.: Challenges & opportunities for blockchain powered healthcare systems: a review. Mara Res J Med Health Sci. **1**, 45–52 (2017)

2. Wang, J., et al.: A cost-benefit analysis of electronic medical records in primary care. Am. J. Med. **114**, 397–403 (2003)
3. Walker, J., Pan, E., Johnston, D., Adler-Milstein, J., Bates, D.W., Middleton, B.: The Value of health care information exchange and interoperability: there is a business case to be made for spending money on a fully standardized nationwide system. Health Aff. (Millwood). **24**, W5–10-W5–18 (2005).
4. Matthias, W., Christian, J., Rainer, R.: Secondary Use of Clinical Data in Healthcare Providers; an Overview on Research, Regulatory and Ethical Requirements. Stud. Health Technol. Inform **180**, 614–618 (2012)
5. Sahama, T., Simpson, L., Lane, B.: Security and Privacy in eHealth: Is it possible? 5 (2013).
6. Ekblaw, A., Azaria, A., Halamka, J.D., Lippman, A.: A case study for blockchain in healthcare:"MedRec" prototype for electronic health records and medical research data. In: Proceedings of IEEE Open & Big Data Conference, p. 13 (2016).
7. Nakamoto, S.: Bitcoin: A peer-to-peer electronic cash system, pp. 1–9 (2008).
8. Antonopoulos, A.M.: Mastering Bitcoin: Programming the Open Blockchain. "O'Reilly Media, Inc.", Sebastopol (2017).
9. Mohan, C.: State of Public and Private Blockchains: Myths and Reality. In: Proceedings of the 2019 International Conference on Management of Data - SIGMOD 2019, pp. 404–411. ACM Press, Amsterdam, Netherlands (2019).
10. Azaria, A., Ekblaw, A., Vieira, T., Lippman, A.: MedRec: using blockchain for medical data access and permission management. In: 2016 2nd International Conference on Open and Big Data (OBD), pp. 25–30. IEEE, Vienna (2016).
11. Fan, K., Wang, S., Ren, Y., Li, H., Yang, Y.: MedBlock: efficient and secure medical data sharing via blockchain. J. Med. Syst. **42**(8), 1–1 (2018). https://doi.org/10.1007/s10916-018-0993-7
12. Xia, Q., Sifah, E.B., Asamoah, K.O., Gao, J., Du, X., Guizani, M.: MeDShare: trust-less medical data sharing among cloud service providers via blockchain. IEEE Access. **5**, 14757–14767 (2017)
13. Crosby, M., Pattanayak, P., Verma, S., Kalyanaraman, V., et al.: Blockchain technology: Beyond bitcoin. Appl. Innov. **2**, 71 (2016)
14. Steichen, M., Fiz, B., Norvill, R., Shbair, W., State, R.: Blockchain-based, decentralized access control for IPFS. In: 2018 IEEE International Conference on Internet of Things, pp. 1499–1506. IEEE, Halifax (2018).
15. Sun, J., Yao, X., Wang, S., Wu, Y.: Blockchain-based secure storage and access scheme for electronic medical records in IPFS. IEEE Access. **8**, 59389–59401 (2020)

Data Minimisation as Privacy and Trust Instrument in Business Processes

Rashid Zaman$^{(\boxtimes)}$, Marwan Hassani , and Boudewijn F. van Dongen

Process Analytics Group, Faculty of Mathematics and Computer Science,
Eindhoven University of Technology, Eindhoven, The Netherlands
{r.zaman,m.hassani,b.f.v.dongen}@tue.nl

Abstract. Data is vital for almost all sorts of business processes and workflows. However, the possession of personal data of other beings bear consequences. Data is prone to abuses through the exposure to adversaries in case of data breaches or insider's illegitimate access and processing, hence adding to customer distrust. The data minimisation principle of the General Data Protection Regulation (GDPR), as a proactive approach, requires the collection of personal data to be limited to what is necessary for the legitimate processing purpose(s). Data degradation advocates for periodic inter-process data minimisation in a multi-process environment. In this context, we are proposing intra-process data degradation as a continuous data minimisation function during the process life. In our solution, the granularity or the information level of the process data is reduced at *suitable* instances in the process life to the *minimum sufficient level* for a successful completion of the remaining process. We devise three effective data degradation policies to realise and guide intra-process data degradation in business processes. We show through a proof-of-concept implementation the applicability of the introduced concept and the effectiveness of one of the policies. Our proposed approach intrinsically reduces privacy infringement damages which contribute to end-users trust in the processes.

1 Introduction

Data is an exploitable asset which organisations process in primary processes for fulfillment of organisational goals and in secondary processes to create added value for the organisation as well as the customers. Process data may contain sensitive personal data of related actors: customers and the organisational resources. Rosemann [12] identifies the consideration for privacy measures in processing their data as one of the end-user *trust concerns*, which negatively impacts their engagement with a process. Long-term possession of data has added risk of either being exposed to adversaries through data breaches[1] or illegitimately accessed by ill-intentioned insiders[2], causing intractable losses[3].

---

[1] www.thesun.co.uk/money/11657383/easyjet-hacked-cyber-attack-customer-data.

[2] https://www.compliancehome.com/dutch-dpa-gdpr-fine-haga-hospital.

[3] www.egnyte.com/blog/2017/06/how-much-does-a-data-breach-cost-a-business.

© Springer Nature Switzerland AG 2020
A. Del Río Ortega et al. (Eds.): BPM 2020 Workshops, LNBIP 397, pp. 17–29, 2020.
https://doi.org/10.1007/978-3-030-66498-5_2

A plethora of security and privacy enabling solutions have been devised but as a matter of fact no solution is foolproof. Data leakages and breaches still happen. One potential damage reduction solution in the context is *data minimisation*. Being one of the General Data Protection Regulation processing principles, data minimisation requires that "personal data should be adequate, relevant and *limited* to what is necessary in relation to the purposes for which they are processed". Data minimisation, implying data minimality at collection stage, inherently minimises any exposure aftermath. Data minimisation is frequently practiced by most of us. For instance, for birthdays greeting and celebration purposes, usually only the day and month of the birth is disclosed to the acquaintances. Maintaining own privacy but not revealing the exact birth year *data* does not spoil the birthday celebration *process*.

Even while collecting *limited* data, data controllers still run into the risk of retaining data which can be of interest to adversaries. The data degradation concept advocates incremental data minimisation in chained processes, where the same data is processed for multiple purposes with varying granularity. In essence, the level of details in data shall be lowered at specific instances in between these processes to the extent that the resulting granularity is exactly sufficient for succeeding process(es) in the processes chain. The term *granularity* is used in place of *precision*, the native term in the related literature, to avoid confusion when using it in process mining field.

Practically, data elements can become granularity-wise fully or partially superfluous even during the execution of a process. In this paper, we are taking data degradation one step further by proposing intra-process data degradation, where data shall be continuously degraded during the course of an individual process. Various *vertical and horizontal* granularity or information reduction techniques for degrading data elements can be realised. Business process(es) may end up starving for data if the data elements are degraded beyond the limit required for successful execution of the remaining process activities. To avoid such *data starvation*, we propose multiple data degradation policies. Each one of these data degradation policies is suitable for specific business process category in terms of usual process life, process activities flux, and data degradation associated costs.

With the larger research goal of realising data degradation in the data perspective of business processes, we are framing our contributions in this paper by answering two concrete research questions. *RQ1*: What are the different realisable intra-process data degradation policies? *RQ2*: What are the open challenges for realising data degradation in the data perspective of business processes?

The rest of this paper is organised as follows. Section 2 provides an overview of the related work on data minimisation and data degradation, followed by a use case example. Section 3 provides the preliminaries for intra-process data degradation and various data degradation techniques. Further, this section details on our introduced data degradation policies. Section 4 provides a proof of concept implementation for one of the introduced policies. Section 5 discusses some of the major foreseen challenges. Finally, Sect. 6 discusses the extent to which the two research questions have been answered.

2   Related Work and Use Case

This section details on the related work for data minimisation and data degradation, along with a use case example to be referenced throughout the paper.

2.1   Related Work

Domain generalisation graphs were introduced by Hilderman et al. [10]. Primarily aimed at large databases, the hierarchical levels of these graphs are utilised for rolling up and drilling down the data in the database. Anciaux et al. [4,5], taking into account the domain generalisation graphs of [10], proposed degrading data elements in case the current level of granularity is not required in future processes. For this purpose, the authors utilise the hard-wired time-based lifecycle policy model of Anciaux et al. [3]. van Heerde et al. [9] attempts to personalise the time-based lifecycle policies to stakeholders having varying preferences.

Our work, progressing upon the initial work in [15], is different from [3–5] on three grounds. First, the approach proposed in [3–5] is time-based recurrent for inter-processes, where multiple processes having distinct and mutually exclusive boundaries use the same data. Granularity or information level of data elements remains the same during the course of an individual process. Our proposed approach is intra-process and has three variants. Second, the approach in [4,5] is database-centric while our data degradation policies are more process-centric. Third, the approach in [3–5] being database-centric, highlights the data degradation implementation related challenges therein. Our process-centric approach highlights the implementation challenges being faced from business processes perspective. To the best of our knowledge no work exists that extends the data degradation approach, especially from the perspective of business processes.

A theoretical approach, that applies data degradation but in a self-triggered way has been proposed by Geambasu et al. [6], where data become unusable after a designated period. The 'Sticky policies' approach of Pearson et al. [11] propose appending allowed usage and associated obligations as machine-readable policies to data transferred outside organisational boundaries. The sticky policy solution is primarily aimed at data access minimisation. k-Anonymity, proposed by Samarati [13], is an example of group-level granularity and retained information reduction technique where a group of at least k data element instances are made indistinguishable. Complete GDPR compliance is non-trivial and different works such as Agostinelli et al. [2] have addressed some aspects of GDPR.

2.2   Running Use Case Example

We consider an example test drive process of a car dealership (cf. Fig. 1). The car dealerships arrange test drive events in first and second quarter of each year.

People interested in one of the upcoming test driving events can register in the car dealership information system through an online portal. Personal data especially date of birth, driving license details, and occupation are provided during registration to be used for eligibility assessment of the test drive candidates.

The three eligibility assessment checks are performed in parallel in this example process to introduce a parallelism scenario. Apart from checking the validity of the driving license, the *Driving License Check* activity compares the license issuance year and the birth year of the candidate to ascertain if the license was issued at an adult age, which otherwise can highlight a fraud. Eligible candidates are assigned to one of the test drive events and later contacted at some time prior to the planned test drive event for scheduling a test drive. Usually, the complete process may span over several months. Also, the duration from registering a test drive request till the activity *Assign to Event* may just be a few weeks, and therefore contributes as a fraction to the total process duration. Note that the example process is represented in BPMN formalism since the implementation platform for the concept currently does not support Petri nets.

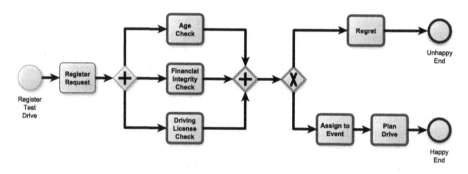

Fig. 1. A car dealership example business process model.

3   Intra-process Data Degradation

This section constitutes the building blocks of our intra-process data degradation approach. We present some preliminaries, followed by data elements degradation techniques. Then we explain our proposed data degradation policies.

3.1   Intra-process Data Degradation Preliminaries

Article 4 of the GDPR abstractly defines personal data of the data subjects.

Definition 1 (Personal Data). *"Personal data" means any information relating to an identified or identifiable natural person ("data subject").*

Definition 2 (Universes). Let:

- V be the universe of all variables,
- U is the universe of all possible values for the universe of variables V,
- $V_p \subset V$ is the set of process variables of a business process,
- $V'_p \subseteq V_p$ is the set of process variables containing personal data.

Definition 3 (Process Model). A process model as a Petri net is represented as a Tuple $N = (P, T, F, M, V_p)$ where P represents a finite set of places, T a finite set of transitions, $F \subset (P \times T) \cup (T \times P)$ a set of flow relations between places and transitions, and M represents the marking i.e., state of the process, as tokens over the places P. $M_I \subset M$ is the set of initial markings and $M_F \subset M$ is the set of the final markings. Transitions T read, write, or manipulate the values of the process variables V_p in order to fulfill their specific purposes.

Definition 4 (Intra-process Data Degradation). The level of details in personal data (i.e. the values $u \subset U$ of all $v'_p \in V'_p$), should always be lowered at suitable points in the process(es) timeline such that the resulting granularity and retained information remain sufficient for the successful completion of the process(es) i.e., at least a state from M_F is still reachable.

Our intra-process data degradation definition highlights two important aspects, namely: *suitable points* and *granularity and retained information*. The techniques presented in the next two sections will cater for these two aspects.

3.2 Data Elements Degradation Techniques

A process variable $v'_p \in V'_p$ can be assigned multiple distinct but related values $\{u_i, u_j, u_k, \cdots, u_n\} \subset U$ such that all these values contain the required information to fulfill the common processing purpose(s). At the same time, these values have the potential to be utilised for fulfilling purposes unique to each value or a subset of these values. For instance, both the cell number and the landline number of a person can be used for correspondence purpose but the cell number can also be used for tracing purposes. Not all potential purposes are legitimate and some may lead or exclusively be meant for privacy infringements.

The values $\{u_i, u_j, u_k, \cdots, u_n\}$ can be ordered $u_i \prec u_j \prec u_k, \cdots, \prec u_n$ such that the information, or in other words the number of purposes a value can serve, in a specific context reduces down the list. Importantly, a valid ordering in one context may not hold valid in a different context. Since our prime concern is *privacy*, we are interested in orderings where the sensitivity of information and therefore the potential of privacy infringements reduces down the list.

The sensitivity-reduction based ordering of possible values of a process variable v'_p can be achieved through two types of techniques: vertical and horizontal. As proposed by Anciaux et al. [5], the domain generalisation graphs of Hilderman et al. [10] provide a mechanism for vertical ordering of the multiple values of v'_p. Values of some data elements can be fitted into a tree-like graph, such that traversing from the leaf node towards the root node reduces the granularity and retained information of the data element. A house address makes a perfect example where the subsequent address parts are have increasing abstraction and fits in to a generalisation graph. The horizontal techniques transform a data element into some alternate representation such that the embedded information is reduced or in other words, the entropy is increased in a specific context. For instance, the data element date of birth in the *dd-mm-yyyy* format can be transformed to {child, teen, adult, elderly} configuration.

Degradation of a data element's value can be realised by replacing the current value u_i of the data attribute v'_p with an alternate value down its sensitivity-reduction based ordered list i.e., u_i should be replaced by some other value in the ordered list $u_i \prec u_j \prec u_k, \cdots, \prec u_n$. A degradation function takes as input the current value of a data attribute and provides as output the next potential value. The degradation instances, for calling the degradation function, are managed by the data degradation policies which is the subject of the next section.

3.3 Data Degradation Policies

Definition 5 (Data Degradation Policy). A data degradation policy defines the (recurrence) criteria for initiation of data degradation during the course of a business process.

A data degradation policy essentially caters the *suitable points* aspect of our intra-process data degradation definition. In this paper, three of such policies are proposed: time-driven, event-driven, and event-driven with margin. These three policies are depicted in Fig. 2 with data degradation triggering events lying on x-axis and degradation levels of data on y-axis. The heatmap colouring depicts the sensitivity of the data in-between degradation instances. Each data degradation policy is intrinsically suitable for specific variants of processes.

Time-Driven Data Degradation: In time-driven data degradation policy, the value degradation of a data element is initiated at predefined time instances $\{T_i, T_j, T_k, \cdots\}$ in the process timeline. Usually, these time instances are equidistant i.e., $\Delta T_{ij} = \Delta T_{jk}$ for all i, j, k values, where $i, j, k, \cdots$ are alternating data degradation instances. Referring to Fig. 2a and the example business process of Fig. 1, a typical time-driven data degradation policy would be *"degrade data element* date of birth *one level every week"*.

The periodicity of data degradation instances, or the *suitable points* aspect of our intra-process data degradation, depends on multiple factors such as data degradation costs, completion times of the process activities, usual process duration, processing nature of the process i.e., single-instance or multi-instance. Single-instance processes usually batch process data related to multiple cases such as the customer segmentation process for instance. While, in multi-instance processes, an individual instance of the business process is initiated for each individual case such as a bank loan application process for example.

From the perspective of data repository, the time-driven data degradation policy is suitable for environments where the data degradation costs are high.

This policy can be used in cases where the unavailability of the repository during data degradation operations highly affects the performance of the core business process. The composition of the data repositories is a further critical factor to be taken into account. If the process data is scattered or replicated over redundant (heterogeneous) repositories then infrequent time-driven data degradation policy is adequate.

For the time-driven data degradation policy to be effective and starvation-safe, activities in the business process should have time-bounds on completion. Business processes, however, are prone to deviations from normal behavior in both control and time perspectives. Therefore, time-driven data degradation policy can potentially lead to data starvation in case of deviations where activities are executed later than expected. Additionally, this policy degrades data at pre-defined time instances and as such data can be unnecessarily retained although it may have become superfluous.

Event-Driven Data Degradation: Event-driven data degradation policy caters for the shortcomings of time-driven data degradation policy. With this policy in place, the value degradation of a data element is initiated as soon as the current granularity or retained information level is no longer required for the successful completion of the process.

Data degradation initiation instances, or the *suitable points* aspect of our intra-process data degradation, are linked to the completion of specific process activities. The completion of these activities marks the current information level of a specific data element to be superfluous and renders it eligible for degradation. Referring to our example process of Fig. 1 and the event-driven data degradation policy in Fig. 2b, the value of data element `date of birth` is degraded one level as soon as activity *Check Age* is executed and later degraded one further level when activity *Assign to Event* is executed.

From the perspective of data repository, the event-driven data degradation policy is suitable for environments where data degradation costs are low and the unavailability of the repository during data degradation operations has a negligible effect on the performance of the core business process. From the business process perspective, the event-driven data degradation policy is suitable for both single-instance and multi-instance categories of business processes. Business processes with evenly or sparsely distributed activities constitute an ideal case for event-driven data degradation policy.

The control-flow deviations are a major threat to event-driven data degradation policy. Activities triggering data degradation can be executed ahead of their predecessor activities which may consequently leave the latter starving for data at pre-degradation information level. In case of deviations with respect to time perspective, processes with event-driven data degradation in place are unlikely to suffer from data starvation.

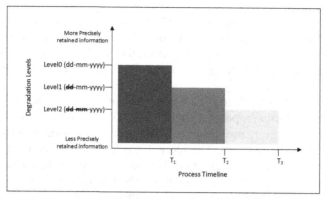

(a) Time-Driven Data Degradation Policy.

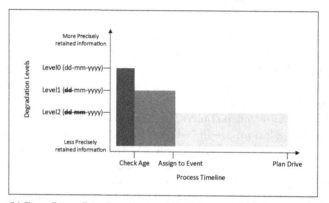

(b) Event-Driven Data Degradation Policy (cf. Figure 1 for the events in the process model).

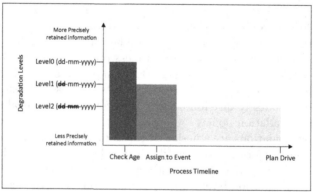

(c) Event-Driven Data Degradation Policy (cf. Figure 1 for the events in the process model).

Fig. 2. Data degradation policies discussed in this work for data element `date of birth` in the context of example process of Fig. 1, adopted from [15].

Event-Driven Data Degradation with Margin: In this policy, the value degradation of a data element is initiated at a *suitable offset* after the execution of the last process activity requiring the data at the current granularity and retained information level.

This policy is suitable for categories of processes where either an applicable regulation or some sort of an expected preemption requires that data should obligatorily be retained for a specified period of time. Different potential scenarios could be: (i) audit teams can randomly pick cases to be analysed for compliance and therefore data should be retained for some specified period, (ii) customers are entitled to object to an automated decision within a specified time frame, or (iii) the management occasionally inspects cases leading towards an undesirable result such as loan applications leading towards rejection.

In all the previously mentioned and many more related scenarios, data should not be degraded promptly but should instead be retained for a specific period of time, even if it is believed to be superfluous. Referring to our example process of Fig. 1 and the event-driven data degradation with margin policy of Fig. 2c, the value of data element `date of birth` is degraded with a suitable offset following the execution of activity *Check Age* and later degraded one additional level with a suitable offset after the execution of activity *Assign to Event*.

4   A Proof of Concept Implementation

We have realised our intra-process data degradation approach in Camunda BPM[4], which is a java-based open-source workflow automation platform. It currently supports BPMN (Business Process Modeling Notation), CMMN (Case Management Model and Notation) and DMN (Decision Model and Notation). Being a powerful workflow platform, the process engine can be bootstrapped in a java application or alternatively java applications can be deployed to the process engine through webapps mechanism. The engine provides Java APIs for providing interfaces. Our implementation as webapps is available on SurfDrive[5].

In our implementation, the example process of Fig. 1 is coupled with the event-driven data degradation policy of Fig. 2b. We are presenting screenshots of the Camunda *cockpit* at different process stages in Figs. (3a–3d). The date of birth, represented as variable `DateOfBirth` in the screenshots, is provided at registration stage (cf. Fig. 3a). Once utilised for checking age eligibility for a test drive by activity *Age Check*, the `date of birth` data element is degraded one level such that only month and birth year are retained (cf. Fig. 3b). Recalling from the process specification in Sect. 2, the assessment activity *Check Driving License* also utilises the year of birth for a check. Even with preceding of the activity *Age Check* and its degradation of the `DateOfBirth`, the mentioned

---

[4] https://www.camunda.com.

[5] https://surfdrive.surf.nl/files/index.php/s/E7mU4UQCLffmyoQ.

check can still be performed (cf. Fig. 3c). If eligibility is proven, the request is further processed through activity *Assign to Event* and `date of birth` is further degraded to contain birth year only (cf. Fig. 3d).

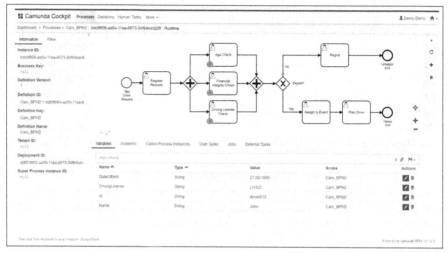

(a) Process Execution without Degraded Process Variables.

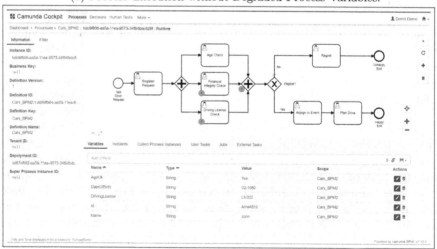

(b) *Level1* Degradation of Data Element `date of birth` (cf. Figure 2b).

Fig. 3. Screenshots of Camunda Process Execution with Data Degradation (cont.)

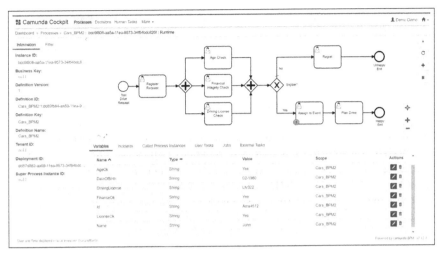

(c) Execution of activity "Driving License Check" even with *Level1* Degradation of Data Element `date of birth` (cf. Figure 2b).

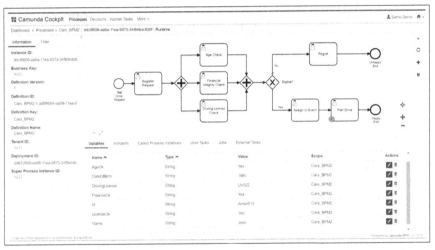

(d) *Level2* Degradation of Data Element `date of birth` (cf. Figure 2b).

Fig. 3. (*continued*)

To illustrate the efficacy and a qualitative analysis of our approach, assume a data exposure incident happening when applying this policy compared to the case of a similar process implemented without data degradation. Usually the *elapsed time* from the start of the process till the activity *Assign to Event* contributes as a fraction to the process completion time. Therefore, during the major portion of the process duration stretching from activity *Assign to Event* till end of the process, the incident will result in revealing far less sensitive information in our data degradation backed approach. The risk in our case is comparably less even before the activity *Assign to Event* as we start degrading `DateOfBirth` very early in the process.

5 Challenges and Future Directions

In spite the initial effectiveness of the introduced concepts, some initial challenges identified in the context of the BPR4GDPR [1] project are still to be addressed. The foremost challenge is the formalisation of the presented concepts. Such formalisation will in turn highlight further challenges. The complexity of the processing environments is considered to be another challenging situation for our intra-process data degradation approach. Having multiple processes operating on the same data concurrently, sequentially or in a cascading pattern and having multiple organisations involved in beyond-boundary processing as part of a single process are examples of this kind of complexity.

The structural complexity of real world business processes is a further important challenge. Process models contain combination of parallelism, choices and the most challenging looping construct. All these constructs introduce behavioral complexity in the incumbent business processes e.g. with sophisticated customer journeys handling sensitive data [14]. Our current version of the data degradation policies are still unable to handle such behavioral diversity which makes data degradation hard to realise. We will therefore investigate much more advanced, probably hybrid, data degradation policies possibly by checking concept drifts [7,8] in the underlying process model to infer suitable degradation points.

6 Conclusion

With the research goal of realising data degradation in the data perspective of the business processes, we introduced inter-process data degradation in this paper. Section 3 introduced vertical and horizontal data degradation techniques as means for reducing granularity and retained information of data elements. Having a mechanism to degrade data, we proposed three data degradation policies for guiding the intra-process data degradation, thus catering for *RQ1*. To answer *RQ2*, we highlighted several potential future challenges for positioning data degradation in the data perspective of business processes in Sect. 5. We demonstrated the viability and damage reduction potential of our proposed approach in cases of data exposure with the help of an example process.

Acknowledgments. The authors of the paper have received funding within the BPR4GDPR [1] project from the European Union's Horizon 2020 research and innovation programme under grant agreement No 787149.

References

1. BPR4GDPR EU H2020 Project. https://www.bpr4gdpr.eu
2. Agostinelli, S., Maggi, F.M., Marrella, A., Sapio, F.: Achieving GDPR compliance of BPMN process models. In: International Conference on AISE, pp. 10–22 (2019)
3. Anciaux, N., Bouganim, L., Van Heerde, H., Pucheral, P., Apers, P.: The life-cycle policy model. Rapport de recherche (2008)
4. Anciaux, N., Bouganim, L., Van Heerde, H., Pucheral, P., Apers, P.M.: Data degradation: Making private data less sensitive over time. In: Proceedings CIKM 2008, pp. 1401–1402 (2008)
5. Anciaux, N., Bouganim, L., Van Heerde, H., Pucheral, P., Apers, P.M.: Instantdb: enforcing timely degradation of sensitive data. In: ICDE, pp. 1373–1375 (2008)
6. Geambasu, R., Kohno, T., Levy, A.A., Levy, H.M.: Vanish: increasing data privacy with self-destructing data. In: USENIX, vol. 316 (2009)
7. Hassani, M.: Concept drift detection of event streams using an adaptive window. In: International ECMS Conference on Modelling and Simulation, pp. 230–239 (2019)
8. Hassani, M.: Overview of efficient clustering methods for high-dimensional big data streams. In: Nasraoui, O., Ben N'Cir, C.-E. (eds.) Clustering Methods for Big Data Analytics. USL, pp. 25–42. Springer, Cham (2019). https://doi.org/10.1007/978-3-319-97864-2_2
9. van Heerde, H., Anciaux, N., Fokkinga, M., Apers, P.M.: Exploring personalized life cycle policies. CTIT Technical Report Ser. (Supplement/TR-CTIT-07-85) (2007)
10. Hilderman, R.J., Hamilton, H.J., Cercone, N.: Data mining in large databases using domain generalization graphs. JIIS **13**(3), 195–234 (1999)
11. Pearson, S., Casassa-Mont, M.: Sticky policies: an approach for managing privacy across multiple parties. Computer **44**(9), 60–68 (2011)
12. Rosemann, M.: Trust-aware process design. In: BPM, pp. 305–321 (2019)
13. Samarati, P.: Protecting respondents identities in microdata release. IEEE TKDE **13**(6), 1010–1027 (2001)
14. Terragni, A., Hassani, M.: Optimizing customer journey using process mining and sequence-aware recommendation. In: SAC 2019, pp. 57–65 (2019)
15. Zaman, R., Hassani, M.: On enabling GDPR compliance in business processes through data-driven solutions. SN Computer Science **1**(4), 1–15 (2020)

Verifiable Multi-Party Business Process Automation

Joosep Simm, Jamie Steiner, and Ahto Truu$^{(\boxtimes)}$

Guardtime, A.H.Tammsaare Tee 60, 11316 Tallinn, Estonia
{joosep.simm,jamie.steiner,ahto.truu}@guardtime.com

Abstract. Lack of trust is one of the main problems preventing multi-party business process automation. Solutions based on smart contracts and distributed ledgers have been proposed, but suffer from scalability issues. We present a more performant alternative approach to enable trust between business partners, based on authenticated data structures.

Keywords: Multi-party business process automation · Distributed ledger technology · Authenticated data structures · Sparse Merkle trees

1 Introduction

Business process automation (BPA) is the use of technology to execute recurring tasks or processes with the goal of replacing manual effort. Sometimes also referred to as digital transformation, its potential benefits include increasing service quality, improving service delivery, and reducing costs.

Most BPA deployments aim to automate a firm's internal operations. However, many business processes are composed of a series of steps taken by different firms. Single-party business processes such as invoice production and processing, which might be handled using a corporate enterprise resource planning (ERP) system, can be viewed as steps in the overarching multi-party business process of two or more firms engaging in a trade of goods or services. Thus, if BPA aims to improve performance of a firm's internal business activities, we may define multi-party business process automation (MPBPA) as the use of technology to optimize and automate firms' interactions with each other.

As an example, we may consider a hypothetical supply chain relationship. A Supplier regularly produces and ships widgets to a Customer. The widgets are transported by a series of third-party Logistics companies. Payment is typical net 30 terms. Since the Supplier is interested in being paid as soon as possible, he uses the services of a Bank, which pays the Supplier upon delivery to the first Logistics provider, in exchange for interest. This arrangement is common and is effectively a loan secured by the Supplier's receivables.

The challenges in realizing MPBPA differ from those solved by extant BPA technology. The most important difference is the inherent lack of trust between

© Springer Nature Switzerland AG 2020
A. Del Río Ortega et al. (Eds.): BPM 2020 Workshops, LNBIP 397, pp. 30–41, 2020.
https://doi.org/10.1007/978-3-030-66498-5_3

firms on matters of value. A firm would be naturally hesitant to allow a computer program to automatically pay invoices; instead, review and approval by a trusted employee in accounts payable is typically required. If it were possible to automate such interactions in a trusted way, many of the same benefits that BPA has yielded within the scope of individual firms could be realized at an inter-company or even systemic level. In our example, the Customer receives the goods, and must reconcile the received amount against the invoice. Then they must make payment to the Bank according to their terms. The Bank receives payment and also reconciles it against the terms of the original invoice before extinguishing the loan to the Supplier.

The existence of current manual reconciliation processes shows that firms are willing to pay a substantial premium to ensure that business rules, designed to protect against fraud or error, are applied correctly.

Distributed ledger technology (DLT) can be described as the process of replicating and synchronizing shared data between several mutually distrusting parties. The data can then be relied on in situations where no single party can be trusted to create and maintain a central database. Byzantine fault tolerant (BFT) consensus protocols are typically used for the replication and synchronization, and are often seen as the underlying technological basis for DLT. These consensus algorithms, however, present a number of challenges.

First, BFT consensus has high network overhead. This has caused many practical implementations to appoint a small subset of nodes in the network as designated validators who receive all transactions. The validators then apply the business rules, which are encoded in the ledger itself as so-called smart contracts, to determine which transactions are valid, and follow the consensus protocol to agree on how the ledger state should be updated.

Second, clients might not want to share their confidential business data. In such cases, only the parties to a contract would receive the transaction details and know the code to be executed. In our example, trade information is commercially sensitive. For example, the supplier may not want other customers to know that they provide widgets to one customer at a discount, because other customers might also ask for a discounted price. Thus, the parties execute the code independently and only post attestations about the current state of the contract to the ledger. The validators follow the consensus protocol to ensure that the attestations are reliably recorded.

In the latter case, the validators do not know the transaction data or the contract code. Thus, they can not possibly know whether the contract has been executed correctly. The attestation of state is still useful, however, since all parties can be assured of having a common, non-repudiable view of the state. We refer to this arrangement as using "privacy limited validation." In a supply chain context, if the validators were also part of the supply chain, by validating their competitors' plain text transactions, they might be able to obtain commercially useful information that they would not otherwise be entitled to.

In summary, several key ingredients are necessary for multi-party business process automation:

- a process definition which describes the possible or acceptable interactions between the parties, i.e. the business rules;
- a reliable record of the information added by different parties, at different times, to contribute to and progress the execution of that process;
- means to prevent parties from repudiating the current process state, claiming an alternative process state, or claiming alternative histories of how the process progressed to the current state.

Smart contracts underpinned by DLT are one way to meet the above requirements. However, implementations that use them have proven difficult to scale. We propose an alternative approach that is more scalable and can be used as the basis for many of the multiparty process enforcement use cases that others address through DLT. Our solution is based on committing process states into an authenticated data structure (ADS) operated by a server whose actions can be independently verified.

2   Related Work

Several DLT-based business process management (BPM) systems have been proposed to support execution of multi-party processes.

Caterpillar [7] is a BPM system that runs on top of the Ethereum blockchain. It keeps all process related data on the blockchain, in order to ensure its reliability. On one hand, no off-chain data is required to read or execute a process. On the other hand, potentially sensitive data must be posted to the blockchain.

Lorikeet [16] is a model-driven engineering tool. It generates smart contract code (in the Solidity language of the Ethereum blockchain) from a Business Process Model and Notation (BPMN) representation. It uses DLT for communication between parties, but sensitive data is not posted to the blockchain.

Both solutions suffer from the inherent limitations of DLT design, namely the overhead required to reach consensus between parties. This limits the throughput of these systems. A comparative overview of these systems is given in [5].

The Laava platform [17] focuses on multi-tenant aspects of BPM and proposes to solve these by creating a private blockchain instance for each tenant and periodically aggregating the states of these private chains into a commitment posted to a public blockchain as an external anchor. A drawback of this approach is that a user would still have to scan a whole private blockchain to extract, prove, or verify the history of one process.

There are other solutions outside of academic research, such as the Proof of Process by Stratumn [14]. Their system is also based on DLT, and therefore inherits its limitations.

Mendling et al. [8] discuss different directions in business processes and blockchain interaction. One of the proposed approaches is based on monitoring:

This provides a suitable basis for continuous conformance and compliance checking and monitoring of service-level agreements. Second, based on monitoring data exchanged via the blockchain, it is possible to verify if a process instance meets the original process model and the contractual obligations of all involved process stakeholders.

Our proposed solution is close to this description. It does not put restrictions on execution of business processes, but provides a secure, performant, and scalable way to monitor the execution of multi-party processes.

Breu et al. [2] introduced the distinction between process orchestration, where the process instance is managed by a central service provider, and process choreography, where the ownership of the instance is transferred from one participant to another. We note that our monitoring-based approach is applicable in both cases and perhaps even more valuable in the context of choreographies where it is easier for the parties to get confused over the current state of the process [11].

3 Preliminaries

A hash function h maps arbitrary-sized inputs to fixed-size outputs.

Hash values are often used as representatives of inputs that are either too large or too confidential to be handled directly. For example, in the hash-then-sign model, a document's hash value is signed instead of the document itself. Likewise, in the hash-then-publish time-stamping model, a hash value is published to establish evidence of the existence of the input without disclosing the input itself.

To facilitate such uses, cryptographic hash functions are required to have several additional properties, such as one-wayness (it is infeasible to reconstruct the input from the output), second pre-image resistance (it is infeasible to change the input so that it still maps to the same output), and collision resistance (it is infeasible to find two distinct inputs mapping to the same output). We refer to [12] for a more extensive treatment.

SHA-256 and SHA-512 are practical hash functions (with 256-bit and 512-bit outputs, respectively) that have all these properties and are expected to withstand cryptoanalytic attacks on them for at least the next ten years [10].

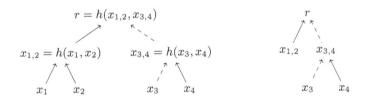

Fig. 1. Merkle tree with 4 leaves and the hash chain linking x_3 to r.

Hash trees [9], also known as Merkle trees (MT), are tree-shaped structures of hash values. Each node is either a leaf with no children or an internal node with two children. The value x of an internal node is computed as $x = h(x_l, x_r)$, where x_l and x_r are the values of the left and right child, respectively. There is one root node r that is not a child of any node.

In order to prove that a value x_i participated in the computation of the root hash r, it is sufficient to present values of all the sibling nodes on the unique path from x_i to the root of the tree, along with the "shape" of the path (defining the concatenation order on each step). For example, to claim that x_3 belongs to the tree shown on the left in Fig. 1, one has to present the values x_4 and $x_{1,2}$ to enable the verifier to compute $x_{3,4} = h(x_3, x_4)$, $r = h(x_{1,2}, x_{3,4})$, essentially re-building a slice of the tree, as shown on the right in Fig. 1.

A standard hash tree can thus be used to prove that a value was in a set (so-called "inclusion proof"). It is, however, not suitable to proving that a value was absent from a set (so-called "exclusion proof"). The value could be in any node, and to verify its absence, the whole tree would have to be scanned.

A way to overcome this limitation is to dedicate a leaf to each potential value of the set. Then the leaves corresponding to values not in the set could be set to some "empty" sentinel value (shown as white in Fig. 2) and absence of a value in the set can be proven by showing that the dedicated leaf is empty.

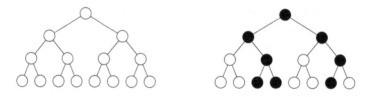

Fig. 2. Sparse Merkle trees: an empty one (left), and a populated one (right).

With 256-bit hash values, such a tree would have 2^{256} leaves, which is clearly infeasible. To resolve this, we amend the rule for populating internal nodes so that a parent of two empty child nodes also gets the "empty" value, and hashing is only used to compute values of parents with at least one non-empty child. Such a tree, called sparse Merkle tree (SMT), is fully defined by its non-empty nodes (shown as black in Fig. 2), the number of which is more tractable.

An SMT can also be used to manage maps of key-value pairs. In this case, the hash of the value is stored in the leaf dedicated to the hash of the key, and the leaves corresponding to absent keys will be left empty [1,4].

4   Our Approach

Verifiable business processes (VBP) is a transaction-based server-centric solution for adding trust to a registry or a service in order to achieve verifiable multi-party business process automation. In the case of traditional blockchains, new transactions are approved by majority agreement; VBP instead is based on single trust domain where it generates proofs of correct operation, which would make incorrect behavior immediately evident. VBP consists of:

- authenticated data structures (Sect. 4.1);
- verifiable state machines (Sect. 4.2);
- verifiable log of registry changes (Sect. 4.3).

4.1 Authenticated Data Structures

An authenticated data structure (ADS) is a data structure whose operations can be carried out by an untrusted service provider, the results of which a verifier can efficiently check as correct [15].

VBP uses an authenticated map of key-value pairs maintained in a sparse Merkle tree (SMT). A VBP server maintains one such tree and populates it incrementally, starting from an empty tree. For better performance, the server operates in rounds, collecting updates within a time period and committing them as a batch at the end of the round. All modifications are public for the participants of the process, which enables them to audit the operation of the server if they wish to do so.

The root of the SMT is the trust anchor for the proof holder. It is critical for a verifier to know that only one root value is produced at the end of each round. This guarantees there exists only one version of the tree for all participants. Our solution uses a set of external auditors who digitally sign the roots of the SMT for that purpose.

4.2 Verifiable State Machines

The SMT, by itself, cannot be used to describe a business process. We assume that the allowed states and transitions of each process are defined in some formalism. This could be based on a general standard, such as BPMN, or some custom formalism. For VBP the main requirements are:

- the process description can be deterministically serialized;
- a process instance's state can be deterministically serialized.

VBP tracks the state of each process instance as a key-value pair where the key is a permanent and unique process identifier and the value represents the current state of the process instance. For example, a single shipment from the Supplier to the Customer might be represented in a single leaf. The hash of the process identifier determines the leaf of the SMT for tracking the evolution of the state of that process instance and the hash of the current state is stored in the designated leaf of the SMT.

It is desirable to use meaningful process instance identifiers, as this reduces the risk of participants mistaking a process instance for another. For that, participants need to agree on the naming convention for process instances, in addition to the verification logic for the business process model they are participating in. For example, the counterparties, invoice number, and date of shipment might be used to derive the process instance identifier. A single SMT can be used to hold multiple instances of the same kind of process, or of different processes.

As a process is executed, participants send hashes of the updated process states to the VBP server. The server will record each update in the corresponding leaf of the SMT. This enables extraction of a proof of state for that process instance relative to the root hash of the SMT. Anyone who holds the process instance and the state proof can verify independently that the process instance is in fact in that state and that other participants must know the same. If a process instance is altered to indicate that the shipment has been delivered to the Customer, the hash value for this process will change. Anyone might see that this has occurred, but without having access to the underlying process instance data that is hashed, it reveals nothing. Thus, the VBP server functions as an independent witness of the process state, but without having to know anything about the transactions or the business rules, in the manner described as the "privacy limited validation" in Sect. 1.

How the process instance moves from one party to another is beyond the scope of VBP. This could be done by any means of transport agreed upon by the participants. In all cases, the integrity of the process instance is guaranteed by the VBP server. Note that the VBP server does not receive any sensitive data to provide that service. Only meta-data on when processes are started and modified is required.

Our current implementation uses a JavaScript based process definition library to describe the process verification rules. Process definitions are modeled as finite state machines (FSM). The same library generates representations of process instances. These are not based on BPMN, although it would be possible to generate these artifacts from a BPMN tool.

4.3 Verifiable Log of Registry Changes

The evolution of process instances can be modeled as a new SMT being constructed by the VBP server for each round. To ensure that all participants have a consistent view of the evolution of a process, we need to prove that the server is not maintaining parallel histories for a process instance and showing different histories to different participants. This general goal can be broken down in following more specific questions:

- How can we prove that a given state for a particular process instance is the current state?
- How can we prove that state was not changed and changed back in secret at some time in the past? (We have termed this the "flip-flop" attack.)
- How can we efficiently traverse the history of state changes of a process instance?

Eijdenberg et al. [6] discuss the properties of verifiable log backed maps (VLBM) where the history of changes to a simple SMT is recorded in a verifiable log for auditing purposes. However, full audit is required to prove correct operation of a VLBM. Since a full audit would be impractical in large trees, it could allow a malicious operator to perform flip-flop attacks with low probability of detection.

The data structure used in VBP improves upon the VLBM in this respect. Instead of just the hash of the state of a process instance, each leaf of the SMT in VBP stores a record (H, C, T) where:

- H is the hash of the current state;
- C is the counter of state changes since the process instance was created;
- T is the time of the last state change (expressed as the number of the round when that change was recorded in the SMT).

This data structure enables efficient auditing of the VBP server in a manner similar to the server auditing process described in [3]. For efficency, the VBP server applies updates to the SMT in batches. An auditor keeps the current value of the root hash of the SMT as its internal state and uses that to verify the correctness of the updates submitted by the server. The correctness of the batch of updates that transforms the SMT with the root hash R in round N to the SMT with the root hash R' in round $N + 1$ as follows:

1. The auditor sets R^* to R, to start from the SMT state as of round N.
2. For each process state update, the server presents
   - the record $X = (H, C, T)$ of the process instance state as of round N;
   - the hash chain A linking X to R^*;
   - the updated record $X' = (H', C', T')$.
3. The auditor processes each of those updates by
   - checking that A indeed links X to R^*; this establishes the correctness of the initial state of the record;
   - checking that $C' = C + 1$ and $T' = N + 1$; this establishes the correctness of the update of the metadata in the record; note that the auditor does not check the state change, because it is auditing the behavior of the VBP server and not of the business process participants;
   - updating R^* to the output value of the hash chain A when the record X is replaced with X'; note that the hash chain A is the same as in the first check, which ensures that the server could not have changed any other records with this update.
4. After processing all updates, the auditor checks that $R^* = R'$; this ensures that the server has presented exactly the set of updates that transformed the SMT from the state with the root R to the state with the root R'.
5. If all the checks pass, the auditor signs the new root hash value R' as approved and the server can use it as a reference relative to which proofs can be delivered to clients and also as the starting state for the next batch of updates.

Crucially, the audit protocol allows multiple auditors to work in parallel and independently sign their approvals, thus avoiding the need for a distributed consensus protocol, which would reintroduce some of the limitations of DLT.

For clients, the VBP server provides an interface for querying the (H, C, T) records, together with the associated hash chain proofs. This enables efficient traversal of records relating to a particular process. The client can query the latest state of the process, then use the time T of the last change to query the previous state at time $T - 1$, and so on until the history of the process instance has been traced back to its creation.

The client can then verify that all the state transitions are indeed valid according to the agreed process model. The client can also verify that the counter of state changes increases by one with each change, which ensures that there are no other changes to the process state outside this reported history.

This mode of operation prevents the possibility that processes could be altered briefly, in collusion with the VBP server operator, for the purpose of producing a fake "proof" of an incorrect state, after which the process would be put back to the correct, unaltered state.

Motivations for doing this attack could vary. For example, our Supplier might wish to obtain a proof attesting more widgets had been shipped than really had, for the purpose of fraudulently recognizing revenue early. Since the shipment process state would be corrected moments later by removing the invalid step from the process data, it would be extremely unlikely that anyone would notice.

This type of fraud mainly affects parties who are not participating in the process, but have some outside interest in it, and rely upon its accuracy for some reason. Without a mechanism to prevent it proactively, this type of fraud can only be detected by full audit.

5 Discussion

5.1 Storage Requirements

Keeping the SMT with full modification history takes storage space. Each internal node of the SMT contains a hash value of k bits. In the leaf nodes the size of the hash value dominates over the two integers, so for simplicity we can assume equal-sized nodes. As only the hashes of the process states are kept in the SMT, the space requirement does not depend on the size of process states, but only on the number of process instances and their modification rates.

When a full SMT would be naively stored for each round, each process instance would fill one leaf node and cause at most k internal nodes to be populated with non-empty values, for a total of about $N \cdot M \cdot k$ nodes for M processes maintained over N rounds.

However, if only a minority of the processes change in each round, most nodes of the SMT remain the same from one round to the next and those nodes can be shared between the two trees, thus reducing the storage to $N \cdot C \cdot k$, where C is the average number of process instances that change per round.

Since the tree is a sparse one, most paths from a non-empty leaf to the root have many empty siblings. Neither these siblings nor the parents computed from them need to be stored, as each such parent can be re-computed on demand from the only non-empty child. With this optimization, only $\log_2 M$ nodes need to be kept on an average path, for a total of $N \cdot C \cdot \log_2 M$ nodes, each containing k bits, or $k/8$ bytes of data.

The process count is not a constant, however. The append-only nature of the SMT means it is always growing. We can assume constant rate of new process instances to get some useful estimates. For example, with 1 000 clients and 1 000 new process instances per client per year, and an average of 10 state modifications per process instance, we get the rate of about 10 million updates per year and the tree with 1 million leaves by the end of the first year, 2 million leaves by the end of the second year, etc. Some example schedules of storage requirements are listed in Table 1.

Table 1. Storage requirements of a VBP server, depending on the number of clients, number of new process instances per client per year, and the number of state updates over the lifetime of a process instance, assuming a 256-bit hash function.

Clients	Processes	Updates	Year 1	Year 2	Year 3	Year 4	Year 5	Total
1 000	1 000	10	6.4 GB	6.7 GB	6.9 GB	7.0 GB	7.1 GB	34.1 GB
10 000	10 000	100	8.5 TB	8.8 TB	9.0 TB	9.1 TB	9.2 TB	44.7 TB

It is up to the application to decide how long historical state should be stored. A rolling window approach (e.g., store only last year's worth of full proof history) could be applicable in many cases.

5.2 Deployment Models

Depending on the business needs, there are multiple ways the VBP service could be deployed. A few examples:

1. Internal organization process auditing: Deploy a VBP server for the organization only. No need for external auditors. Root publication does not have to be global.
2. Coordinate processes within a consortium: Deploy one VBP server for the consortium. Appoint auditors trusted by the consortium members for the VBP server validation and root attestation.
3. Need to prove small number of processes in global scale: Use a global VBP service provider. Service provider handles VBP service auditing and root publication for many clients. Processes are provable independently from the infrastructure of any one organization.
4. Need to prove many processes, some locally, some globally: Use a layered federated architecture, as shown on Fig. 3. Intra-organizationally, short proofs up to the root of the local tree suffice. Cross-organizationally, full proofs up to the root of the global VBP publishing service are needed.

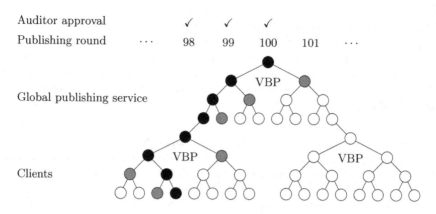

Fig. 3. VBP root publication scheme.

6   Conclusions

We have proposed an alternative to distributed ledgers for adding integrity to multi-party business process automation. This solution uses a "trust, but verify" model, which allows us to create a highly scalable solution. In addition, our solution does not require each participant to install new infrastructure. The central piece—the VBP server—could be shared between many participants in a software-as-a-service (SaaS) model.

Our main contribution is adding efficient auditability of modification history to the sparse Merkle tree (SMT) based authenticated data structure, which protects clients against malicious behavior by the service provider, including advanced threats, such as the "flip-flop" attack (cf. Sect. 4.3).

The other contribution is showing how such an authenticated data structure could be used in multi-party business process automation use cases.

7   Future Work

A known limitation of VBP is that it can handle only linear processes. Extending the model to support branching, parallel execution, and merging is an interesting avenue of future research.

Another direction would be to develop a root publication service, so it would be usable by other VBP server instances. The layered deployment model proposed in Fig. 3 is conceptual and needs more research.

The solution is not yet integrated into any established BPM systems. This is a possible direction of future practical development effort. As the VBP server is agnostic to what data it contains, multiple such integrations could be supported even by a single instance of the VBP service.

Also, it would be beneficial to implement more real-life use cases with this model, to find shortcomings that could be addressed in the future. Currently, a previous version of VBP is used in one production system for monitoring the business process execution and easing the auditing process of the Certus service [13].

Acknowledgements. This research was partly supported by the EU H2020 project PRIViLEDGE (grant 780477). The authors are also grateful to the anonymous reviewers who pointed out additional related work.

References

1. Bauer, M.: Proofs of zero knowledge. https://arxiv.org/abs/cs/0406058 (2004)
2. Breu, R., et al.: Towards living inter-organizational processes. In: 2013 IEEE 15th Conference on Business Informatics, pp. 363–366. IEEE (2013)
3. Buldas, A., Laanoja, R., Truu, A.: A blockchain-assisted hash-based signature scheme. In: Gruschka, N. (ed.) NordSec 2018. LNCS, vol. 11252, pp. 138–153. Springer, Cham (2018). https://doi.org/10.1007/978-3-030-03638-6_9
4. Dahlberg, R., Pulls, T., Peeters, R.: Efficient sparse merkle trees. In: Brumley, B.B., Röning, J. (eds.) NordSec 2016. LNCS, vol. 10014, pp. 199–215. Springer, Cham (2016). https://doi.org/10.1007/978-3-319-47560-8_13
5. Di Ciccio, C., et al.: Blockchain support for collaborative business processes. Informatik Spektrum **42**, 182–190 (2019)
6. Eijdenberg, A., Laurie, B., Cutter, A: Verifiable data structures. https://www.continusec.com/static/VerifiableDataStructures.pdf (2015)
7. López-Pintado, O., García-Bañuelos, L., Dumas, M., Weber, I., Ponomarev, A.: Caterpillar: a business process execution engine on the Ethereum blockchain. Softw. Pract. Exerience **47**(7), 1162–1193 (2019)
8. Mendling, J., et al.: Blockchains for business process management–challenges and opportunities. ACM Trans. Manag. Inf. Syst. **9**(1), 4:1–4:16 (2018). Article no. 4. https://doi.org/10.1145/3183367
9. Merkle, R.C.: Secrecy, Authentication and Public Key Systems. PhD thesis, Stanford University (1979)
10. NIST. Recommendation for key management. SP 800–57, Part 1, Rev. 5 (2020)
11. Prybila, C., Schulte, S., Hochreiner, C., Weber, I.: Runtime verification for business processes utilizing the Bitcoin blockchain. Future Generation Comput. Syst. **107**, 816–831 (2020)
12. Rogaway, P., Shrimpton, T.: Cryptographic hash-function basics: definitions, implications, and separations for preimage resistance, second-preimage resistance, and collision resistance. In: Roy, B., Meier, W. (eds.) FSE 2004. LNCS, vol. 3017, pp. 371–388. Springer, Heidelberg (2004). https://doi.org/10.1007/978-3-540-25937-4_24
13. SICPA. Certus. https://www.certusdoc.com [2020-06-13]
14. Stratumn. Proof of process. https://www.proofofprocess.org [2020-05-20]
15. Tamassia, R.: Authenticated data structures. In: Di Battista, G., Zwick, U. (eds.) ESA 2003. LNCS, vol. 2832, pp. 2–5. Springer, Heidelberg (2003). https://doi.org/10.1007/978-3-540-39658-1_2
16. Tran, A.B., Lu, Q., Weber, I.: Lorikeet: a model-driven engineering tool for blockchain-based business process execution and asset management. In: BPM 2018, volume 2196 of CEUR Workshop Proceedings, pp. 56–60. CEUR-WS.org (2018)
17. Weber, I., Lu, Q., Tran, A.B., Deshmukh, A., Gorski, M., Strazds, M.: A platform architecture for multi-tenant blockchain-based systems. In: ICSA 2019, pp. 101–110. IEEE (2019)

The Thirteenth Workshop on Social and Human Aspects of Business Process Management (BPMS2'20)

The Thirteenth Workshop on Social and Human Aspects of Business Process Management (BPMS2'20)

Social software [1, 2] is a new paradigm that is spreading quickly in society, organizations, and economics. It enables social business that has created a multitude of success stories. More and more enterprises use social software to improve their business processes and create new business models. Social software is used both in internal and external business processes. Using social software, the communication with the customer is increasingly bi-directional. E.g., companies integrate customers into product development to capture ideas for new products and features. Social software also creates new possibilities to enhance internal business processes by improving the exchange of knowledge and information, to speed up decisions, etc. Social software is based on four principles: weak ties, social production, egalitarianism and mutual service provisioning.

Social software is part of social information systems [3] that comprise a large variety of software used by organizations including social networking platforms, collaborative project management tools, or online/content communities. They differ from traditional information systems by enabling emergent interactions [4]. Emergent interactions are interactions that are defined during run-time by two or more stakeholders. No plan or approval from a supervisor or management is necessary. Emergent interactions enable the articulation of personal into collective knowledge thus representing mechanisms for harnessing collective intelligence in the digital age [5, 6].

Up to now, the interaction of social and human aspects with business processes has not been investigated in depth. Therefore, the objective of the workshop is to explore how social software interacts with business process management [7], how business process management has to change to comply with weak ties, social production, egalitarianism and mutual service, and how business processes may profit from these principles [8]. The workshop discussed the three topics below.

1. **Social Business Process Management (SBPM),** i.e., the use of social software to support one or multiple phases of the business process lifecycle
2. **Social Business: Social software supporting business processes**
3. **Human Aspects of Business Process Management**

Based on the successful BPMS2 series of workshops since 2008 [9], the goal of the 13th BPMS2 workshop is to promote the integration of business process management with social software and to enlarge the community pursuing the theme.

Four of eight submitted papers were accepted in the workshop:

In their study "A Classification of Digital-Oriented Work Practices", Pooria Jafariand Amy Van Looy take the perspective of employees who are agents for executing and digitalizing work. The authors investigate current work practices related to DI and BP and present five types of current digital-oriented work practices, which they translate into a gradual adoption model. An importing finding is that the more advanced groups of digital workers also had a lower work satisfaction.

Olivier Hotel, Lilia Gzara, Hervé Verjus, and Wafa Triaa describe in their paper with the title "Competency Cataloging and Localization to Support Organizational Agility in BPM" a competency-based framework to support organizational agility within the context of business process management. The authors also develop a generic methodology to integrate competency management that enables locating knowledgeable actors among the corporate employees and finding the right actors to perform the process tasks.

Johannes Tenschert, Willi Tang, and Martin Matzner address that most process support systems turn to cloud solutions, and that the integrated systems are used as one consolidated enterprise system (ES). Applying the perspective of an ES vendor to enable sustained support within the same ES, the authors identify integral and supporting capabilities. The authors also derive corresponding features in an ES. In this way potential users are able to analyze vendors in regard to capabilities for expected change.

In their paper "SentiProMo: A Sentiment Analysis-enabled Social Business Process Model Tool" Egon Lüftenegger and Selver Softic introduce SentiProMo as a novel social business process modeling tool based on sentiment analysis. SentiProMo improves the BPM lifecycle in the process analysis phase by capturing and processing stakeholder's opinions. It takes a social information systems perspective to transform these opinions and classifiy them.

We wish to thank all the people who submitted papers to BPMS2 2020 for having shared their work with us, the many participants creating fruitful discussion, as well as the members of the BPMS2 2020 Program Committee, who made a remarkable effort in reviewing the submissions. We also thank the organizers of BPM 2020 for their help with the organization of the event.

Selmin Nurcan
Rainer Schmidt

Organization

Program Committee

References

1. Schmidt, R., Nurcan, S.: BPM and Social Software. In: Ardagna, D., et al. (eds.) BPM 2008. LNBIP, vol. 17, pp. 649–658. Springer, Heidelberg (2009). https://doi.org/10.1007/978-3-642-00328-8_65
2. Bruno, G., et al.: Key challenges for enabling agile BPM with social software. J. Softw. Maintenance Evol.: Res. Pract. 23, 297–326 (2011). https://doi.org/10.1002/smr.523
3. Schmidt, R., Alt, R., Nurcan, S.: Social Information Systems. In: Proceedings of the 52nd Hawaii International Conference on System Sciences, pp. 2642–2646, Hawaii (2019)
4. Alter, S.: Information systems. Addison-Wesley Longman Publishing Co., Inc. (1998)
5. Bhatt, G.D.: Management strategies for individual knowledge and organizational knowledge. J. Knowl. Manage. 6, 31–39 (2002)
6. Schmidt, R., Kirchner, K., Razmerita, L.: Understanding the business value of social information systems – towards a research agenda. In: 2020 53rd Hawaii International Conference on System Sciences (HICSS), pp. 2639–2648 (2020)
7. Schmidt, R., Nurcan, S.: Augmenting BPM with Social Software. In: Business Process Management Workshops, pp. 201–206, Ulm (2010)

8. Erol, S., et al.: Combining BPM and social software: contradiction or chance? J. Softw. Maintenance Evol.: Res. Pract. **22**, 449–476 (2010). https://doi.org/10. 1002/smr.460
9. Nurcan, S., Schmidt, R.: Introduction to the first international workshop on business process management and social software (BPMS2 2008). In: Ardagna, D., et al. (eds.) BPM 2008. LNBIP, vol. 17, pp. 647–648. Springer, Heidelberg (2009). https://doi.org/10.1007/978-3-642-00328-8_64

A Classification of Digital-Oriented Work Practices

Pooria Jafari[(⊠)] [iD] and Amy Van Looy[(⊠)] [iD]

Faculty of Economics and Business Administration, Department of Business Informatics and Operations Management, Ghent University, Tweekerkenstraat 2, B-9000 Ghent, Belgium
{Pooria.Jafari,Amy.VanLooy}@UGent.be

Abstract. Digital innovation (DI) profits from new IT opportunities to affect the internal and external interactions of organizations. This impact is also present on business process management (BPM), which is an important managerial approach, resulting in digital process innovation projects to acquire more reliable business processes for all stakeholders. While most studies focus on the innovation outcome of better serving end customers, this study takes the perspective of employees who are crucial agents for executing and digitalizing work. Based on a representative European dataset, we statistically investigated current work practices related to DI and BPM, and linked them to work satisfaction. Our classification presents five types of current digital-oriented work practices, which we translated into a gradual adoption model. Remarkably, the more advanced groups of digital workers also had a lower work satisfaction. We encourage organizations to launch efforts for affecting employees' intrinsic and extrinsic motivation to help realize digital work more efficiently.

Keywords: Digital innovation · process innovation · Empowerment · Work satisfaction · Classification

1 Introduction

Advancements in emerging technologies have the potential to impact on how organizations work. For instance, artificial intelligence (AI) and data mining methods offer more intelligent and agile solutions to design, execute and improve business processes [1]. Such new ways for managing business processes result in more explorative business process management (BPM) solutions during which digital innovation (DI) is brought into the work practices of organizations [2]. The digitalization of work has already taken place during recent decades by the fast evolution of internet-based technologies, and is now increasingly stimulated by new technologies such as blockchains, AI or internet-of-things [3]. Consequently, digital process innovation projects have been launched to assist organizations in the injection of technologies to BPM in order to become more financially sustainable and to offer excellent value to customers [1, 4].

While the literature stresses the importance of digital process innovation in organizations [5], many of these studies focus on particular technologies affecting BPM

© Springer Nature Switzerland AG 2020
A. Del Río Ortega et al. (Eds.): BPM 2020 Workshops, LNBIP 397, pp. 49–59, 2020.
https://doi.org/10.1007/978-3-030-66498-5_4

methods and techniques [6, 7], and how customers benefit from innovation [8]. While such studies mainly focus on the technical aspects related to digital process innovations [9], more attention is needed to the employees involved in a process change and their features (e.g., cultural, behavioral, motivational) to examine ways in which they interact [10, 11]. An employee focus is essential because digital innovation projects are likely to be unsuccessful due to employee resistance [12]. More specifically, relatively little attention has been paid to how employees experience their current working conditions, although observing work conditions is important for diverse reasons, such as producing quality. For this purpose, we address two research questions:

- **RQ1. Which types of digital-oriented work practices by employees exist?**
- **RQ2. What are the differences in work satisfaction between the types of RQ1?**

We were granted access to a large-scale European dataset comprising various aspects of digital-oriented work practices and work satisfaction. Based on statistical classification techniques and ANOVA-based testing, we distinguish various types of current digital-oriented work practices that currently exist in organizations. Besides enriching the knowledge base about specific aspects that contribute to both BPM and digital innovation, our work is valuable for practice. Business executives, managers and consultants are offered a gradual adoption model for digital work, resulting in practical advice to pay sufficient attention to the well-being of employees.

This article continues by providing the research background in Sect. 2, followed by describing our quantitative research design in Sect. 3. The findings are then presented (Sect. 4) and discussed (Sect. 5). We conclude in Sect. 6.

2 Background

This section describes the three central constructs to our work, namely BPM, DI and digital process innovation.

2.1 Business Process Management

We rely on the definition of [13] to describe BPM as *"the art and science of overseeing how work is performed in an organization to ensure consistent outcomes and to take advantage of improvement opportunities"* (p. 1). Each business process is expected to advance through iterations in a lifecycle similar to Plan-Do-Check-Act (PDCA) [14]. This iterative lifecycle emphasizes the notion of iterative process improvements, in which process optimizations can be categorized ranging from smaller or incremental adjustments to larger or more radical improvements [14]. Regular process improvements have been supported by quality management philosophers for decades (e.g., total quality management or TQM) [15]. TQM characteristics that also apply to BPM are: process-centered, continuous improvement, integrated systems, strategic thinking, effective communication, fact-based decision making, employee commitment and customer satisfaction [15]. Alternatively, the idea of radical process improvements gained momentum in the 1990s due to IT and globalization pressures. The predominant delegates of those radical process changes were [16], with the notion of "business process reengineering" (BPR), and [17] for introducing the term "process innovation". Similarly, in the early 2000s, the related triggers (i.e., IT and globalization) raised up more powerful with novel internet-based technologies. Nowadays, emerging technologies cause a similar innovation wave and trigger a BPM adoption because of the expected performance outcomes.

Prior surveys have demonstrated that BPM can lead to higher business (process) performance and competitiveness [18, 19]. Alternatively, the success factors to advance in BPM have been quantitatively calculated in various maturity models, which point to continuously reinforcing organizations with gradual advice [20–22]. To enhance the maturity of business processes and BPM, organizations can choose between maturity models of different types. Examples of different approaches for applying BPM may focus on process innovation (i.e., which is typically based on lower BPM maturity levels) [19] or on customized BPM (i.e., which is mostly based on higher BPM maturity levels) [23]. Such BPM approaches emphasize the increasing role of digitalization in applying BPM [24].

2.2 Digital Innovation

Innovation means the creation and adoption of a new idea in a product or service, a business process or a business model [25]. A digital innovation is an innovation enabled by means of an information technology (IT) [26], and can therefore be seen as the combination of innovation and digitalization influences (e.g., emerging technologies and IT infrastructures) to transform from non-digital artifacts to digitized one, and which help enhance the traditional BPM approaches to more intelligent and agile BPM approaches.

DI has become increasingly important because organizations implement most of their activities by digital solutions and they are keen to use innovative solutions to realize more efficient business processes and to increase financial performance [27–29]. To assess whether a DI is usable and affordable, organizations can look at key factors like: (1) user centricity and value added, (2) the applicable features of actual solutions in the market environment, and (3) skills to work (e.g., empowerment). These three items help reinforce and sustain a DI [11].

2.3 Digital Process Innovation

Digital innovation has assisted the PDCA cycle by applying new IT-based solutions. For instance, IT offers an infrastructure for collaborating, coordinating and communicating within and across teams [30], and for managing knowledge for a new product/service/process development [4].

Incremental digital process innovation mainly focuses on continually improving existing products or services by concentrating on cost reduction and improving or adding features. On the other hand, radical digital process innovation explores new technologies, and is more risky to meet the organization's mission and vision [31]. In both approaches, digital process innovations have enabled organizations to implement their strategic plans and to better align their organizational layers (i.e., operational, tactical and strategic).

Hence, digital process innovation projects intend to improve an organization's business (process) performance, and this especially for delivering an excellent quality to customers.

An organization's business environment is one of the main decisive factors in a process innovation adoption [32]. E.g., the combination of BPM and DI focuses on enriching organizational data. To access high-quality data, digital process innovation uses environmental data (e.g., business plans) and employee-related features (i.e., cultural, behavioral, motivational) [10]. Alternatively, while technology developments and innovation opportunities have empowered new BPM solutions to become more efficient, agile and smart [1, 2], preparing users for accepting new technologies has been another challenge which we will address [33].

3   Methodology

3.1   Dataset

The European Foundation for the Improvement of Living and Working Conditions has been taken a periodical survey since the 1990s. This survey consists of different human resource and work-related aspects throughout the European zone. We focused on the European working conditions to reveal the effects of digital-oriented work practices on employees' work satisfaction. We selected the most recent round belonging to 2015, which comprised 35 countries. The initial dataset contained 43,850 respondents, of which 49.59% female and 50.41% male. During the data preparation step, we cleaned the data to improve data quality, resulting in 30,556 respondents without missing values.

3.2   Variables

The dataset contained 343 questions related to working conditions, of which we extracted nineteen as independent variables related to BPM and DI (or digital process innovation). Four of them consisted of individual and organizational characteristics (i.e., current status of attendance, gender, age and organization size). The other 15 variables covered different aspects of digital-oriented work characteristics, more specifically: (1) IT use, (2) quality control (QT), (3) pace of work, (4) frequency of work interruptions, (5) effect of work interruptions, (6) problem-solving tasks, (7) task complexity, (8) task novelty, (9) task repetitiveness, (10) multi-skilled tasks, (11) employee involvement, (12) decision autonomy, (13) skills match, (14) on-the-job training, and (15) payment for team performance.

Since the statistical analyses for answering the research questions required at least ordinal variables, nominal variables were first converted to dummy variables.

Regarding work satisfaction (i.e., acting as the dependent performance outcomes), we selected seven ordinal variables from 18 questions in total, namely: (1) perceived job performance, (2) job usefulness, (3) perceived working conditions, (4) career advancements, (5) job motivation, (6) job satisfaction, and (7) job security.

3.3 Statistical Tests for RQ1

To classify the European employees along their digital-oriented work practices, we conducted a cluster analysis [34] (i.e., for exploratory classification) and a discriminant analysis [35] (i.e., for confirmatory classification) based on the 15 digital-oriented work variables.

3.4 Statistical Tests for RQ2

To scrutinize any difference in work satisfaction outcomes among the clusters of RQ1, ANOVA-based testing was conducted [36]. Before being able to investigate RQ2, a principal factor analysis was used to indicate scale validity and reliability among the combined work satisfaction variables, and to identify which factors served as index for the ANOVA-based tests. When the means of work satisfaction were supposed to be unequal across the clusters, post-hoc testing [37] was required to distinguish which cluster differs from which other cluster in terms of work satisfaction.

4   Statistical Tests for RQ1

4.1   Classification

The K-means solution turned out to have the best fit with our data by resulting in five equal clusters (i.e., with 4,308, 5,856, 6,062, 7,157, and 7,173 employees, respectively). Also the mean values of all variables related to the digital-oriented work practices were close to the predicted centroid per cluster.

To give additional evidence for the five clusters, a discriminant analysis was applied to predict which digital-oriented work variables best fit to which cluster. The independent variable was the categorical membership variable, coming from the cluster analysis. There are two discriminant methods in SPSS (version 26): regular and stepwise. In both methods, the discriminant functions were highly significant ($P < 0.001$). The discriminant functions in the regular method respectively explained 96.9% and 69.7% of the total variability between the clusters (R-Squared), whereas those in the stepwise method respectively explained 93.4% and 66.4%. The discriminant analysis confirmed the clustering findings by presenting all employees with a score near to the respective cluster centroid.

4.2   Feature Analysis of the Obtained Classification

We opted to extract the typical or differentiating characteristics per cluster by comparing the descriptive statistics (percentages) of each variable option through the five clusters. Our criterion to keep or eliminate a particular variable focused on a difference in the percentages across the 15 variables related to digital-oriented work. More specifically, if we observed that multiple options had a similar behavior across the five clusters and that major percentages were repeated through all clusters, we eliminated the related variable for not being a differentiator. In total, we kept five out of 15 variables (i.e., IT use or working with IT devices, employee involvement, decision autonomy, task repetitiveness, and task complexity) because only they had diverse cluster percentages and thus suggesting patterns.

Next, we did an internal evaluation of the clusters by reviewing the cluster percentages among the five remaining variables. This means that we turned towards a more content-wise approach by translating the quantitative percentages to textual labels to acquire a qualitative cluster comparison (Fig. 1). As such, it appeared that the five remaining variables could be grouped along three domains or groups of characteristics, which formed the central features of our classification. The three groups were: (1) digital characteristics (i.e., IT use), (2) people characteristics and empowerment (i.e., employee involvement and decision autonomy), and (3) process characteristics and task types (i.e., task repetitiveness and task complexity).

Clusters	Cluster A	Cluster B	Cluster C	Cluster D	Cluster E	
IT use	Superior IT use	Moderate IT use	Little IT use	Little IT use	High IT use	IT use (digital characteristics)
Employee involvement	Superior involvement	High involvement	Superior involvement	Little involvement	Little involvement	Empowerment (people characteristics)
Decision autonomy	Superior autonomy	High autonomy	Superior autonomy	Little autonomy	Little autonomy	
Task repetitiveness (i.e., monotonous tasks)	Little monotonous	Little monotonous	Frequently monotonous	Frequently monotonous	Frequently monotonous	Task type (process characteristics)
Task complexity	Highly complex	Highly complex	Frequently complex	Little complex	Frequently complex	

Fig. 1. Turning the quantitative feature analysis to a qualitative comparison. (Color figure online)

We summarized the derived features per cluster as follows.

- Cluster A: greatly digitized and greatly empowered work with challenging tasks
- Cluster B: much digitized and much empowered work with challenging tasks
- Cluster C: little digitized and greatly empowered work with difficult tasks
- Cluster D: little digitized and little empowerment work with straightforward tasks
- Cluster E: highly digitized and little empowered work with difficult tasks

When looking for a gradation among the three central features of our classification, we made use of a color scheme introducing three levels ranging from lower (i.e., red), average (i.e., orange) to higher (i.e., green) values regarding digital-oriented work. This coloring exercise (as is frequently done in management tools like the Balanced Scorecard) better visualized the patterns among the clusters. Consequently, Fig. 1 suggests a gradation among existing digital-oriented work practices: **D < E < C < B < A**.

5   Statistical Tests for RQ2

5.1   Factor Analysis of the Performance Outcomes

In order to combine the seven performance variables, a principal factor analysis resulted in two latent constructs for work satisfaction: "Future Job Prospect Perception" (FJPP) and "Actual Job Perception" (AJP). No rotations were required.

- For "Future Job Prospect Perception (FJPP)", the factor analysis extracted one factor, which explained 37.14% of the total variance of four process performance statements (i.e., job security, perceived working condition, career advancement and job motivation). Cronbach's alpha was 0.792.
- For "Actual Job Perception (AJP)", the factor analysis extracted one factor, explaining 55.46% of the total variance of three organizational performance statements (i.e., perceived job performance, job usefulness and job satisfaction). Cronbach's alpha was 0.850.

5.2   ANOVA-Based Tests with Regard to the Obtained Classification

Given both factors' non-normality and variance inequality, we conducted the ANOVA Welch's F test [38, 39], followed by the Games-Howell post-hoc test [37]. The ANOVA results were similar for both factors:

- for FJJP: $F(4, 4655.962) = 283.631, P = 0.000$
- for AJP: $F(4, 4594.180) = 265.560, P = 0.000$

For FJJP, all clusters had significantly different cluster means, although not between cluster B and cluster C. A gradual increase in work satisfaction was observed as follows: $A < (B + C) < E < D$.

For AJP, again, all clusters had significantly different cluster means, but not between cluster D and cluster E. Likewise, a gradual performance was presented: $C < A < B < (E + D)$.

6   Discussion

Section 4 and Sect. 5 have both worked towards a gradation among the identified clusters of existing digital-oriented work practices. The discussion section combines both insights to propose a phased approach (Fig. 2) by translating them in a metaphorical way similar to the idea of organizational maturity levels.

6.1   Towards a Gradual Adoption Model for Digital-Oriented Work Practices

Organizational maturity models typically assess and improve an organization's readiness in organizational capabilities related to its people, processes, data and/or technologies. For example, maturity models can encourage to advance in certain capabilities in order to make the business more efficient, and provide advice on what organizations can do to improve [40]. Maturity models are a recognized way to boost the business processes and the BPM capabilities of an organization to bring higher efficiency for all stakeholders (e.g., manager, employees and costumers) [20, 21]. In recent decades, advancements in IT and BPM have helped organizations to change their business processes faster and cheaper than in a traditional organization (i.e., with less digitization, less employee empowerment in their governance bodies and with less optimized processes). Since

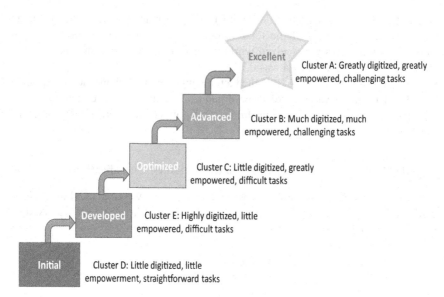

Fig. 2. A gradual adoption model for digital-oriented work practices. (Color figure online)

increasing work satisfaction is crucial for concretizing an organization's mission and vision, we have dug deeper into a gradual adoption of digital-oriented work practices.

Since the survey data merely describe existing work practices, we aim to be descriptive rather than prescriptive, and this by understanding the implementation levels that organizations currently follow. From the perspective of an organizational maturity model, the five clusters covering aspects of digital-oriented work (i.e., IT use with digital characteristics, empowerment with people characteristics, and task type with process characteristics) can be related to five maturity levels (i.e., from lower to higher levels in digital work), depending on their degree of digital-oriented practices (RQ1).

We have placed two clusters (i.e., cluster A and cluster B) that included higher values of digital, people and process characteristics on higher maturity levels to show an organization's ambition of progressing towards the most advanced situations related to digital-oriented work practices. The clusters with average values (i.e., cluster C and cluster E) show an organization's smooth ambition. The cluster with the lowest values on digital, people and process characteristics (i.e., cluster D) represents more traditional organizations which are less open to digitization or empowerment, and which processes are less challenging.

The qualitative assessment from a maturity point of view is summarized in Fig. 2, which presents the five clusters via more diverse colors and symbols (i.e., initial (red), developed (orange), optimized (yellow), advanced (green) and excellent (star)). Although the cluster gradations resulting from the performance outcomes were similar for both work satisfaction factors (FJJP: $A < (B + C) < E < D$; AJP: $C < A < B < (E + D)$), they contrast to the gradation from the feature analysis (Sect. 4.3: $D < E < C < B < A$). Hence, RQ2 has shown that lower work satisfaction is situated in the more advanced clusters of digital-oriented work practices. This means that digitalization should rather focus on tackling task complexity to increase job satisfaction, as well as that organizations need to pay more attention to the wellbeing of their employees.

While more research is needed on how organizations can boost their employees' work satisfaction, we already make some suggestions. Organizations should especially focus on their employees' intrinsic motivation, for instance, by regular communication about the organization's mission and vision, by non-financial incentives that stimulate team spirit, or by letting employees do bottom-up suggestions for process improvements. Alternatively, manager can work on their trustworthiness by working towards goals derived from strategies. Employees' extrinsic motivation can be triggered by human resource management policies, including mandatory training and formal rewards for employees adapting to digital work.

In sum, the statistical clusters allowed us to derive practical implications for climbing from lower to higher digital-oriented work practices. When an organization's maturity level shows a gap between its ideal level (e.g., a green or star level) and its actual status of maturity, then the organization now understands its actual digital-oriented work practices and can start planning the next iterative rounds to reach higher levels of the ladder [40]. We acknowledge that not all organizations should strive for the highest level, but for an optimal level that is contingent upon its business environment. Future research elaborates on such contingencies to progress along Fig. 2.

6.2 Research Limitations

The study profits from internal validity, which has been statistically ensured. For instance, in RQ1, we repeated the classification exercise during exploratory cluster analysis and confirmatory discriminant analysis. Also the use of a factor analysis in RQ2 improved data reliability. Additionally, an expert panel could add interesting insights for testing Fig. 2 and to come up with more concrete practical recommendations concerning the related practical implications. Although the dataset profits from a large number of respondents, we acknowledge that external validity is still reduced to European digital-oriented work practices, and that a worldwide generalization needs to be justified in more detail.

6.3 Future Research

We will continue working on a longitudinal study by adding the previous working condition datasets. We will first add a chronological perspective across the datasets to verify the extent to which the employees' use of technology has evolved (i.e., between the 1990s, 2000s and 2010s). Afterwards, the clustering exercise will be repeated to find trends across the decennia.

In future work, we will also position the findings into other disciplines, like applied psychology and human resource management in which work satisfaction is more deeply ingrained. We also intend to adopt an additional research method for taxonomy development in order to build and test the statistical classification as a taxonomy, resulting in deeper theoretical implications than the current empirical findings.

7   Conclusion

This study has revealed a classification of five clusters (types) related to digital-oriented work practices, for which strong evidence has been found for the respective differences in work satisfaction. Although we acknowledge limitations regarding our work (e.g., with many binary variables), scholars and practitioners can benefit from the classification and the related 'star model'. Interestingly, we uncovered that the more organizations work digitally, the lower their work satisfaction among staff members, whereas the opposite is more desirable in a digital economy. This study encourages organizations to pay sufficient attention to prepare their employees while advancing in their digital-oriented work practices.

References

1. Clemons, E.K., Dewan, R.M., Kauffman, R.J., Weber, T.A.: Understanding the information-based transformation of strategy and society. J. Manag. Info. Sys. **34**(2), 425–456 (2017)
2. Pontieri, L.: Extending Process Mining techniques with additional AI capabilities to better exploit incomplete/low-level log data. In: AI4BPM Proceedings (2019).
3. Kirchmer, M., Franz, P., Gusain, R.: Digitalization for agile business process management. In: BPM-D Proceedings, London, Philadelphi (2018).
4. Huesig, S., Endres, H.: The role of functionality for the adoption of innovation management software by innovation managers. Eur. J. of Innov. Manag. **22**(2), 302–314 (2018)
5. vom Brocke, J., Zelt, S., Schmiedel, T.: On the role of context in business process management. Int. J. Info. Manag. **36**(3), 486–495 (2016)
6. Hussein, D.M.E-D.M., Taha, M.H.N., Khalifa, N.E.M.: A blockchain technology evolution between business process management (BPM) and Internet-of-Things (IoT). Int. J. Adv. Comp. Science and Applications, 9(8), 442–45, (2018).
7. Mendling, J., Weber, I., Van der Aalst, W., vom Brocke, J., Cabanillas, C., Daniel, F., Debois, S., Di Ciccio, C., Dumas, M., Dustdar, S., Gal, A., Garcia-Banuelos, L., Governatori, G., Hull, R., La Rosa, M.: Blochchains for business process management. ACM Trans. Manag. Info. Sys. **9**(1), 1–16 (2018)
8. Trkman, P., Mertens, W., Viaene, S., Gemmel, P.: From business process management to customer process management. BPM J. **21**(2), 250–266 (2015)
9. Mikalef, P., Krogstie, J.: Big Data Analytics as an Enabler of Process Innovation Capabilities: A Configurational Approach. In: Weske, M., Montali, M., Weber, I., vom Brocke, J. (eds.) BPM 2018. LNCS, vol. 11080, pp. 426–441. Springer, Cham (2018). https://doi.org/10.1007/978-3-319-98648-7_25
10. Schmiedel, T., vom Brocke, J.: Business Process Management: Potentials and Challenges of Driving Innovation. Springer, Cham (2014)
11. Van Looy, A.: A Quantitative Study of the Link Between Business Process Management and Digital Innovation. In: Carmona, J., Engels, G., Kumar, A. (eds.) BPM 2017. LNBIP, vol. 297, pp. 177–192. Springer, Cham (2017). https://doi.org/10.1007/978-3-319-65015-9_11
12. Venkatesh, V., Morris, M.G., Davis, G.B., Davis, F.D.: User acceptance of information technology: toward a unified view. MIS Q **27**(3), 425–478 (2003)
13. Dumas, M., La Rosa, M., Mendling, J., Reijers, H.A.: Fundamentals of BPM. Springer, Berlin (2018)
14. Hasan, Z., Hossain, M.S.: Improvement of effectiveness by applying PDCA Cycle or Kaizen. J. of Scientific Research **10**(2), 159–173 (2018)

15. Yousif, A.S.H., Najm, N.A., Al-Ensour, J.A.: Total quality management (TQM), organizational characteristics and competitive advantage. J. Econ. Financ. Studies **5**(4), 12–23 (2017)
16. Hammer, M., Champy, J.: Reengineering the Corporation. HarperCollins, NY (2003)
17. Davenport, T.: Process Innovation. Harvard Business School, Boston (1993)
18. Bronzo, M., de Resende, P.T.V., de Oliveira, M.P.V., McCormack, K.P.: Improving performance aligning business analytics with process orientation. Int. J. Inf. Manag. **33**(2), 300–307 (2013)
19. Dijkman, R., Lammers, S.V., de Jong, A.: Properties that influence BPM maturity and its effect on organizational performance, *Info. Sys., Front.,* **18**(4), 717–734 (2016).
20. McCormack, K., Johnson, W.C.: Business process orientation. St. Lucie, Florida (2001)
21. Hammer, M.: The process audit. Harvard Bus. Rev **85**(4), 111–123 (2007)
22. De Bruin, T., Rosemann, M.: Using the Delphi study technique to identify BPM capability areas. ACIS Proceedings **42**, 642–653 (2007)
23. Bucher, T., Winter, R.: Taxonomy of business process management approaches. In: vom Brocke, J., Rosemann, M. (Eds.). *Handbook on BPM 2* (pp. 93–114). Berlin: Springer (2010).
24. vom Brocke, J., Rosemann, M.: Handbook On Business Process Management 2. Springer, Berlin (2010)
25. Adomavicius, G., Bockstedt, J.C., Gupta, A., Kauffman, R.J.: Making sense of technology trends in the information technology landscape. MIS Q **32**(4), 779–809 (2008)
26. Yoo, Y., Lyytinen, K.J., Boland, R.J., Berente, N.: *The next wave of digital innovation,* Accesssed 1/2020: https://papers.ssrn.com/sol3/papers.cfm?abstract_id=1622170 (2010).
27. Chavan, M.: The balanced scorecard: a new challenge. J. Manag. Dev. **28**(5), 393–406 (2009)
28. Kane, G.C., Palmer, D., Philips, A.N., Kiron, D.: Is your business ready for a digital future? MIT Sloan Manag. Rev. **56**(4), 37–44 (2015)
29. Mithas, S., Tafti, A., Mitchell, W.: How a firm's competitive environment and digital strategic posture influence digital business strategy. MIS Q. **37**(2), 511–536 (2013)
30. Attaran, M., Attaran, S., Kirkland, D.: The need for digital workplace. Intern. J. Enterpr. Inform. Sys. **15**(1), 1–23 (2019)
31. Paunov, C., Planes-Satorra, S.: How are digital technologies changing innovation, OECD Science, Technology and Industry Policy (74) (2019).
32. Trkman, P.: The critical success factors of BPM. Int. J. Inf. Manag. **30**(2), 125–134 (2010)
33. von Rosing, M., Scheer, A.-W., von Scheel, H.: The Complete Business Process Handbook. Morgan Kaufmann, Waltham, MA (2015)
34. Everitt, B.S., Landau, S., Leese, M., Stahl, D.: Cluster Analysis. Wiley, London (2011)
35. Klecka, W.: Discriminant analysis. Sage Publications, California (1980)
36. Fiss, P.C.: A set-theorectic approach to organizational configurations. Acad. of Manag. **32**(4), 1180–1198 (2007)
37. Shingala, M.C., Rajyaguru, A.: Comparison of post hoc tests for unequal variance. *Int. J. of New Techn. in Science and Engin.,* **2**(5), 22–33 (2015).
38. Box, G.E.: Non-normality and tests on variances. Biometrica **40**(3–4), 318–335 (1953)
39. Vickers, A.J.: Parametric versus non-parametric statistics in the analysis of randomized trials with non-normally distributed data. BMC Med. Res. Methodol. **5**(35), 1–2 (2005)
40. Röglinger, M., Pöppelbuss, J., Becker, J.: Maturity Models in Business Process Management. BPM Journal **18**(2), 328–346 (2012)

Competency Cataloging and Localization to Support Organizational Agility in BPM

Olivier Hotel[1], Lilia Gzara[1(✉)], Hervé Verjus[2], and Wafa Triaa[1]

[1] G-SCOP, Technology Institute of Grenoble, Grenoble, France
{olivier.hotel,lilia.gzara,wafa.triaa}@grenoble-inp.fr
[2] LISTIC, Université Savoie Mont-Blanc, Annecy, France
herve.verjus@univ-smb.fr

Abstract. In this article, we describe a competency-based framework to support organizational agility within the context of business process management. We develop a generic methodology to integrate competency management to locate knowledgeable actors among the corporate employees and to find the right actors to perform the process tasks. This article exploits a competency catalog to register the corporate employees competencies and proposes a methodology to retrieve the knowledgeable actors from the database.

Keywords: BPM · Organizational agility · Human-centric process · Competency management · Competency modeling

1 Introduction

Considering the constant evolution of the work environment, corporate process must be monitored and continuously improved. These aspects are addressed by Business Process Management (BPM) methodologies which combine information technology and management science knowledge to manage operational business processes [1]. Business process can be classified into two main categories. The first one concerns stable or simple processes; whereas the second concerns processes that can be impacted by unplanned events or changes.

This article deals with this second category of process for which the concept of agility is primordial [2]. We therefore consider agility as a characteristic of processes subject to unplanned events or changes. Agility in BPM approaches has often been considered from a technological perspective. We propose to investigate the organizational agility which is little studied in the field of BPM; in this context we propose an approach centered on the agile management of competencies applied to business processes facing unplanned events or changes.

A. Del Río Ortega et al. (Eds.): BPM 2020 Workshops, LNBIP 397, pp. 60–69, 2020.
https://doi.org/10.1007/978-3-030-66498-5_5

2 Related Works and Proposal

2.1 Agility in BPM

Agility is the company's ability to adapt continuously in response to a complex, turbulent and uncertain environment, it is defined as the ability to react quickly to change by mastering it with significant anticipation, innovation and learning skills [3]. Therefore, agility is based on both:

- organizational agility: the capability of an organization to renew itself, adapt, change quickly, and succeed in a rapidly changing environment; and
- technical agility: the capability of an organization to readjust its information system.

Despite the variety of approaches developed to help companies improve their efficiency and become more agile, the agility of business processes remains a real challenge today. In the literature, the concept of agility has been considered from a particular perspective without approaching it in its entirety. Most of the work has focused on the technical or IT dimension [20]. The consideration of the organizational dimension of agility is still little studied today as shown by Fig. 1. The authors of [16] and [19] proposed a social recommender system to support organizational agility during the design phase.

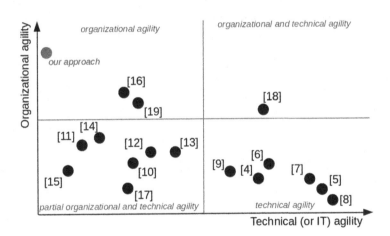

Fig. 1. Organizational and technical agility in the literature.

2.2 Proposal

We state that BPM should encompass competency management methodologies to fully support organizational agility. Many definitions of competency have been proposed in the literature. In this article, we share the definition of the HR-XML Open Standards Consortium Competencies Schema [21]: *a competency is a specific, identifiable, definable, and measurable knowledge, skill, ability and/or other deployment-related characteristics (e.g., attitude, behavior, physical ability) which a human resource may possess and which is necessary for the performance of an activity within a specific business context.* There are two main categories of competency derived from the literature [22]:

1. hard competency; which refers to two types:
   (a) *know*: it concerns everything that can be learned from educational or formative systems. It represents the theoretical understanding of something such as a new or updated method or procedure, etc.; and
   (b) *know-how*: it is related to personal experiences and working conditions. It is learned by doing, by practice and by experience. It is the practical knowledge consisting of how to get something done.
2. soft competency: which consists of relational know-whom, cognitive know-whom and behavior. It is referred to individual characters, talents, human traits or qualities that drive someone to act or react in a certain way under certain circumstances.

In the rest of this paper, we do not make any distinction between hard and soft competency.

We focus on competency management because it was established that competencies and organization's performance are correlated [23], and that a human-centric approach can enhance industrial systems and factories to their full potential [24].

To that end, we introduce a competency catalog (Sect. 3.1) and a methodology (Sect. 3.2) to locate knowledgeable actors among the corporate employees and to find the right actors that can respond to an unplanned event at runtime. The typical workflow, as shown in Fig. 2, is:

1. the execution of a task failed, the execution context changed, or a new activity emerged (Fig. 3.1: the task E failed); and
2. the process manager and the stockholders take corrective actions (Fig. 3.2: task F).

Our approach makes it possible to determine which actors must be assigned to the different corrective actions:

1. the process manager identifies the required competencies; and
2. he queries the catalog to automatically retrieve the list of appropriate actors (Fig. 2.3).

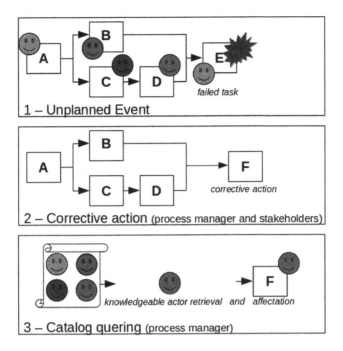

Fig. 2. Application of competency management for supporting organizational agility at runtime.

3   Competency Cataloging

3.1   Competency Modeling

Many studies evaluate different methods of competency modeling [22]. The authors of [13] show that ontologies have a greater expressiveness and are able to describe them despite their diversity. Ontologies are collections of concepts, instances of concepts and relation among them that are expressed at the desired level of formality that are deemed to be important in characterizing the knowledge domain under consideration at the desired level of detail [25].

In our approach, the central concept is *Competency*. A *Competency* is described by its name and by a set of keywords which provide a semantic description of the competency as shown in Fig. 3.

We consider two types of ontological relationship:

1. the requirement *(requires)* relation which appears when the execution of a competency requires the presence of other ones; and
2. the semantic relation *(isSimilarTo)* which is a relation between two names of the same concept.

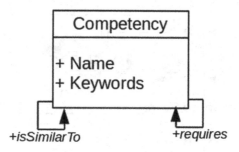

Fig. 3. Competency model.

In the rest of this article, C_i denotes a certain competency. The profile of each employee is formally represented as a list combining each acquired competency with the corresponding level of mastery:

$$E_n = \begin{bmatrix} (C_1, & k_{n,1}) \\ (C_2, & k_{n,2}) \\ \vdots \\ (C_i, & k_{n,i}) \end{bmatrix};$$

where E_n denotes the profile of the employee n and where $k_{n,i}$ represents the level of competency i by the employee n. Thereby, the catalog of the employees' competencies is the set containing the profile's vector of each of them.

Creating a catalog can be performed declaratively: the manager or the employees manually populate the catalog; or deductively: the manager uses employees' evaluation or observations to create catalog, it may also involve the use of process execution's traces. Such traces can be obtained manually or automatically: it can be the traces of executions recovered in a BPM software but also in other business software used by employees such as calculation, forecasting, simulation software or even emails and corporate social networks. The mastery level is dynamic and depends on two factors: learning (it increases) and degradation (it decreases). Hence it should be updated periodically.

3.2 Competency Localization

In this section, we tackle the issue of finding the corporate employees having the required competency to respond to the unplanned events. We define a catalog request as a set of competency; for instance, requesting the competencies $\{C_1, C_3, C_4\}$ returns all the employees having these competencies. Then the manager will filter them according to their mastery level and its objective.

Three situations may arise when solving the problem of finding the right actor to perform a task whereas the required competency already exists in the catalog or not, and whereas some employees have all the requested competencies or not.

Locating Existing Competencies

This situation occurs when the requested competencies have already been registered in the catalog. In this situation, one simply has to iterate over the employee's profile and retrieve the ones having the requested competencies.

Locating Nonexisting Competencies

This situation occurs when stakeholders are searching for a competency which does not exist in the catalog or when the desired competency was registered with a different name in the catalog.

These issues can be tackled by looking in the catalog for similar competencies. Two approaches may be considered:

- The first approach uses the requirement relation to find similar competencies. Let C_r and C_o respectively denote the set of the required competencies of the requested competency and of a given competency o in the data structure. Then, one can use a semantic proximity measure between C_r and C_o to extract the similar competencies present in the catalog: the semantic proximity measures between the requested competency and those of the catalog are computed, the most similar one is identified, and the associated employees are returned.
- The second approach is similar to the first one: the sets of the keywords providing the semantic description of the competencies are used to compute the semantic proximity measure instead of the sets C_r and C_o.

The semantic proximity could be evaluated using different measures, among them, the Dice measure has gained a certain interest because of its simplicity [26]. The Dice measure is given by:

$$Dice(C_r, C_o) = \frac{2card(Cr \cap Co)}{card(C_r) + card(C_o)}.$$

The similarity values are real numbers between 0 and 1 where:

- $Dice(C_r, C_o) = 0$: the competencies are totally different; and
- $Dice(C_r, C_o) = 1$: the competencies are identical.

For instance, let C_1, C_2, and C_3 be threes competencies whose semantic descriptions are respectively:

$$\{keyword\ 1;\ keyword\ 2;\ keyword\ 3\},$$

$$\{keyword\ 1;\ keyword\ 2;\ keyword\ 4\};\ and$$

$$\{keyword\ 1;\ keyword\ 5;\ keyword\ 6\}.$$

Then, one has: $Dice(C_1, C_2) = \frac{2}{3}$ and $Dice(C_1, C_3) = \frac{1}{3}$: C_2 is more similar to C_1 than C_3.

Partial Request This situation occurs when no employee has all the requested competencies. In this situation, one can find the employees that match most of the initial request by building a lattice of the original request. Such structure provides a formal and efficient classification process to discover and represent hierarchies of concepts [27,28]. A lattice is a directed acyclic graph representing the structure of a partially ordered set. Each element of the set is represented as a node and the vertices correspond to the ordering relation: there exists a vertex between the node A and the node B if $A > B$. In the context of a partial request, the considered ordered set is the power set of the different requested competencies and the ordering relation is the inclusion relation: there exists a vertex between the node A and the node B if $B \subset A$. The Fig. 4 represents the lattice associated with the request $\{C_1, C_3, C_4\}$.

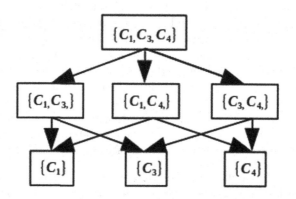

Fig. 4. Lattice of the request $\{C_1, C_3, C_4\}$.

Such structure allows to quickly retrieve the employees that partially match a request. For instance, if one wants to find the corporate employees having the competencies C_1, C_3, C_4; one should:

1. build the associated lattice;
2. starting from its root node, locate its children $\{C_1, C_3\}$, $\{C_1, C_4\}$, $\{C_3, C_43\}$;
3. and return the employees associated with them.

If no employees are associated with the children, the search process should be recursively applied to them until employees are found.

4   Conclusion

This article discusses the different approaches to support agility within the context of BPM, it shows that such approaches especially support technical agility whereas organizational agility is partially taken into account.

Our work is part of a complementary perspective to work on technical agility by providing an approach and mechanisms dealing with the organizational agility of business processes. We state that BPM should encompass competency management methodologies in order to locate knowledgeable actors among the corporate employees. To that end, we presented a generic competency management based framework for BPM using competencies management features to support changing processes and rapidly find the relevant actor. This article highlights that the assignment of a level of mastery to each couple *(employee - competency)* allows building a catalog of the corporate employee competency. Such database can be queried to locate the knowledgeable that can respond to an unplanned event actors whereas the requested competency exist in the catalog or not. This approach is currently applied to a real life industrial process.

In this paper, our approach deals with unplanned events that handle process changes at runtime, but it is generic enough to also deal with unanticipated situations in general, and may be used to respond to non (or partially) determinist business processes, emergent processes [29] or highly collaborative processes [30].

References

1. van der Aalst, W.M.P., ter Hofstede, A.H.M., Weske, M.: Business process management: a survey. In: Business Process Management, pp. 1–12 (2003)
2. Verjus, H., Pourraz, F.: A formal framework for building, checking and evolving service oriented architectures. In: Fifth European Conference on Web Services (ECOWS 2007), pp. 245–254 (2007)
3. Dove, R.: Agile enterprise cornerstones: knowledge, values, and response ability. In: Baskerville, R.L., Mathiassen, L., Pries-Heje, J., DeGross, J.I. (eds.) TDIT 2005. IIFIP, vol. 180, pp. 313–330. Springer, Boston, MA (2005). https://doi.org/10.1007/0-387-25590-7_20
4. Nurcan, S., Etien, A., Rim, K., Zoukar, I., Rolland, C.: A framework for context-aware adaptable web services. In: Advances in Database Technology (EDBT04), pp. 826–829 (2004)
5. van der Aalst, W.M.P., Gunther, G.W., Recker, J., Reichert, M.: Using process mining to analyze and improve process flexibility. In: BPMDS Workshop at the 18th International Conference on Advanced Information Systems Engineering, pp. 168–177 (2006)
6. Boukhebouze, M., Amghar, Y., Benharkat, A.N., Maamar, Z.: A rule-based modeling for the description of flexible and self-healing business processes. In: 13th East-European Conference on Advances in Databases and Information Systems (ADBIS09), pp. 15–27 (2009)
7. Pourraz, F., Verjus, H.: Managing service-based EAI architectures evolution using a formal architecture-centric approach. In: Enterprise Information Systems, pp. 269–280 (2008)
8. Chaâbane, M.A., Andonoff, E., Bouaziz, R., Bouzguenda, L.: Versions to address business process flexibility issue. In: Grundspenkis, J., Morzy, T., Vossen, G. (eds.) ADBIS 2009. LNCS, vol. 5739, pp. 2–14. Springer, Heidelberg (2009). https://doi.org/10.1007/978-3-642-03973-7_2

9. Lezoche, M., Missikoff, M., Tininini, L.: Business process evolution: a rule-based approach. In: 9th Workshop on Business Process Modeling, Development, and Support (BPMDS 2008) (2008)
10. Qiu, L., Chang, L., Lin, F., Shi, Z.: Context optimization of AI planning for semantic Web services composition. Serv. Oriented Comput. Appl. **1**(2), 117–128 (2007)
11. Koschmider, A., Song, M., Reijers, H.: Social Software for Modeling Business Processes. In: Lecture Notes in Business Information Processing, vol. 17, pp. 666–667 (2007)
12. Vanderhaeghen, D., Fettke, P., Loos, P.: Organizational and technological options for business process management from the perspective of web 2.0. Bus. Inf. Syst. Eng. **2**(1), 15–28 (2010)
13. Brambilla, M., Fraternali, P.: Combining social web and BPM for improving enterprise performances. In: 21 international conference companion on Wolrd Wide Web, pp. 223–226 (2012)
14. Silva, A.R., Meziani, R., Magalhães, R., Martinho, D., Aguiar, A., Flores, N.: AGILIPO: embedding social software features into business process tools. In: Business Process Management Workshops, pp. 219–230 (2010)
15. Dollmann, T., Houy, C., Fettke, P., Loss, P.: Collaborative business process modeling with comomod - a toolkit for model integration in distributed cooperation environments. In: IEEE 20th International Workshops on Enabling Technologies: Infrastructure for Collaborative Enterprises, pp. 217–222 (2011)
16. Qu, H., Sun, J., Jamjoom, H.T.: SCOOP: automated social recommendation in enterprise process management. In: 2008 IEEE International Conference on Services Computing, pp. 101–108 (2008)
17. Hauder, M.: Bridging the gap between social software and business process management: a research agenda: Doctoral consortium paper. In: IEEE 7th International Conference on Research Challenges in Information Science (RCIS), pp. 1–6 (2013)
18. Rangiha, M.E., Karakostas, B.: Process recommendation and role assignment in social business process management. In: 2014 Science and Information Conference, pp. 810–818 (2014)
19. Pesic, M., Schonenberg, H., van der Aalst, W.M.P.: DECLARE: full support for loosely-structured processes. In: 11th IEEE International Enterprise Distributed Object Computing Conference (EDOC07), pp. 287–297 (2007)
20. Triaa, W., Gzara, L., Verjus, H.: Organizational agility key factors for dynamic business process management. In: 18th IEEE Conference on Business Informatics (CBI 2016), pp. 64–73 (2016)
21. HR Open Standards: hropenstandards.org
22. Hammer, M., Champy, J.: Reengineering the Corporation - A Manifesto for Business Revolution. Harper Business Essentials (2003)
23. Tutu, A., Constantin, T.: Understanding job performance through persistence and job competency. Soc. Behav. Sci. **33**(1), 612–616 (2012)
24. Panetto, H., Weichhart, G., Pinto, R.: Special section Industry 4.0: challenges for the future in manufacturing. Ann. Rev. Control **47**, 198–199 (2019)
25. Schmidt, R., Nurcan, S.: BPM and social software. In: Ardagna, D., Mecella, M., Yang, J. (eds.) BPM 2008. LNBIP, vol. 17, pp. 649–658. Springer, Heidelberg (2009). https://doi.org/10.1007/978-3-642-00328-8_65
26. Martinez-Gil, J.: An overview of textual semantic similarity measures based on web intelligence. Artif. Intell. Rev. **42**(4), 935–943 (2012). https://doi.org/10.1007/s10462-012-9349-8
27. Moor, A.: Community memory activation with collaboration patterns. In: 2009 International Community Informatics Conference, pp. 9–11 (2009)

28. Koschmider, A., Song, M., Reijers, H.A.: Social software for modeling business processes. In: Ardagna, D., Mecella, M., Yang, J. (eds.) BPM 2008. LNBIP, vol. 17, pp. 666–677. Springer, Heidelberg (2009). https://doi.org/10.1007/978-3-642-00328-8_67

29. Debenham, J., Simoff, S.: Managing Emergent Processes. Lecture Notes in Computer Science, vol. 4251, pp. 228–235 (2006)

30. Ariouat, A., Andonoff, E., Andonoff, C.: Do process-based systems support emergent, collaborative and flexible processes? Comparative analysis of current systems. Procedia Comput. Sci. **96**, 511–520 (2016)

Enterprise System Capabilities for Organizational Change in the BPM Life Cycle

Johannes Tenschert[✉], Willi Tang, and Martin Matzner

Chair of Digital Industrial Service Systems, Friedrich-Alexander-Universität
Erlangen-Nürnberg (FAU), Nuremberg, Germany
{johannes.tenschert,willi.tang,martin.matzner}@fau.de

Abstract. In recent years, the frequency of organizational change
increased due to internal innovation and external forces. Classical enterprise life cycle theories expect infrequent, large-scale implementations
that lead to costly and risk-intensive shakedowns after implementation.
Currently, most process support systems turn to cloud solutions, and
integrated systems are used as one consolidated enterprise system (ES).
Processes and organizational routines are the unit of analysis for organizational change, and the BPM life cycle captures the phases involved in
changing them. We apply the perspective of an ES vendor to enable sustained support within the same ES, despite inevitable change. Hence, for
each phase, we identify integral and supporting capabilities, and derive
corresponding features in an ES. This view is also useful to potential
users to analyze vendors in regard to capabilities for expected change.

Keywords: Enterprise systems · Social software · BPM life cycle ·
Change management · Organizational change

1 Introduction

Today, the increasingly complex and dynamic business world demands that organizations can adapt to changing business environments, and their systems' agility
has become a critical component of organizational agility [14]. Often, changing
requirements of actual processes entail changing the systems in use. Tailored
information or enterprise resource planning (ERP) systems may be replaced
completely, and problems are greatest at the shakedown phase right after implementation [23], ranging from dips in job performance to abandonment of the
enterprise system [18]. Classical enterprise life cycle theories expect infrequent,
large-scale implementations, but the wide range of requirements on today's organizations also led to scattered process support, i. e. to many systems that are
used as one consolidated ES. Even for larger subsystems, avoiding large-scale
replacements is useful both for organizations applying the system as well as the
vendors that want to keep their users as long as possible.

© Springer Nature Switzerland AG 2020
A. Del Río Ortega et al. (Eds.): BPM 2020 Workshops, LNBIP 397, pp. 70–82, 2020.
https://doi.org/10.1007/978-3-030-66498-5_6

Since most process support systems nowadays gradually turn to cloud solutions, the typical mode for changes will likely turn from customer-specific, local adaptations to customer-specific support in configuration, and integration with external solutions. An ES has to cope with varying and evolving requirements from a wide range of organizations, and vendors can expect change as inevitable. Hence, we take the vendor perspective to identify integral and supporting capabilities, and derive what types of features could enable these capabilities for each BPM life cycle phase. Hence, we investigate the following research question: *Which supporting capabilities should ES vendors provide to support sustained organizational change?* Our unit of analysis is that of individual enterprise systems. Since ESs support the processes and routines of an organization, we establish a connection between the BPM life cycle and the ES life cycle. Due to the high risk potential of introducing a new ES, this focus is useful for vendors that intend to keep their users for a long time, and organizations using an ES that supports their organizational change in the long run.

The following sections first introduce ESs and the BPM life cycle, and the methods we apply for the analysis. In Sect. 4, we investigate each phase in the BPM life cycle in regard to capabilities and features. Section 5 discusses our methods and results, and Sect. 6 outlines related work. Finally, we conclude the paper.

2 Background

Enterprise systems are large-scale software applications geared towards supporting an organization's information needs [8]. They provide companies with a broad range of organizational and inter-organizational functionalities, e. g. customer service, supply chain optimization, sales order processing, or product life cycle management [8,11]. ESs are packaged *off-the-shelf solutions*, usually embedded with standardized processes that are marketed by vendors and consultants as *best practices* [30]. This standardization is expected to bring several business improvements, e. g. clearer communication about how a business operates or smooth hand offs across process boundaries [9]. As a result, organizations purchasing a new ES need to engage in business process reengineering and usually enter into a long-term relationship with software vendors. The organizations depend on these vendors for *continued enhancement of the package* [18].

ESs usually follow an experience cycle consisting of four phases: project chartering, the project (or configure and roll-out), shakedown, and onward and upward [18]. The *project chartering* phase consists of decisions about the business case and the identification of constraints. The *project/configure and roll-out* phase involves all activities that prepare the system for *going live*, e. g. software configuration, system integration, documentation, and roll-out and startup. The *shakedown* phase follows the implementation and characterizes the period where the system steadily evolves into a mode of *normal operation* and where people are concerned with learning how to use the system for productive work. Finally, the *onward and upward* phase refers to the routine operation of the system.

These four ES life cycle phases can conceptually be grouped into two overarching *phases of implementation*: problem-based and innovation-based implementation [20]. The first two phases, i.e. project chartering and the project, characterize a *problem-based implementation* phase, where the organization invests in the ES to address current problems and constraints and to overcome existing disadvantages against competitors. After the ES has been implemented, the organization will be preoccupied with adapting and innovating selected processes and activities: this is the *innovation-based implementation* phase, consisting of shakedown and onwards and upwards. This is the phase where new capabilities and *innovations* can come into fruition. Particularly in the onward and upward phase, the organization will exploit the system following an iterative cycle, where configuration and implementation will alternate cyclically. During this period, organizations should engage users selectively and allocate resources to do so [30]. Also, the organization requires the ability for organizational change and technical expertise to capitalize on the power of their ES for productive use [20]. The ES life cycle is depicted in Fig. 1.

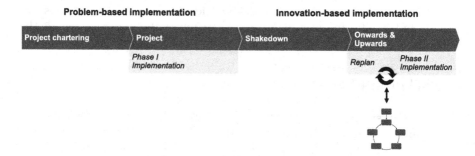

Fig. 1. The enterprise system life cycle (adapted from [18,20])

Enterprise Systems and Organizational Change. During the innovation-based phase (sometimes called the *post-adoption* phase), organizations usually experience a period where productivity is depressed [18]. This dip in organizational performance is typical during shakedown. Misalignments in information and functionality [32] of the system and the business requirements can increasingly surface. Users, who are the dominant drivers of *change* during shakedown and onward and upward, will demand change. Organizations that fail to manage such forces during this *second implementation phase* risk contributing to discontinuance intentions of the users, i.e. the rejection of the ES by the employees [13]. During the life cycle of an existing ES, organizations will therefore need to actively engage in organizational change to prolong the time where the existing system is productive. From a software vendor's perspective, they need to equip their system with capabilities that foster organizational adaptiveness and the organization's (i.e. their customer's) ability to change within the system.

Multiple forces determine whether an organization will be subject to change [27]. Prior studies in organizational change suggested to use organizational routines as a unit of analysis [19,29]. Organizational routines are "repetitive, recognizable patterns of interdependent actions, carried out by multiple actors" [12]. They can be distinguished between an ostensive (i. e. structural) and a performative (i. e. agentic) aspect [12]. The concept of organizational routines is closely related to business processes [5,6]: These can be considered a subset of organizational routines and represent "end-to-end work across an enterprise that creates customer value" [3,15]. These processes are forces that initiate change in an organization, but they will also be subject to change themselves. Changes in the processes will immediately affect how well ESs are equipped to aid organizational operations. To understand how organizations and their ESs change, they need to understand the mechanics of how processes change. The BPM life cycle covers exactly the change process of business processes. It characterizes BPM as a continuous cycle with the following phases: process identification, process discovery, process analysis, process redesign, process implementation, and process monitoring [10], cf. Fig. 2. Section 4 outlines each phase.

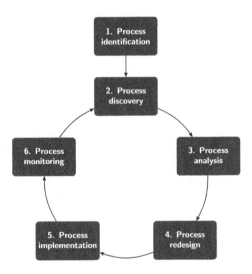

Fig. 2. The BPM life cycle [10]

3   Methods

Motivated by inevitable organizational change and based on the perspective of a system vendor, we want to identify 1) which capabilities an ES needs to support to facilitate organizational change within the system, and 2) which concrete features of a software system can enable these capabilities. Since our questions revolve around the processes of an organization that are changed and supported by an overall ES, we draw on the BPM life cycle [10] to illustrate and structure process change. We assume that each phase requires different capabilities not only by stakeholders of the process change, but also in regard to the ES. For each phase, based on Dumas et al. [10] and related literature, e. g. in regard to the ES life cycle and change management, we synthesize capabilities to be supported by the ES.

Guided by the principles suggested by Webster and Watson [31], we first searched for literature about ES in the context of change management through the Association for Information Systems' (AIS) basket of eight journals. We later extended our search for publications in related areas, e. g. computer science and organization theory. For each of the synthesized ES capabilities, we derive concrete software features that can support them. These lists are non-exhaustive and exemplary, and ES vendors may set a different focus.

The scope of the analysis is restricted to single ES vendors that need to facilitate all capabilities, and subjective in regard to which capabilities and challenges are emphasized in reviewed literature as well as to which exemplary features we derive for support them.

4   Enterprise System Capabilities for Organizational Change

In this section, we investigate all phases of the BPM life cycle (cf. Fig. 2) in regard to capabilities an ES should support, and derive concrete features. This information is consolidated in Table 1. During **process identification** [10], a business problem is posed and related to relevant processes. The outcome of this phase is a collection of processes and links, and typically, a collection of performance measures. These outcomes facilitate finding suitable processes to improve or initially support at all.

There are many drivers for performing processes, e. g. they might be core processes based on the business strategy and goals of an organization, derived from the financial, customer, internal, or learning & growth perspective, or based on regulatory requirements. Processes in this realm may also cover projects that became repetitive or activities around projects. *Organizational transparency* facilitates assessing the source of processes, i. e. over the abstract activities, stakeholders, strategy, external requirements, and infrastructure. Some processes may not be documented at all, e. g. smaller sets of activities requested by customers alongside other processes, but driving their decision for repeated business.

Van de Ven and Poole [27] introduced a taxonomy on process theories of organizational development based on mode and unit of change. Units of change distinguish changes for one or multiple organizations, e. g. based on internal innovation, regulatory requirements or common improvements. Modes of change distinguish whether the change is prescribed or constructive, i. e. emergent. While prescribed changes are fairly easy to detect by predetermined actions, constructive changes need to be identified, e. g. by actual performance of actions, dissatisfaction and conflicts or rather symptoms of the changes. As the change journeys may differ, finding the appropriate theory can avoid breakdowns [28], and ES could support in this regard. These characteristic entail needs in regard

Table 1. Capabilities and features facilitating them for BPM life cycle phases

Stage	Outcomes	Capabilities	Stakeholders	Example Features
Process Identification	Collection of processes and links Collection of performance measures	Organizational Transparency	MT, PO, PA, PM	- Role management / org. charts - Process landscape - Product / service catalog - Data management - Document management (manual EA documentation)
		Internal Feedback	MT, PO, PA, PM	- Forums - Annotations (Artifacts, Processes) - Mailing lists (cross-org) - Reference models
		Import External Knowledge	MT, PO, PA, PM	- Forums - Mailing lists (cross-org) - Reference models
		Change Detection Analysis	PO, PA	- Process mining (variations) - System of record (data availability)
Process Discovery	Collection of as-is process models Intermediate documentation	Instance Transparency	PA	- System of record (data availability) - Data management - Role management ($\rightarrow$ stakeholders) - Event log extraction
		Communication	All	- Groupware / social software - Document management - Project management (transparency for stakeholders)
		Knowledge Management & Modeling	PA	- Document management (traceability, connections) - (Collaborative) modeling
		Process Mining	PA, PM	- System of record (data availability) - Event log extraction - Integrated analysis methods
Process Analysis	Unnecessary steps Sources of waste Process issues KPIs (snapshot)	Knowledge Management	PO, PA	- Role management - Issue register - Document management - Modeling / diagramming tools
		Instance Transparency	PA	- Dashboards - System of record (data availability)
Process Redesign	Collection of to-be process models	Communication	PA	- Groupware / Social software - Project management - Document management
		Knowledge Management	PA	- Wikis - Reference models - Mailing lists (cross-org)
		Collaboration	PA	- Groupware / social software - Modeling tools - Real-time sharing (models, documents, screen)
Process Implementation	Executable process models	Communication	All	- Social software
		Knowledge Management	PO, PA, system engineer	- Role management - Issue register - Document management - Modeling / diagramming tools
		Support Structures	PA, process participants, BPM group	- Extensive system manual - Access to knowledge bases - Fluid user access control - Social software

(*continued*)

Table 1. (*continued*)

Stage	Outcomes	Capabilities	Stakeholders	Example Features
Process Monitoring	KPIs (continuously) Bottlenecks Recurrent errors Deviations	Instance Transparency	MT, PA	- System of record (data availability) - Dashboards - Data integration
		Process Mining	PA, PM	- System of record (data availability) - Event log extraction - Integrated analysis methods

(Stakeholders MT = Management Team, PO = Process Owner, PA=Process Analyst, PM = Process Methodologist)

to internal and external *communication* for sharing of knowledge and providing feedback, and for analysis methods for *change detection* to unveil symptoms of emergent changes. In **process discovery** [10], the current state of each relevant process is documented and consolidated to a set of as-is process models. Process discovery consists of defining the setting, gathering information, modeling, and assuring process model quality. Hence, the outcomes are a collection of as-is process models and potentially intermediate documentation gained by discovery methods.

The main challenges for process discovery are fragmented process knowledge, domain experts primarily thinking on case level, and potential lack of modeling experience. Different methods are primarily based on evidence, interviews, and workshops. Evidence-based discovery can be manual, e.g. through document review and observation, or automated, e.g. via process mining [26]. Hence, the main capabilities are *communication* support for multiple stakeholders with complementary skills and different settings of interviews and workshops, *instance transparency*, i.e. availability of information and stakeholders, *knowledge management* to capture and connect information, *modeling*, and *process mining*.

The reason behind process discovery for existing processes is that available process models and actual practice may diverge. *Process orientation* depends on BPM experience of stakeholders and application of process models in some regard. If process models are restricted to automation and quality management manuals, this may not be a driver to establish BPM experience. Rather, easy applicability in e.g. low-code settings of process models for smaller automation, or visible models that guide through the process may establish familiarity with notations, and applying documented models in real-world processes. **Process Analysis** [10] is a structured collection of issues of the *as-is process*. The phase involves a qualitative and quantitative analysis of business processes. The *qualitative process analysis* identifies unnecessary steps and reveals sources of waste. The *quantitative process analysis* considers performance measures, e.g. cycle time, waiting time, or cost. Indicators from flow analysis, queueing analysis, and simulation provide additional criteria for decisions [10].

System vendors can facilitate ease of access to relevant data and knowledge. Hence, we see knowledge management and data accessibility as crucial capabilities for process analysis. *Knowledge management* can for example be supported via role management, issue registers, document management, and diagramming tools. Role management helps in identifying key stakeholders of the process.

An issue register provides details about problems and their impact to facilitate resolution. Document management can give insights to the intricacies of processes and help in finding inefficiencies *by design*, e. g. regulatory requirements. With *data accessibility* refers to ease of accessing and analyzing measures. Dashboards provide relevant metrics for quantitative analysis.

The goal of **process redesign** [10] is to address and analyze the identified issues in regard to change options and consequence for performance measures. Promising options are translated and combined to a redesigned process, to *to-be process model*, and potentially change management for implementation gets prepared. While there are many redesign methods, their characteristics in regard to necessary support are quite similar.

Typically, redesign is an emergent, creative method in the sense that methods like 7FE, heuristic process design, business process reengineering, or product-based design impose a varying degree of structure, but the focus is on communication, knowledge management, collaborative modeling and ideation, and document management of results. *Communication* is important in regard to managing meetings, guiding the discussion, and potentially for evaluation or validation steps of respective methods. *Knowledge management* for heuristical or pattern-based methods may require capturing best practices from own or external sources, e. g. (anti-)patterns and reference models. *Collaboration* in regard to modeling and ideation may result in tailored modeling tools and collaborative document management. In all derived capabilities, characteristics of social software facilitate discussion, documentation, and exchange of information. In **process implementation** [10], the changes for turning *as-is* to *to-be* processes are performed. The results of completing this phase are executable process models. The process implementation phase is not only concerned with automating processes, but is also tied to active organizational change. During this change initiative, organizational change management needs to explain the planned business process changes to process participants, devise a change management plan, and train and prepare users for the new work procedures [10].

The goal is that changes will not impede the whole ES. We assume that required capabilities to avoid common problems and for successful completion are similar to those of ES implementation projects. While not explicitly covered in the BPM life cycle, organizations experience a shakedown between *going live* and *normal operation* [18]. We identified communication, knowledge management, and support structures a capabilities for minimizing sources of friction. *Communication* refers to exchange of information, active communication for feedback e. g. on remaining errors, and is facilitated by characteristics of social software [7]. For *knowledge management* in this phase, issue registers, document management, or diagramming tools can aid the creation of knowledge [1].

Support Structures refer to organizational mechanisms that help employees overcome challenges encountered when adopting new software or workflows [2]. Traditional support structures (e. g. help desk) are provided by the organization. ES can influence the degree of efficiency of support structures in regard to resources and tools. Peer advice ties can also act as a support structure and foster exchange of information among employees. Embeddedness in these ties has a positive impact on job performance during this phase [24]. Vendors could facilitate such ties by easing access to or applying features of social software.

Finally, in **process monitoring** [10], process-related data is collected and analyzed to determine performance in regard to measures and objectives. This way, bottlenecks, recurrent errors, and deviations can be identified to trigger corrective actions. After implementation, process monitoring should be applied continuously. Process monitoring utilizes process instance data, e. g. event logs, and additionally depends on available analysis methods as well as on the experience of the analyst. Typical tools for process monitoring are *dashboards* for certain key performance indicators and *process mining* for interactive drill-downs into process peculiarities based on event logs in regard to conformance and performance. While process discovery happens again after process monitoring, automated process discovery may unveil differences between the currently modeled process and the actual as-is process instances.

In regard to capabilities, this phase requires transparency, process mining experience, and integration. In *instance transparency*, a clear system of record facilitates that the right information is processed and available, and dashboards reveal an aggregated view. For *process mining*, at least an event log should be available to be analyzed in external systems, but process mining techniques could be directly integrated into an ES.

5   Discussion

Our objective was to take on the perspective of a software vendor to identify capabilities needed by ES that ease process and organizational change. We conceptualized how the ES life cycle relates to the BPM life cycle and how business processes serve as a unit of analysis for organizational change. Following the BPM life cycle, we identify and synthesize capabilities as well as supporting features based on literature, primarily on Dumas et al. [10]. While their contents are widely applied in academia and practice, our analysis still needs to be evaluated empirically in regard to impact and completeness. Our results are generally consistent with the socio-technical perspective in information systems [22] and they highlight how technical and social components within an organization need to jointly be optimized. In case of the continued ES use, software vendors can extend the life cycle of their ES by facilitating capabilities to react to change within the system. Social software and social software characteristics support communication channels for feedback during process implementation and allow the exchange of advice for support for the new workflows or features. By balancing these technical and social components, software vendors can extend exploitation of the system during the onward and upward phase, where configuration and implementation alternate cyclicallly. Despite the importance of these social components to the change-enabling capabilities of an ES, these social aspects are severely underrepresented in practice.

Many capabilities, especially in process identification, process discovery, process analysis, and process monitoring, require that instance artifacts are captured within a *system of record*, i. e. that artifacts are accessible, available, and related data is connected. This characteristic is not only useful during the BPM life cycle. Adding *knowledge management* features to change processes, or reusing existing ES features to support project and knowledge management activities during change is already applied in knowledge-intensive processes. Especially for adaptive case management systems, where planning is typically part of the work, this is already common [25].

The list of capabilities and features is not only useful to software vendors, but also to organizations searching for an ES, as we explicitly consider characteristics useful for change management and in regard to evolutionary information systems [17]. Even if vendors not explicitly capture these characteristics, selection teams can compare single or consolidated solutions to prevent unnecessary high-risk and cost-intensive shakedown phases.

Our analysis is relevant to ES vendors especially if they provide holistic cloud solutions. Here, organization-specific adaptations relate to (low-code) configuration instead of manual programming and consulting activities. This increases the importance of change within the same environment. Facilitating support structures can enable cross-organizational learning, as many stakeholders still apply a system with little differences between concrete deployments.

6   Related Work

A mapping between theoretical constructs and software features has been conducted in other contexts. Karahanna et al. [16] analyzed how social media features provide affordances (e. g. self presentation or group management) that motivate their use. The need-affordances-features perspective on social media use posits that social media features fulfill individual psychological needs (e. g. autonomy, relatedness, competence [21]) via affordances. This fulfillment of needs motivates use of social media applications. Karahanna et al. followed an integrative approach that synthesized affordances and features of popular social media. We adapted this approach and used the BPM life cycle, corresponding activities, and expected outcomes for each phase as sources of information to synthesize integral and supporting capabilities.

Brehm and Schmidt [4] examined how social software features (weak ties, social production, egalitarianism, and mutual service provision) can support design and improvement of business processes in the ES life cycle. While we also apply the ES life cycle within our analysis, we primarily used it to relate areas of business processes that lead to or are affected by organizational change. We concentrated on the second, *innovation-based* implementation phase, and used the BPM life cycle as central theory to characterize process and organizational change.

Bruno et al. [7] linked social software features to agile BPM requirements (organisational integration, semantic integration, and responsiveness). They highlight the ability of social software to enable organizations in quickly reacting to change. Our analysis confirmed that ES and organizations could benefit from social software, at least in regard to organizational change.

7 Conclusion

We investigated integral and supporting capabilities for each BPM life cycle phase, and derived features of ES that support these capabilities. We took the perspective of a system vendor with the clear intention of enabling organizational change within the same ES. Expecting and supporting organizational change in regard to support structures, project management, processes, and deployment allows sustainable usage of the system, and reduces the number of costly and risk-intensive new ES implementations. In cloud environments, capabilities are not only of technical nature and ES are socio-technical systems. There, adaptions are mostly configuration instead of individual deployments. Hence, support structures change and deployments of processes become configuration.

Still, there are open questions. We investigated the situation for a single, large-scale ES, but actual implementations may also be a consolidated ES integrating several BPMS and (tailored) information systems. Then, a vendor may emphasize further on integration and loose coupling from the technical perspective, and restricting the scope of required capabilities and phases to the scope of the individual aspect of the system, while a consolidated ES has to capture all capabilities. We based our results on a literature review, but future work should also investigate the impact and further classify required and desirable capabilities empirically. Finally, we expect that the rise of knowledge work as well as organizations primarily centered around informational production demonstrates different impact factors of capabilities of ES than in classical industrial settings.

References

1. Alavi, M., Leidner, D.E.: Review: Knowledge management and knowledge management systems: conceptual foundations and research issues. MIS Q. **25**(1), 107 (2001)
2. Ann Sykes, T.: Support structures and their impacts on employee outcomes: a longitudinal field study of an enterprise system implementation. MIS Q. **39**(2), 473–A11-A11 (2015)
3. Beverungen, D.: Exploring the interplay of the design and emergence of business processes as organizational routines. Bus. Inf. Syst. Eng. **6**(4), 191–202 (2014)
4. Brehm, L., Schmidt, R.: Potential benefits of using social software in ERP-based business process management. In: Piazolo, F., Felderer, M. (eds.) Multidimensional Views on Enterprise Information Systems. LNISO, vol. 12, pp. 71–83. Springer, Cham (2016). https://doi.org/10.1007/978-3-319-27043-2_6
5. Breuker, D., Matzner, M.: Statistical sequence analysis for business process mining and organizational routines. In: ECIS 2013 Completed Research (2013)
6. Breuker, D., Matzner, M.: Performances of business processes and organizational routines: similar research problems, different research methods - a literature review. In ECIS 2014 Proceedings (2014)
7. Bruno, G., et al.: Key challenges for enabling agile bpm with social software. J. Softw. Maintenan. Evol. Res. Pract. **23**(4), 297–326 (2011)
8. Davenport, T.H.: Mission critical: realizing the promise of enterprise systems. Harvard Business School Press Books, p. 1 (2000)

9. Davenport, T.H.: The coming commoditization of processes. Harvard Business Rev. **83**(6), 100–108 (2005)

10. Dumas, M., La Rosa, M., Mendling, J., Reijers, H.A.: Fundamentals of Business Process Management. Springer, Heidelberg (2018). https://doi.org/10.1007/978-3-662-56509-4

11. Elmes, M.B., Strong, D.M., Volkoff, O.: Panoptic empowerment and reflective conformity in enterprise systems-enabled organizations. Inf. Organ. **15**(1), 1–37 (2005)

12. Feldman, M.S., Pentland, B.T.: Reconceptualizing organizational routines as a source of flexibility and change. Adm. Sci. Q. **48**(1), 94 (2003)

13. Furneaux, B., Wade, M.: An exploration of organizational level information systems discontinuance intentions. MIS Q. **35**(3), 573–598 (2011)

14. Goodhue, D.L., Chen, D.Q., Boudreau, M.C., Davis, A., Cochran, J.D.: Addressing business agility challenges with enterprise systems. MIS Q. Executive **8**(2), 73–87 (2009)

15. Hammer, M.: What is business process management? In: vom Brocke, J., Rosemann, M. (eds.) Handbook on Business Process Management 1. International Handbooks on Information Systems, pp. 3–16. Springer, Heidelberg (2015). https://doi.org/10.1007/978-3-642-00416-2

16. Karahanna, E., Xin Xu, S., Xu, Y., Zhang, N.: The needs–affordances–features perspective for the use of social media. MIS Q. **42**(3), 737–756 (2018)

17. Lenz, R.: Evolutionäre Informations systeme. Habilitation thesis (2005)

18. Markus, M.L., Tanis, C.: The enterprise system experience - from adoption to success. In: Zmud, R.W., Price, M.F. (eds.) Framing the domains of IT management, pp. 173–207. Pinnaflex Educational Resources, Cincinnati (2000)

19. Nelson, R.R.W., Sidney, G.: An evolutionary theory of economic change. Harvard University, Cambridge (1982)

20. Peppard, J., Ward, J.: Unlocking sustained business value from it investments. Calif. Manag. Rev. **48**(1), 52–70 (2005)

21. Ryan, R.M., Deci, E.L.: Self-determination theory and the facilitation of intrinsic motivation, social development, and well-being. Am. Psychol. **55**(1), 68 (2000)

22. Sarker, S., Chatterjee, S., Xiao, X., Elbanna, A.: The sociotechnical axis of cohesion for the is discipline: its historical legacy and its continued relevance. MIS Q. **43**(3), 695–719 (2019)

23. Sykes, T.A.: Support structures and their impacts on employee outcomes: a longitudinal field study of an enterprise system implementation. MIS Q. **39**(2) (2015)

24. Sykes, T.A., Venkatesh, V., Johnson, J.L.: Enterprise system implementation and employee job performance: understanding the role of advice networks. MIS Q. **38**(1), 51-A4-A4 (2014)

25. Tenschert, J.: Speech-Act-Based Adaptive Case Management. Ph.D. thesis, Friedrich-Alexander-Universität Erlangen-Nürnberg, Erlangen (2019)

26. van der Aalst, W., et al.: Process mining manifesto. In: Daniel, F., Barkaoui, K., Dustdar, S. (eds.) BPM 2011. LNBIP, vol. 99, pp. 169–194. Springer, Heidelberg (2012). https://doi.org/10.1007/978-3-642-28108-2_19

27. Van de Ven, A.H., Poole, M.S.: Explaining development and change in organizations. Acad. Manag. Rev. **20**(3), 510–540 (1995)

28. Van de Ven, A.H., Sun, K.: Breakdowns in implementing models of organization change. Acad. Manage. Perspect. **25**(3), 58–74 (2011)

29. Volkoff, O., Strong, D.M., Elmes, M.B.: Technological embeddedness and organizational change. Organ. Sci. **18**(5), 832–848 (2007)

30. Wagner, E., Newell, S.: Exploring the importance of participation in the post-implementation period of an ES project: a neglected area. J. Assoc. Inf. Syst. **8**(10), 508–524 (2007)
31. Webster, J., Watson, R.T.: Analyzing the past to prepare for the future: writing a literature review. MIS Q. **26**(2), xiii-xxiii (2002)
32. Wei, H.L., Wang, E.T.G., Ju, P.H.: Understanding misalignment and cascading change of ERP implementation: a stage view of process analysis. Eur. J. Inf. Syst. **14**(4), 324–334 (2005)

SentiProMo: A Sentiment Analysis-Enabled Social Business Process Modeling Tool

Egon Lüftenegger[✉][iD] and Selver Softic[iD]

CAMPUS 02 University of Applied Sciences, IT and Business Informatics,
Graz, Austria
{egon.lueftenegger,selver.softic}@campus02.at

Abstract. SentiProMo is a novel social business process modeling tool enabled by sentiment analysis. We designed and developed SentiProMo for supporting social business processes management to enhance the business process management (BPM) lifecycle. In particular, we socially improve the BPM lifecycle in the process analysis stage by capturing and processing stakeholder's opinions regarding the tasks within a business process. By taking a social information systems perspective, SentiProMo transforms these opinions with sentiment analysis and classifies them into positive and negative feedback. The aim is to support the business analysts for redesigning a business process by considering the sentiment-analyzed opinions for designing the to-be process. We illustrate the current SentiProMo's capabilities with a simple process.

Keywords: Social BPM · BPM · Sentiment analysis · Social IS

1 Introduction

Social BPM (SBPM) focuses on providing new tools for activities within the BPM lifecycle by adding social features to conventional BPM systems [2,4,9]. In [8], the authors define SBPM as the involvement of all relevant stakeholders in the BPM lifecycle by applying social software and its underlying principles. Therefore, social software can support the creation, operation, and adaptation of the abstract business process in design-time. In the SBPM paradigm, social software brings new possibilities to enhance business processes by increasing the exchange of knowledge and information [4].

Data science is a field that is gaining importance due to the availability of large data sets. Currently, only process mining has achieved a connection between data science and business processes for analysis and discovery. Process mining seeks to confront event data captured from observed behavior and hand-made or automatically discovered process models [11]. Nevertheless, such a bridge does not exist in SBPM and social information systems (social IS).

© Springer Nature Switzerland AG 2020
A. Del Río Ortega et al. (Eds.): BPM 2020 Workshops, LNBIP 397, pp. 83–89, 2020.
https://doi.org/10.1007/978-3-030-66498-5_7

In this paper, we establish a bridge between business processes and sentiment analysis for the advancement of social BPM and social IS. With SentiProMo, we seek the confrontation between hand-made as-is business processes and stakeholders' feedback to assist analysts as a starting point for business process redesign.

The structure of this paper is as follows: First, in Sect. 2, we present the relationship between SentiProMo and the BPM lifecycle and social IS. Then, in Sect. 3, we present the conceptual architecture specification of our tool. Next, in Sect. 4, we present our SentiProMo software tool [1]. Finally, we end this paper with discussion and next steps.

2   SentiProMo in Context: BPM Lifecycle, Social BPM and Social Information Systems

Researchers and practitioners adopted data mining in the field of BPM for process mining. However, process mining is focused on processes at run-time for re-creating a business process from systems logs. Opinion mining is a sub-discipline of data mining and computational linguistics for extracting, classifying, understanding, and assessing opinions. An opinion is just a positive or negative sentiment, attitude, emotion, or appraisal about an entity or an aspect of the entity from an opinion holder [7]. In opinion mining, the sub-field of sentiment analysis focuses on the extraction of opinions expressed in a text. However, the current research focus on sentiment analysis lies in e-business and e-commerce like social media and social networks like Twitter and Flickr rather than BPM [1]. Hence, we identify this development as an opportunity for bringing it into SPBM and social IS.

Social IS supports four paradigms: weak ties, social production, egalitarianism, and mutual service provisioning [10]. With our tool, we aim to implement the egalitarianism aspect of social IS by involving different stakeholders. Egalitarianism in social IS intends to encourage a maximum of contributors and get the best solution combining a high number of contributions, consequently empowering the wisdom of the crowds. Social IS highly relies on egalitarianism and accordingly aims to give all participants the same rights to contribute. Furthermore, Social IS is used in business processes to provide additional functionality and to improve the integration of stakeholders. For supporting and egalitarianism, we need to identify the relevant stakeholders. We use the following stakeholders identified in [5] for knowledge acquisition in the BPM lifecycle: Customer (C), process designer (PD), process owner (PO), activity performer (AP), and, superior decision-makers (SDM).

We support egalitarianism by capturing the stakeholders' opinion regarding a task for a given process with the aim to capture their wisdom for improving such process. Therefore, we narrowed the problem by focusing in the highlighted

---

[1] The software tool, training data, and, testing data is available at https://sites.google.com/view/sentipromo.

area within the BPM lifecycle [3] presented in Fig. 1. We define as requirement the following method of use for SentiProMo: First, we import the as-is process into the tool. Second, the stakeholders insert their personal opinions into the process' tasks. Third, SentiProMo calculates the opinions' sentiment for each task and overall score. Finally, the analyst use this information for producing a to-be process.

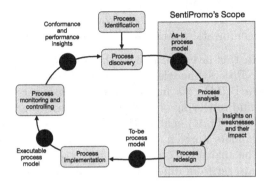

Fig. 1. SentiProMo and the business process Lifecycle. Adapted from [3].

3   SentiProMo's Conceptual Architecture Specification

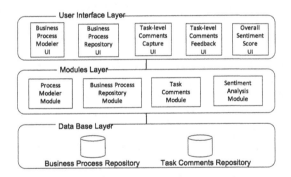

Fig. 2. SentiProMo's architecture as components with an aggregated level of detail.

As shown in Fig. 2, SentiProMo uses a three-layer architecture approach: User interface (UI) layer, modules layer and data base (DB) layer. We describe them as follows:

UI Layer. In this layer, we define the user interface for interacting with the functionalities defined in the Modules Layer. In this layer, we define the following UI components: Business process modeler UI (BPModeler-UI) suitable for redesigning business processes; Business process repository UI (BPRepo-UI) for storing and retrieving business processes into the system; task-level comments capture (TCC-UI) for capturing of comments in the form of text written by stakeholders; task-level comments feedback UI (TCF-UI) that shows the sentiment of the stakeholders' comments in the form of positive or negative; and, overall sentiment score UI (OSC-UI) that provides a number of the calculated sentiment of whole comments within a business process.

Modules Layer. In this layer, located in the middle, we distinguish the following components: Process modeler module (PMM) enables the modeling of business processes in BPMN 2.0; business process repository module (BPRM) enables the storing and retrieving of business processes stored in the DB Layer; task comments module (TCM) enables the storing and retrieving of comments from different stakeholders stored in the DB Layer; and, sentiment analysis module (SAM) enables the analysis of comments with a trained opinion mining algorithm that uses a data set from different social networks.

DB Layer. In this layer, we use the business process repository database to store BPMN 2.0 files imported in the tool and the comments repository DB to store the comments made from different stakeholders.

4   SentiPromo Software Tool

In this paper, we are focusing our progress regarding sentiment analysis of stakeholders' comments and the resulting feedback with an illustrative example. In Fig. 3, we show a cropped picture of the tool for highlighting the BPModeler-UI at the top, TCC-UI in the middle, and the TCF-UI at the bottom. By using the defined SentiProMo's usage method, we go through an airport boarding process as a running example: First, we import the as-is airport boarding business process in BPMN 2.0 within the tool by using the business process modeler UI: The resulting process shown in the Fig. 3. Second, the stakeholders insert their personal opinions into the process' tasks by using the TLC UI. For instance, as shown in Fig. 3, within TLC UI implemented as Windows Form, the process owner insert into the "Check-In at Self Service Counter" task the following comment: "We need more self-service counter because sometimes its gets too crowded in from of them". Then, the sentiment of this comment is calculated and displayed in the table below. In this case a negative value for representing a negative comment from the process owner regarding the task. Hence, an opportunity for the analyst for improving the business process.

In Fig. 4, we present a deeper look to the TCF-UI that shows the resulting sentiments. In the columns: First, we have the task name. Then, we have the stakeholders category. Next, we have the comment made by the stakeholders. Finally, we have the calculated sentiment score: Negatives in red color and

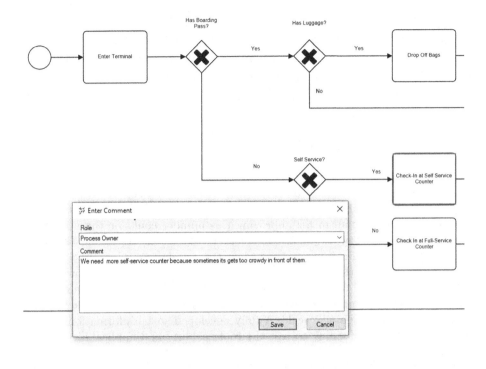

Fig. 3. Inserting and analyzing tasks commentaries with SentiProMo.

Task	Stakeholder Perspective	Comment	Score
Check-In at Self Service Counter	Customer	Check-In at Self Service terminals is very fast, con...	3.851823
Check-In at Self Service Counter	Process Owner	We need more self-service counter because some...	-0.2897689
Check-In at Self Service Counter	Activity Performer	Some of the cusomers need assistance on the pro...	-1.468892

Task	Stakeholder Perspective	Comment	Score
Check In at Full-Service Counter	Activity Performer	Domestic flights are overbooked. We need more counters open wit...	-11.39...
Perform Security Check	Customer	Opening task in ticket system is not intuitive.	-10.31...
Perform Security Check	Activity Performer	We need more security lanes open at rush hours.	-4.488...
Check In at Full-Service Counter	Customer	Waiting times at check in are too long. Do something about this.	-2.0221
Enter Terminal	Activity Performer	We need to distinguish between priority and common customers.	-1.96525
Check-In at Self Service Counter	Activity Performer	Some of the cusomers need assistance on the process. We need s...	-1.468...
Check-In at Self Service Counter	Process Owner	We need more self-service counter because sometimes its gets to...	-0.289...
Perform Security Check	Process Owner	Automation in ticket dispensing is working fine.	1.930...
Drop Off Bags	Customer	Line of people is a lot . Still weiting time is acceptable.	2.037...
Perform Security Check	Activity Performer	Ticket forwarding works but we need simpler user interface for use...	2.951...
Check-In at Self Service Counter	Customer	Check In at Self Service terminals is very fast, convenient and inst...	3.851...
Check In at Full-Service Counter	Superior Decision Maker	Desk in person is friendly.	4.105...

Fig. 4. Results of the sentiment analysis module applied to comment analysis from different tasks and stakeholders. (Color figure online)

positives in green. The TCF-UI uses the sentiment analysis module that we implemented with the Group-Instance Cost Function algorithm. We trained implemented modules with thousands of real-world comments from Amazon, IMBD, Yelp, and Tweeter: 1000 product reviews from amazon, 1000 IMDB movie reviews, 1000 Yelp reviews, and 639580 Tweets. These data sets[2,3] are used for training algorithms in sentiment analysis at the sentence level with the group-instance cost function algorithm [6].

5   Discussion and Next Steps

As a result of our experimental application, our tool identifies the sentiment of the comments as positive or negatives with an accuracy of 80%. We can explain this result by pointing out the way we trained the algorithm with existing datasets. A mechanism for improving this number could be the use of the captured and processed comments for further training an algorithm. However, we need thousands of comments to improve this number.

We implemented the current SentiProMo version as an MS-Windows application. Nevertheless, as the next steps, we will extend the task commenting features through a complementary Web-Application for capturing stakeholders' feedback on a more massive scale: A fully developed social information system for SBPM. We are planning to test the tool usability and applicability in real settings as future work. Nowadays, we are working with an airline company that has been affected by the pandemic caused by the SARS-CoV-2 virus, for studying the feasibility to use the next version of SentiProMo in a real setting scenario for business process redesign.

References

1. Chen, H., Zimbra, D.: Ai and opinion mining. IEEE Intell. Syst. **25**(3), 74–80 (2010)
2. Das, M., Chow, L.: Next generation bpm suites: social and collaborative. Social BPM-Work, Planning and Collaboration Under the Impact of Social Technology, pp. 193–208 (2011)
3. Dumas, M., La Rosa, M., Mendling, J., Hajo, R.: Fundamentals of Business Process Management, 2nd edn. Springer-Verlag, Berlin (2018)
4. Erol, S., et al.: Combining bpm and social software: contradiction or chance? J. Softw. Maintenance Evol. Res. Pract. **22**(6–7), 449–476 (2010)
5. Hrastnik, J., Cardoso, J.S., Kappe, F.: The business process knowledge framework. In: ICEIS (3), pp. 517–520 (2007)
6. Kotzias, D., Denil, M., De Freitas, N., Smyth, P.: From group to individual labels using deep features. In: Proceedings of the 21th ACM SIGKDD International Conference on Knowledge Discovery and Data Mining, pp. 597–606. ACM (2015)
7. Liu, B., Zhang, L.: A survey of opinion mining and sentiment analysis. In: Aggarwal, C., Zhai, C. (eds.) Mining Text Data, pp. 415–463. Springer, Boston (2012)

---

[2] https://archive.ics.uci.edu/ml/datasets/Sentiment+Labelled+Sentences.
[3] http://help.sentiment140.com/for-students/.

8. Pflanzl, N., Vossen, G.: Human-oriented challenges of social bpm: an overview. Enterprise Modelling and Information Systems Architectures (EMISA 2013) (2013)
9. Pucher, M.J.: How to link bpm governance and social collaboration through an adaptive paradigm. Social BPM-Work, Planning and Collaboration Under the Impact of Social Technology, pp. 57–76 (2011)
10. Schmidt, R., Alt, R., Nurcan, S.: Introduction to the minitrack on social information systems. In: Proceedings of the 52nd Hawaii International Conference on System Sciences (2019)
11. Aalst, W.: Data Science in Action. Process Mining, pp. 3–23. Springer, Heidelberg (2016). https://doi.org/10.1007/978-3-662-49851-4_1

4th International Workshop on Business Processes Meet Internet-of-Things (BP-Meet-IoT 2020)

4th International Workshop on Business Processes Meet Internet-of-Things (BP-Meet-IoT 2020)

The arrival of the Internet of Things (IoT) has put into play a huge amount of interconnected and embedded computing devices with sensing and actuating capabilities that are revolutionizing our way of living and doing business. The incorporation of this technology into the Business Process Management field has the potential to create business processes with higher levels of flexibility, efficiency, and responsiveness, providing as a result a better support to the evolving business requirements. In addition, the proper combination of these two fields can foster the development of innovative solutions not only in the business domain where the BPM emerged, but also in many different application areas in which IoT is applied (e.g., smart cities, smart agro, or e-health).

To ensure that Business Process Management and IoT can mutually benefit from each other, still several challenges must be tackled. Particularly, it has to be understood how processes can improve the IoT and how to exploit IoT for BPM. For further discussion on how to combine these two paradigms we refer to the publication "The Internet-of-Things Meets Business Process Management. A Manifesto."

The objective of this workshop was to attract novel research at the intersection of these two areas by bringing together practitioners and researchers from both communities that are interested in making IoT-enhanced business processes a reality.

The 5th edition of this workshop attracted four international submissions, each of them reviewed by three members of the Program Committee. Based on these reviews, the two following papers have been accepted to take part in the workshop program: "Using Physical Factory Simulation Models for Business Process Management" authored by L. Malburg, R. Seiger, R. Bergmann, and B. Weber, and "Modelling Notations for IoT-Aware Business Processes: a Systematic Literature" authored by I. Compagnucci, F. Corradini, F. Fornari, A. Polini, B. Re, and F. Tiezzi.

The first paper presents a physical Fischertechnik factory simulation model as testbed for conducting research in the context of Industry 4.0 based on the combination of BPM and IoT. The factory simulation model is used for three exemplary generic research topics. The second paper reports on the results of a systematic literature review with the aim of developing a map on modelling notations for IoT-aware business processes. The authors identified that modelling notations for IoT-aware business processes are a hot topic from the BP-meet-IoT community. However, a notation suitable to support all the IoT related requirements is still missing.

In addition to these two papers, the program of the workshop included the invited talk "Data Interaction for IoT-Aware Wearable Process Management" given by Stefan Schönig. Finally, the workshop concluded with a discussion about the challenges of integrating the BP and IoT fields. Around 50 attendees were present during the workshop presentations, talk, and discussion.

Similarly, to previous editions of this workshop, the organizers of this event hope that the reader finds this selection of papers and talks interesting and useful to get a better insight at the integration of these two fields from a theoretical and practical point of view.

October 2020   BP-meet-IoT Workshop Organizers

Organization

Workshop Organizers

Agnes Koschmider   Christian-Albrechts-Universität zu Kiel, Germany

Massimo Mecella   Sapienza Università di Roma, Italy

Francesco Leotta   Sapienza Università di Roma, Italy

Estefanía Serral   KU Leuven, Belgium

Victoria Torres   Universitat Politècnica de València, Spain

Program Committee

Andrea Marrella   Sapienza Università di Roma, Italy

Zakaria Maamar   Zayed University, Dubai, UAE

Vicente Pelechano   Universitat Politècnica de València, Spain

Manfred Reichert   University of Ulm, Germany

Andreas Oberweis   Karlsruhe Institute of Technology, Germany

Matthias Weidlich   Humboldt-Universität zu Berlin, Germany

Pnina Soffer   University of Haifa, Israel

Felix Mannhardt   SINTEF Digital, Norway

Udo Kannengießer   Compunity GmbH, Germany

Mathias Weske   Hasso-Plattner-Institut at the University of Potsdam, Germany

Sylvain Cherrier   University Marne-la-Vallée, France

Jianwen Su   University of California at Santa Barbara, USA

Christian Janiesch   University of Wurzburg, Germany

Data Interaction for IoT-Aware Wearable Process Management

Stefan Schönig

Institute for Management Information Systems,
University of Regensburg, Germany

Abstract. Business processes are frequently executed within application systems that involve humans, computer systems as well as objects of the Internet of Things (IoT). IoT as well as Cyber-Physical Systems (CPS), denoting the internetworking of all kinds of physical devices, have become very popular these days. Process execution, monitoring and analytics based on IoT data can enable a more comprehensive view on processes. Embedding intelligence by way of real-time data gathering from devices and sensors and consuming them through BPM technology helps businesses to achieve cost savings and efficiency. This talk presents an approach[1] that implements an IoT-aware business process management system, comprising an integrated architecture for connecting IoT data to a business process management system. Furthermore, a wearable process user interface is presented that allows process participants to be notified in real-time at any location in case new tasks occur. In many situations operators additionally must be able to directly influence data of IoT objects, e.g., to control industrial machinery or to manipulate certain device parameters from arbitrary places. However, a BPM controlled interaction and manipulation of IoT data has been neglected so far. Therefore, the approach is further extended towards a framework for IoT data interaction by means of wearable process management. Here, BPM technology provides a transparent and controlled basis for data manipulation within the IoT. The introduced techniques have evaluated extensively in different use cases in production industry.

Keywords: Internet of Things • Process execution • Data interaction

---

[1] Screencast of the latest prototype of *iot2flow* - https://youtu.be/gt9aJwTto2E.

Using Physical Factory Simulation Models for Business Process Management Research

Lukas Malburg[1]([✉]) [iD], Ronny Seiger[3] [iD], Ralph Bergmann[1,2] [iD],
and Barbara Weber[3] [iD]

[1] Business Information Systems II, University of Trier, 54296 Trier, Germany
{malburgl,bergmann}@uni-trier.de
[2] German Research Center for Artificial Intelligence (DFKI),
Branch University of Trier, Behringstraße 21, 54296 Trier, Germany
ralph.bergmann@dfki.de
[3] Institute of Computer Science, University of St. Gallen, 9000 St. Gallen,
Switzerland
{ronny.seiger,barbara.weber}@unisg.ch

Abstract. The production and manufacturing industries are currently transitioning towards more autonomous and intelligent production lines within the *Fourth Industrial Revolution* (Industry 4.0). *Learning Factories* as small scale physical models of real shop floors are realistic platforms to conduct research in the smart manufacturing area without depending on expensive real world production lines or completely simulated data. In this work, we propose to use learning factories for conducting research in the context of *Business Process Management* (BPM) and *Internet of Things* (IoT) as this combination promises to be mutually beneficial for both research areas. We introduce our physical Fischertechnik factory models simulating a complex production line and three exemplary use cases of combining BPM and IoT, namely the implementation of a BPM abstraction stack on top of a learning factory, the experience-based adaptation and optimization of manufacturing processes, and the stream processing-based conformance checking of IoT-enabled processes.

Keywords: Cyber-physical production systems · Factory simulation models · Business process management · Industry 4.0 · Digital twins

1 Introduction

The production and manufacturing industries are undergoing major changes with machines, products, materials, and humans becoming increasingly interconnected via information technology to form the industrial *Internet of Things* (IoT)–a process known as Industry 4.0 [11]. Among others, these developments promise more efficient and flexible production processes, optimized supply chains, reduced downtimes and maintenance efforts for machines as well as cost

© Springer Nature Switzerland AG 2020
A. Del Río Ortega et al. (Eds.): BPM 2020 Workshops, LNBIP 397, pp. 95–107, 2020.
https://doi.org/10.1007/978-3-030-66498-5_8

reductions [23]. To enable the development of new concepts and prototypes in the context of Industry 4.0, openly available repositories and interfaces for accessing machine data and control functionality in real world settings are required. However, the majority of current machines and production lines are closed systems not allowing to conduct research due to high costs of downtimes and setup processes as well as safety and security concerns [18]. To remedy this situation, related work usually resorts to simulated artificial data from production environments (*Digital Twins* [5]) or to expensive high-end laboratory setups with real production machines (e. g., the *SmartFactoryKL*[1]). While the latter is infeasible for most research institutions in academic contexts, working with artificial data often does not completely reflect the actual physical properties of a production environment, especially w.r.t. runtime behavior and ad-hoc interactions with the physical world [6]. *Learning Factories* are emerging as suitable platforms for future oriented research and education [1] combining the advantages of both approaches in a *Cyber-Physical Production System* (CPPS) [18]. Being small scale physical models with a sufficient number of sensors and actuators for simulating real world industrial IoT environments, learning factories allow for flexibly conducting research on Industry 4.0 concepts and running experiments at much lower costs while maintaining the transferability of results to real smart factories.

In this paper, we present three use cases using a learning factory as physical simulation model of a CPPS to conduct research in the context of BPM. The application of concepts and technologies from the BPM domain in industrial IoT promises various advantages, among others, an easy and flexible integration and programming of hardware, events, services and humans on a process-oriented level as well as the usage of a wide range of process analysis techniques developed by the process mining community. However, apart from mutual benefits also new challenges arise with the combination of BPM and IoT [8]. With this work, we will address a subset of these challenges linked to the combination of process and event-based systems, the adaptation of processes to deal with new situations, and the application of IoT for process analysis–all in the context of smart factories.

2   The Fischertechnik Factory Simulation Model

We use a physical simulation model consisting of components developed by *Fischertechnik* (FT)[2] as testbed for research in BPM and Industry 4.0. Such physical models are referred to as *Learning Factories* [1] and used for education, training, and Industry 4.0 research (e. g., in [9,20,26]) enabling the development and evaluation of research artifacts in a protected environment before moving to real world production scenarios. The custom factory model we use[3] simulates a complete production line at low costs (<15.000 EUR) consisting of two shop floors that are linked for the exchange of workpieces as shown in Fig. 1.

---

[1] https://smartfactory.de/.

[2] https://www.fischertechnik.de/en/simulating/industry-4-0.

[3] https://iot.uni-trier.de.

Each shop floor consists of 5 identical machines: a sorting machine with color detection, a multi-processing station with an oven and a milling machine connected by a workstation transport, a high-bay warehouse, and a vacuum gripper robot. Additionally, the first floor has a punching machine and a human workstation and the second floor a drilling machine including stations for pickup and delivery. Each shop floor is equipped with 13 light barriers, 16 switches, and 3 capacitive sensors for control of the actuators comprising 16 motors, 4 compressors, and 8 valves. The machines are enhanced with sensors mounted on moving parts, motors, and compressors for condition and pressure monitoring for predictive maintenance [9]. Moreover, RFID and NFC readers/writers are integrated into the stations resulting in 28 communication points. This allows each workpiece to be tracked and required manufacturing operations and parameters to be retrieved and adjusted during production. Furthermore, a camera is placed above the two shop floors to track the workpieces. An additional environmental sensor provides climate data (e. g., room temperature, humidity, illuminance, air pressure). The workpieces used for simulating the production are small cylindric blocks (height = ~1.4 cm, diameter = ~2.6 cm) of varying colors each equipped with an NFC tag, which contains information regarding the individual workpiece such as an identifier, the type (i. e., color), the current production state, and timestamped production history. The sensors and actuators of the processing stations are connected to Fischertechnik TXT controllers; 6 Raspberry PIs and 2 Arduinos are used for managing the additional sensors and the camera, which are all linked via Ethernet to a central network switch. The embedded controllers

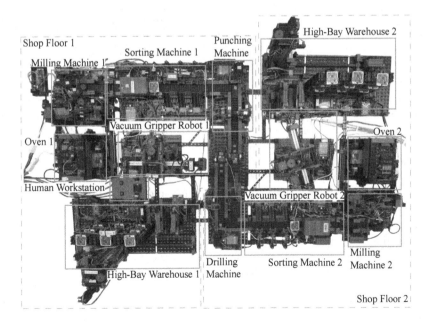

Fig. 1. The physical factory simulation model. (Color figure online)

run C/C++ or Python code to control the sensors and actuators. An integrated MQTT server publishes high-level factory data (e. g., machine states, order and production states, environment and NFC readings). An external Apache Kafka server provides more fine grained access to sensor data.

3   Related Work

Physical factory models are increasingly used to carry out Industry 4.0 research and for education purposes. Primarily, two types of research environments can be distinguished: small scale physical simulation models and full scale physical production lines. The *SmartFactoryKL* and the *LPS* learning factory [20] at the University of Bochum are examples of full scale physical manufacturing environments that are used for research and education. The learning factory *AutFab* of the University of Applied Sciences Darmstadt [26] is another example for using real production machines in this context. Disadvantages of these kinds of production line setups are that basic experimental research is much more difficult and expensive to carry out, since it requires profound knowledge about the machines; the costs for acquisition, networking, maintenance, equipment, and operation are high; and the simulation of errors is difficult and could lead to high costs if the machines are damaged in this process. Therefore, small scale physical factory simulation models are increasingly gaining attention in research and education as alternatives, e. g., the *DBISFactory* at the University of Ulm[4], the Lego factory at the University of Vienna[5], or the DFKI-Smart-Lego-Factory [21], all of which are also partially used for conducting BPM research. In contrast to relying on completely simulated data, these types of simulation models provide much more realistic data and behavior, especially in a highly dynamic CPPS.

Related research addresses the application of BPM in smart environments such as smart logistics [2,16], smart health [7] and emergency management [14], smart homes [25] as well as smart factories [13,15,24,28]. The work by Mangler et al. presents a general discussion of applying BPM technologies in the context of Industry 4.0 [13]. An approach for IoT-aware process execution of industrial maintenance processes is presented by Schönig et al. in [24]. They propose an architecture to integrate IoT data into business processes to determine how and when certain work steps should be carried out by production workers. Baumgraß et al. present an architecture for event-driven process execution and monitoring in smart logistics [2]. This is complemented by work of Meroni et al. showing an artifact-driven approach to monitor business processes through real-world objects [16]. Marrella et al. present the *SmartPM* system in [14], which is able to detect deviations between physical and virtual environments during process execution and resolve them using automated planning techniques. The system is motivated by emergency management scenarios with structured processes and corresponding ad-hoc exceptions. The *PROtEUS* system follows similar goals and approaches for enabling self-healing of processes in the smart

---

[4] https://www.uni-ulm.de/in/iui-dbis/forschung/laufende-projekte/dbisfactory/.
[5] https://wst.cs.univie.ac.at/research/projects/project/292/.

home domain [25]. Wieland et al. discuss an approach for using situation-aware adaptive workflows in the manufacturing domain with a corresponding Workflow Management System (WfMS) called *SitOPT* [28]. In addition to the normal workflow model, situational workflow fragments are constructed that define which actions should be performed in certain real world contexts to adapt the process.

Only a few related approaches discuss the integration of BPM and IoT in the context of Industry 4.0. These works propose new concepts without providing comprehensive evaluations, especially not based on real world experiments (e. g., [13–15,28]). If evaluations are conducted, mostly simulated data from IoT is used, presumably due to high costs and effort w. r. t. hardware, setup, and maintenance for running real world experiments in production environments. At this point, learning factories may help to mitigate some of these issues and facilitate research in the Industry 4.0 domain. Our research is based on real world data obtained from interactions with the physical world via the physical simulation models introduced in Sect. 2. These make it possible to identify and discuss more realistic problems associated with the challenges presented in the BPM-IoT Manifesto [8]. Furthermore, our work puts more focus on industrial processes and their intelligent automation from a BPM point of view.

4   Use Cases for BPM-IoT Research

In this section, we describe three use cases for research in BPM based on physical factory simulation models that we are currently investigating. The presented custom FT factory model thereby serves as testbed for research while the research questions and new concepts are targeted to be more generic and also applicable to other factory configurations and IoT settings.

4.1   Implementation of a Business Process Abstraction Stack

Problem Statement: Many complex IoT environments–including the FT factory described in Sect. 2–consist of a multitude of sensors, actuators, and computing units. These components are usually programmed in low-level languages and controlled by software close to the hardware with code (e. g., G code or C code) running on embedded controllers (e. g., Programming Logic Controllers (PLC) in industry), which limits flexibility and interoperability of components to create more complex processes [18]. As with the FT factory only a few static processes are "hardwired" in C code to demonstrate the functionality. Remote access and a flexible composition of functionality into new (business) processes or to adapt existing ones is not possible. An additional software stack is required on top of the existing IoT components to raise the programming and research to the abstraction level of business processes supported by a WfMS and with that to exploit the potential of integrating BPM with IoT [8].

Research Challenges: Enabling the programming of a smart factory on the level of business processes involves the selection of suitable hardware components (sensors/actuators) to detect events relevant for the process execution, e. g., to monitor the production steps, workpieces, stock, and resources (*C1*: Placing Sensors in a Process-aware Way [8]). It also requires the investigation of the micro-processes at the individual machines and stations as well as their inter-connections to achieve an efficient and flexible production line (*C6*: Managing the Link between Micro-Processes [8]). This also means that static and coarse grained "hardwired" processes have to be relaxed and detailed to achieve a more flexible composition of smaller processes (*C7*: Breaking Down End-to-End Processes [8]). The FT factory is a perfect example of a complex IoT system, which is both event-driven due to large amount of sensors and process-based (*C13*: Bridging the Gap between Event-based and Process-based Systems [8]).

Approach: With the configuration of the FT factory described in Sect. 2, we invested a significant amount of work to create a comprehensive and realistic simulated production line with partial redundancy regarding machines as well as a large number of sensors to monitor production stages, workpieces, machines, and the environment. Figure 2 shows our approach of introducing additional software layers to conduct BPM research using the factory. The individual hardware components are controlled by software written in a low-level programming language running on embedded controllers. We analyzed the data and functionality of these devices, refined, grouped, and abstracted them from an Object-Oriented Programming (OOP) point of view to create software components in an OOP library that can be used for developing more complex programs in higher level OOP languages. An excerpt of a machine class and the general controller interface including relevant attributes and methods can be found in Fig. 2. Furthermore, we added a web service layer on top of this OOP layer to make the functionality (i. e., the machines' methods) and data remotely accessible in a service-oriented (RESTful) architecture and via messaging systems such as Apache Kafka. The developed web services were semantically enriched and integrated into the domain ontology FTOnto [12], which contains formal manufacturing knowledge tailored to the FT factory [10]. These web services are the basis for implementing business processes on top of the factory to model, enact, and monitor processes with the help of a WfMS such as Camunda[6] and thereby enabling us to conduct BPM research in the context of Industry 4.0.

Example: An example for the business process-oriented abstraction is the implementation of the mill functionality of a milling machine in the FT factory. This machine consists of multiple actuators and sensors that have to be activated in low-level function calls to the individual actuators (e. g., for the transport to and from the machine). We abstracted these calls into a sub-routine, which is now available as the *mill* method on the OOP layer and exposed via a service on the Web Service layer. A *Service Task* modeled in BPMN 2.0 and executed by the WfMS can be used to invoke this *mill* method via a web service.

Discussion: One of the most fundamental research activities associated with creating such an abstraction stack is analyzing the functionality of available

---

[6] https://camunda.com/.

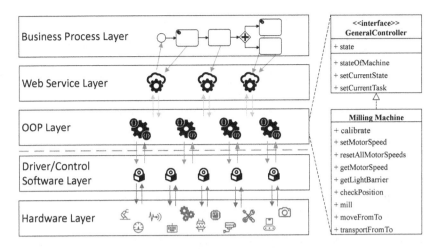

Fig. 2. Hardware and software layers for BPM research on learning factories.

sensors and actuators and grouping and abstracting them at a BPM-oriented level *(C1, C13)*. As there is no generalizable approach for IoT environments here, this process of abstraction should be done based on the actual requirements of the users and domain considering trade-offs between high flexibility of processes with a very fine grained abstraction level and loosing flexibility and expressiveness by being more coarse grained *(C6, C7)*. In the first case, modeling and configuration of the individual processes requires higher efforts, in the latter case, process modeling and implementation is simpler. The degree to which the proposed software stack can be applied in real production environments has to be further investigated as the hardware components of the factory (e. g., the embedded FT controllers as simplified PLCs) are only partially suitable for industrial use and may not fulfill safety and security related requirements.

4.2 Experience-Based Adaptation and Optimization of Processes

Problem Statement: Production processes are often implemented in a rigid manner lacking the flexibility to adapt to dynamic situations such as changed customer demands and individualization or machine breakdowns [11]. When processes cannot be executed as previously planned, constant re-planning and optimization is required [22]. The integration of sensors and other IoT resources in production environments opens new opportunities for process adaptations but it needs to be investigated to what extent this data can be used for production planning and adaptation as well as optimization of processes in real time.

Research Challenges: IoT-aware processes are highly context-sensitive and IoT environments unstructured and dynamic (*C5*: Dealing with Unstructured Environments [8]). New situations can emerge in an ad-hoc manner that lead to unanticipated exceptions during process execution requiring the currently executed process instance to be adapted to the new context (*C12*: Dealing with New

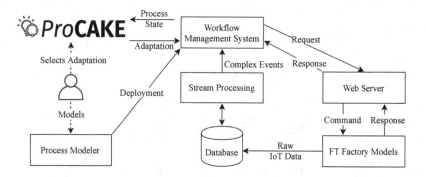

Fig. 3. Architecture of the integration of ProCAKE with a WfMS.

Situations [8]). Furthermore, adaptations can also affect other process instances that may then have to be changed, too. Regarding IoT resources, several autonomy levels exist from full supervision by a central unit to complete independence of central control (cf. *Edge Computing* [7]). Thus, it must be decided what autonomy level the individual resources in the FT factory should have, e. g., if microprocesses of a machine can be performed independently of supervised control by a central unit (*C9*: Specifying the Autonomy Level of Things [8]).

Approach: For adaptation and optimization of manufacturing processes, we investigate the use of Artificial Intelligence (AI) by combining widely used planning techniques [15,22] with other experience-based learning methods. By combining these methods, we expect a reduction of the computational and knowledge acquisition efforts that are often very high for planning-based approaches due to their use of comprehensive real world domain models (full observability assumption). Process-oriented Case-based Reasoning (POCBR) [4,17] is examined as an experience-based learning method that deals with the reuse of procedural experiential knowledge. We use the Camunda WfMS in combination with the *Process-oriented Case-based Knowledge Engine* (ProCAKE) [3], a system that is tailored for developing POCBR applications (cf. Figure 3). Production processes can be modeled in BPMN 2.0 (see Sect. 4.1) using Camunda Modeler. During execution of a process instance, the service tasks invoke the corresponding web services that are semantically enriched to allow for verification of preconditions before the activity is performed. ProCAKE detects state changes of the process and adapts it according to the currently available resources and other executed instances if necessary. Here, users can choose in an interactive way from several adaptation options. The adapted process instances are sent back to Camunda and continued. If several process instances require the same adaptations repeatedly, a migration to an evolved process schema may be necessary. To determine the successful execution or failure of activities, the Complex Event Processing (CEP) platform Siddhi[7] is used, which processes IoT data from Apache Kafka and deduces higher level events for the executed activities (cf. Section 4.3).

---

[7] https://siddhi.io/.

Example: An example scenario could be a malfunction of the drilling machine, which can be determined via sensors and inferred from the FTOnto domain ontology [10]. As only one drilling machine is available in the two shop floors, a process adaptation must be found. ProCAKE determines via the ontology that one of the two milling machines could also be used to drill a hole due to the semantic equivalence of the provided operations, which would also be chosen as an alternative by a production worker. Another option may be that other activities are executed first and the affected process step is relocated to a later stage. In these cases, further process steps must be changed to transport the workpiece to the respective machine. Other examples for this use case include the resource-optimized allocation of orders to the individual machines to achieve a specified goal such as energy, time, or cost optimization.

Discussion: How to capture and formalize the knowledge of production workers and their unstructured environment is one of the fundamental challenges to be addressed *(C5)*. Using past successful process executions to learn possible adaptations of process instances automatically to deal with new or similar already experienced situations *(C12)* seems to be promising [19]. In general, the integration of humans both in the production process itself and in the application of AI-based methods is challenging, e. g., explanations of automatically executed adaptations should be meaningful and transparent for users. Another research question is about the adaptation of simultaneously executed processes within the factory. Production processes may concurrently access the same physical resources and are interrelated among each other. Thus, it is necessary that adaptations of one process instance do not have negative effects on other processes or that users are aware of these impacts. A fundamental technical question in the FT factory is finding the right level of autonomy of IoT devices, which are often resource constrained and therefore not capable of exhaustive computations. The calculation of possible adaptations is rather complex and in addition to the execution of control commands not always feasible on such devices *(C9)*.

4.3   Stream Processing-Based Conformance Checking

Problem Statement: A WfMS may not always exist for monitoring and controlling processes and individual activities. In addition, the implementation of a BPM stack as described in Sect. 4.1 may not be feasible or possible due to high costs and closed hardware/software interfaces of the individual devices. Thus, third party engineers and machine setup workers have no influence on data produced and interfaces provided to program the machines at a BPM-oriented level. However, production processes, machines, and various quality-related aspects still have to be monitored and checked to guarantee the correct execution of processes, also at a BPM level (cf. *Conformance Checking* [27]).

Research Challenges: Here, we investigate ways of integrating IoT sensor data, CEP, and BPM technologies for analysis of process execution (*C3*: Connection of Analytical Processes with IoT; *C13* [8]) even in settings without a WfMS monitoring the execution of processes and activities. The data from IoT

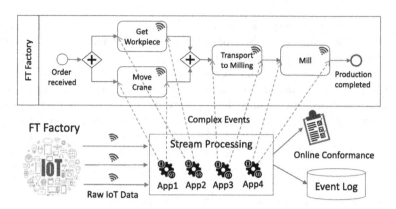

Fig. 4. Architecture of the envisioned conformance checking system.

sensors is used to check the conformance of process executions (*C4*: Integrating IoT into the Correctness Check of Processes [8]) both in offline settings and also at runtime (*C14*: Improving Online Conformance Checking [8]).

Approach: Specific patterns and combinations of IoT data can be used to identify the start, stop, progress, and various other aspects and key performance indicators related to the execution of business processes and activities. Despite the absence of a WfMS for the standard configuration of the FT factory, we are still able to identify the production processes at a BPM level–either based on knowledge or on observations about the individual processes–and to model the processes including activities, gateways, events, messages, resources, etc. The activities can then be enriched by domain experts with IoT related event patterns/queries that are used to detect execution related aspects (e.g., the start and the successful completion of an activity). We use the Camunda WfMS to model the production processes in BPMN 2.0 and associate Siddhi apps with the individual activities. The CEP platform Siddhi is connected to the specified IoT data sources and used for deriving higher level activity-related events based on the event queries within the Siddhi apps. That way, the execution of individual process and activity instances can be monitored and checked for conformance in an online setting but also used for complementing process event logs.

Example: Figure 4 shows the envisioned architecture of the conformance checking system and its correlation with an example process from the FT factory. Each activity is associated with a dedicated event processing app running on the stream processing platform (here: Siddhi), which is connected to the FT factory's sensors via Kafka and/or MQTT. The apps contain one or multiple queries that link sensor data from the factory using logical operators, mathematical functions, aggregations, filters, time-based operations, etc. to identify the start and end of an activity. Based on the raw IoT data (e.g., from light barriers, NFC/RFID readers, production machines, cameras) and the queries defined in the apps, complex events related to the execution of the process are derived, which are then added to the process event log for the executed case or used to provide direct feedback about conformance of the process instance.

Discussion: Among the more fundamental questions of correlating IoT data with process executions is finding the appropriate sensors, algorithms, and domain knowledge to create the correlation patterns, especially regarding activity detection *(C3, C4, C13)*. Available sensors may not provide sufficient information to clearly identify activity executions resulting in uncertainties that need to be dealt with during log creation and to enable conformance checking *(C14)* by looking at the larger process context (e. g., previously executed steps). Also the correlation of a detected activity to a specific process instance is challenging in such a complex IoT environment with multiple instances running in parallel. We expect this correlation of IoT data with process executions to be complementary to existing techniques for (online) conformance checking and used for enriching process event logs with aspects that can be measured through IoT.

5   Conclusion and Future Work

We presented our physical Fischertechnik factory simulation model as testbed for conducting research in the context of Industry 4.0 based on the combination of BPM and IoT. We introduced three exemplary generic research topics that we are currently investigating: 1) the implementation of a business process abstraction stack on top of the factory simulation model; 2) the experience-based adaptation and optimization of manufacturing processes; and 3) the stream processing-based conformance checking of IoT-based processes. By using physical factory models as testbeds for evaluations, research is more realistic–but also more challenging– than using artificial data in this kind of highly dynamic CPPS. The physical factory models enable the validation and demonstration of developed research artifacts in a protected environment. At the same time, this close-to-reality simulation of a real production line facilitates the transfer of developed concepts into practice. In future work, we will further investigate the more fundamental research questions associated with the use cases and implement them. We will also examine how the developed research artifacts can be applied to and evaluated in large scale simulation models and real world shop floors.

References

1. Abele, E., et al.: Learning factories for future oriented research and education in manufacturing. CIRP Ann. **66**(2), 803–826 (2017)
2. Baumgrass, A., et al.: GET controller and UNICORN: event-driven process execution and monitoring in logistics. In: Proceedings of the Demo Session at 13th International Conference on BPM, vol. 1418, pp. 75–79. CEUR-WS.org (2015)
3. Bergmann, R., Grumbach, L., Malburg, L., Zeyen, C.: ProCAKE: a process-oriented case-based reasoning framework. In: Workshop Proceedings of ICCBR 2019, vol. 2567, pp. 156–161. CEUR-WS.org (2019)

4. Bergmann, R., Müller, G.: Similarity-based retrieval and automatic adaptation of semantic workflows. In: Nalepa, G.J., Baumeister, J. (eds.) Synergies Between Knowledge Engineering and Software Engineering. AISC, vol. 626, pp. 31–54. Springer, Cham (2018). https://doi.org/10.1007/978-3-319-64161-4_2

5. Boschert, S., Rosen, R.: Digital twin—the simulation aspect. In: Hehenberger, P., Bradley, D. (eds.) Mechatronic Futures, pp. 59–74. Springer, Cham (2016). https://doi.org/10.1007/978-3-319-32156-1_5

6. Broy, M., Cengarle, M.V., Geisberger, E.: Cyber-physical systems: imminent challenges. In: Calinescu, R., Garlan, D. (eds.) Monterey Workshop 2012. LNCS, vol. 7539, pp. 1–28. Springer, Heidelberg (2012). https://doi.org/10.1007/978-3-642-34059-8_1

7. Chang, C., Srirama, S.N., Buyya, R.: Mobile cloud business process management system for the internet of things: a survey. ACM Comput. Surv. 49(4), 70:1–70:42 (2017). https://doi.org/10.1145/3012000

8. Janiesch, C., et al.: The internet-of-things meets business process management: a manifesto. IEEE Syst. Man Cybern. Mag 6(4), 34–44 (2020). https://doi.org/10.1109/MSMC.2020.3003135

9. Klein, P., Bergmann, R.: Generation of complex data for AI-based predictive maintenance research with a physical factory model. In: 16th International Conference on Informatics in Control Automatic and Robotics, pp. 40–50. SciTePress (2019). https://dblp.uni-trier.de/rec/conf/icinco/KleinB19.html?view=bibtex

10. Klein, P., Malburg, L., Bergmann, R.: FTOnto: a domain ontology for a Fischertechnik simulation production factory by reusing existing ontologies. In: Proceedings of the Conference LWDA, vol. 2454, pp. 253–264. CEUR-WS.org (2019)

11. Lasi, H., et al.: Industry 4.0. BISE 6(4), 239–242 (2014). https://doi.org/10.1007/s12599-014-0334-4

12. Malburg, L., Klein, P., Bergmann, R.: Semantic web services for AI-research with physical factory simulation models in industry 4.0. In: Proceedings of the 1st International Conference on Innovation Intelligent Industrial Production and Logistics (IN4PL), SciTePress (2020). https://www.scitepress.org/PublicationsDetail.aspx?ID=sGMzXE6q768=&t=1

13. Mangler, J., Pauker, F., Rinderle-Ma, S., Ehrendorfer, M.: Centurio work - industry 4.0 integration assessment and evolution. In: Proceedings Industry Forum at BPM 2019, vol. 2428, pp. 106–117. CEUR-WS.org (2019)

14. Marrella, A., Mecella, M., Sardiña, S.: Intelligent process adaptation in the SmartPM system. ACM Trans. Intell. Syst. Technol. 8(2), 25:1–25:43 (2017). https://doi.org/10.1145/2948071

15. Marrella, A., Mecella, M., Sardiña, S.: Supporting adaptiveness of cyber-physical processes through action-based formalisms. AI Commun. 31(1), 47–74 (2018). https://doi.org/10.3233/AIC-170748

16. Meroni, G., Di Ciccio, C., Mendling, J.: An artifact-driven approach to monitor business processes through real-world objects. In: Maximilien, M., Vallecillo, A., Wang, J., Oriol, M. (eds.) ICSOC 2017. LNCS, vol. 10601, pp. 297–313. Springer, Cham (2017). https://doi.org/10.1007/978-3-319-69035-3_21

17. Minor, M., Montani, S., Recio-García, J.A.: Process-oriented case-based reasoning. Inf. Syst. 40, 103–105 (2014)

18. Monostori, L.: Cyber-physical production systems: roots expectations and R&D challenges. Procedia CIRP 17, 9–13 (2014)

19. Müller, G.: Workflow Modeling Assistance by Case-based Reasoning. Springer, Wiesbaden (2018)

20. Prinz, C., et al.: Learning Factory Modules for Smart Factories in Industrie 4.0. Procedia CIRP **54**, 113–118 (2016)
21. Rehse, J.R., Dadashnia, S., Fettke, P.: Business process management for Industry 4.0 - three application cases in the DFKI-smart-Lego-factory. It - Inf. Technol. **60**(3), 133–141 (2018). https://doi.org/10.1515/itit-2018-0006
22. Rossit, D.A., Tohmé, F., Frutos, M.: Production planning and scheduling in cyber-physical production systems: a review. Int. J. Computer Integr. Manuf. **32**(4–5), 385–395 (2019). https://doi.org/10.1080/0951192X.2019.1605199
23. Rüßmann, M., et al.: Industry 4.0: The future of productivity and growth in manufacturing industries. Boston Consult. Group **9**(1), 54–89 (2015)
24. Schönig, S., Ackermann, L., Jablonski, S., Ermer, A.: IoT meets BPM: a bidirectional communication architecture for IoT-aware process execution. Softw. Syst. Model. **19**(6), 1443–1459 (2020). https://doi.org/10.1007/s10270-020-00785-7
25. Seiger, R., Huber, S., Heisig, P., Aßmann, U.: Toward a framework for self-adaptive workflows in cyber-physical systems. Softw. Syst. Model. **18**(2), 1117–1134 (2019)
26. Simons, S., Abé, P., Neser, S.: Learning in the AutFab - the fully automated industrie 4.0 learning factory of the university of applied sciences darmstadt. Procedia Manuf. **9**, 81–88 (2017)
27. van der Aalst, W., et al.: Process Mining Manifesto. In: Daniel, F., Barkaoui, K., Dustdar, S. (eds.) BPM 2011. LNBIP, vol. 99, pp. 169–194. Springer, Heidelberg (2012). https://doi.org/10.1007/978-3-642-28108-2_19
28. Wieland, M., Schwarz, H., Breitenbucher, U., Leymann, F.: Towards situation-aware adaptive workflows: SitOPT - a general purpose situation-aware workflow management system. In: International Conference on Pervasive Computing and Communications Workshops, pp. 32–37. IEEE (2015). https://doi.org/10.1109/PERCOMW.2015.7133989

Modelling Notations for IoT-Aware Business Processes: A Systematic Literature Review

Ivan Compagnucci, Flavio Corradini, Fabrizio Fornari$^{(\boxtimes)}$, Andrea Polini, Barbara Re, and Francesco Tiezzi

Computer Science Division, University of Camerino, Camerino, Italy
{ivan.compagnucci,flavio.corradini,fabrizio.fornari,andrea.polini, barbara.re,francesco.tiezzi}@unicam.it

Abstract. The term *IoT-aware business processes* refers to the interplay of business processes and Internet of Things concepts. Several studies have been carried out on such a topic, so a better awareness of the current state of knowledge can be beneficial. In particular, in a given application domain, this can help the choice of the most suitable modelling approach. This paper reports on the results of a systematic literature review with the aim of developing a map on modelling notations for IoT-aware business processes. It includes 48 research works from the main computer science digital libraries. We first present a description of the systematic literature review protocol we applied, then we report a list of available notations, discussing their main characteristics. A focus has been devoted to modelling tools and application scenarios. Finally, we provide a discussion on the capability of the identified modelling notations to represent requirements of scenarios enriched by IoT adequately.

1 Introduction

Disruptive innovation introduced by the large adoption of Internet of Things in several sectors (e.g., smart agriculture, smart industry, smart environment) received much attention in recent years. This is mainly due to the capability of the Internet of Things to fill the gap between the physical and the digital world, enhancing physical objects with electronic devices. As a side effect, we also observe that complex business processes need to be adapted when electronic devices support foreseen activities. As underlined by the *Business Processes Meet the Internet-of Things* group, and as reported in the manifesto entitled "*The Internet-of-Things Meets Business Process Management: Mutual Benefits and Challenges*" [24], novel research challenges have to be considered [47].

To face such challenges the research community looks for a suitable modelling language to be used as a lingua franca both for documenting and engineering pre-defined models and for representing models discovered by mining logs of IoT systems. Much effort has been devoted to this topic. However, a better understanding of the current state of knowledge can be helpful to select the most suitable approach to a considered purpose.

© Springer Nature Switzerland AG 2020
A. Del Río Ortega et al. (Eds.): BPM 2020 Workshops, LNBIP 397, pp. 108–121, 2020.
https://doi.org/10.1007/978-3-030-66498-5_9

This paper reports on the results of a Systematic Literature Review (SLR)[1] to organize and synthesize the knowledge about the available notations for modelling IoT-aware business processes. We conducted an SLR to present a fair evaluation of the research works on this topic by using a trustworthy, rigorous, and auditable methodology [25].

The paper is organized following the phases of the SLR. Sect. 2 reports on the planning phase. Sect. 3 describes the conducting phase providing some insights on the retrieved research works. Sect. 4 answers to the research questions, while Sect. 5 compares our work with related surveys on the topic. Finally, Sect. 6 provides some conclusive remarks.

2   Planning the Systematic Literature Review

This section illustrates the steps done to plan the SLR, such as the definition of research questions, the search query, and the inclusion/exclusion criteria.

Research Questions. During the planning of our study, we defined one main research question (RQ) and two secondary research questions (SRQ1 and SRQ2). Answers for the secondary research questions are based on scientific works retrieved for answering the main research question.

RQ. Which are the notations used to model IoT-aware business processes? The objective of this question is to gather a list of the available notations that can be used to model business processes integrating IoT concepts. The main characteristics of each notation will be explained with a focus on the actual graphical representation.

SRQ1. Which are the available tools supporting the IoT-aware business process modelling? The objective of this question is to gather a list of available modelling tools that allow designing graphical representations of business processes including IoT concepts. This aims at investigating the possibility of using the notations in practice.

SRQ2. Which are the target application domains for modelling IoT-aware business processes? The objective of this question is to identify which application domains have been already considered in the literature as a target for modelling IoT-aware scenarios. This aims at understanding whether there are domains that will benefit more than others from the integration of IoT concepts in a modelling notation.

Search Query. After defining the research questions, we carefully planned a search query to find suitable works to answer the research questions. The definition of the query has been based on consolidated recommendations as reported in [4]. In particular, the search query was generated by terms related to three different areas: business process, IoT, modelling. It is worth clarifying that among the terms chosen for the query in the IoT area we have also included Cyber Physical System (CPS). This choice is motivated by a NIST research [21], which confirms that the terms IoT and CPS are interchangeably used. The resulting search query is reported in Table 1.

---

[1] All the details of our SLR are available at http://pros.unicam.it/BP-meet-IoT-2020.

Table 1. Search query.

(BPM **OR** "business process management" **OR** "business process") **AND** (IoT **OR** "Internet of Things" **OR** "Cyber Physical Systems" **OR** CPS **OR** Smart **OR** WSN **OR** "Wireless Sensor Network") **AND** (model **OR** behavior **OR** model driven)

We applied such a query over five main digital libraries: IEEE Xplore, ACM Digital Library, Scopus, Science Direct, and Web of Science. We queried the mentioned digital libraries to search the terms of the query into title, abstract, and keywords. Also, since the combination of business process and IoT concepts is relatively recent, we restricted the query to the last nine years, thus considering all the research works published between January 2011 and September 2019 (when the retrieval process took place). We were confident, anyway, that those paper returned by the query would have referenced possibly relevant research works published before 2011, and then included, if relevant, as result of the snowballing step. Our choice about the time frame is also supported by [21], where 2011 is considered as the starting point for the research activity in the IoT area.

Inclusion/Exclusion Criteria. As part of the protocol implementation, a set of inclusion/exclusion criteria has been defined to guarantee the selection of only relevant research works. We defined two criteria for inclusion: *IC.1* - the research work is a primary study; *IC.2* - the research work proposes or uses a graphical notation to model IoT-aware business processes. We also considered two criteria for exclusion: *EC.1* - the research work is not written in the English Language; *EC.2* - the research work does not propose and does not refer to any notation to model IoT-aware business processes.

3   Conducting the Systematic Literature Review

This section describes how we performed the review. In particular, we provide details related to the identification and selection activities of the research works, and we report the categories used to classify them.

Identification and Selection of the Research Works. From the application of the search query over the digital libraries, we identified 1321 research works potentially relevant for the research topic. The selection steps we applied are illustrated in Fig. 1. Through a careful analysis of the title and abstract of these research works, we identified 92 relevant works. After the removal of 53

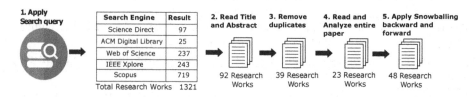

Fig. 1. Research works retrieval process

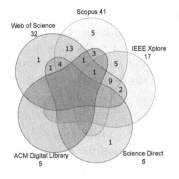

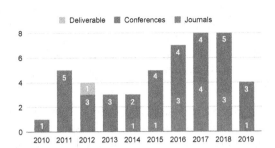

Fig. 2. Research works distribution in digital libraries

Fig. 3. Research works publication venue per year

duplicates, the remaining 39 research works were read and analyzed for a final selection. The final number of research works obtained by applying the search query, and after having analyzed them, is 23.

To improve the accuracy of our review, we also applied *Snowballing* [23]. It consists of analysing all the research works that mention or are mentioned by the research works already selected. The snowballing allows finding research works potentially relevant for our study that have not been found with the query technique previously used. We used both *backward* and *forward* snowballing: we analysed in the first case, all the referenced research works, while in the second case, all the works that refer to those we selected. This step was not limited to the considered time frame, to have a comprehensive picture on related research, possibly including also contributions antecedent 2011. With the first execution of the snowballing, we found 15 new research works; we stopped at the fourth iteration of the snowballing. The entire snowballing procedure added 25 research works to the 23 selected with the search query for a total of 48 relevant research works retrieved.

Figure 2 reports how many research works out of the total 48 can be found querying each digital library. From this analysis, we can say that Scopus and Web of Science appear to be the most suitable digital libraries for such a topic. They include 41 (85%) and 32 (67%) of the total 48 research works, respectively. The results for the other digital libraries are IEEE Xplore 17 (35%), Science Direct 7 (14.5%), ACM Digital Library 5 (10.4%). It is worth mentioning that 2 works (one conference research works and a deliverable) (4.2%) could not be found searching in those libraries.

Figure 3 reports a diagram showing a classification of the different venues from which the research works were selected during the literature review, grouped by year of publication. The majority of the works have been presented at Conferences 34 (70.8%) or published in Journals 13 (27.1%). Among the selected works, we also found a Deliverable (2.1%) worth to be considered, since it was referenced by several of the reviewed scientific works. Moreover, from 2010 to 2019 there has always been at least one contribution related to the analysed topic, with a peak of 8 papers in 2018. This testifies that the research community has been active on the selected research topic and that the interest seems to increase in recent years. As we can see, more than 50% of the works have been published from 2016 and 2019. It is worth noticing a possible decrease in interest in 2019. However, this could be because, at the time of the search, some of the research works published (or under publication) in late 2019 were not indexed, yet.

Data Extraction and Synthesis. To retrieve the necessary data to answer the research questions, we defined the structure of Table 2 to guide the extraction of data out of the retrieved research works. It presents the following columns: *Source* reports the reference to the research work; *Modelling Notation* reports the name of the modelling notation used/proposed or the name of the authors if no name has been assigned to the notation; *Modelling Tool* refers to the tool used to design a process model; *Notation Usage* tells whether the notation is used just to design a business process model related to an IoT application scenario or whether it is also used to guide the implementation and execution of such a process; *Application Domain* refers to the domain of the process represented by the modelling notation. To categorize the application domains we referred to the well-known classification proposed in [2], which groups IoT domains in: healthcare, environmental, smart city, commercial, industrial, and general aspects.

4   Results of the Systematic Literature Review

This section illustrates the data we extracted to answer our research questions.

Answer to RQ. To answer to research question RQ - *Which are the notations used to model IoT-aware business processes?* - we grouped the notation that emerged from the literature in three classes: Not-BPMN, BPMN, and BPMN extension. As it can be seen in Table 2, only 4 (8.3%) out of the 48 identified research works presented a notation to model IoT-aware business processes that are not related to BPMN; 14 (29.2%) of the identified research works make usage of the BPMN 2.0 notation as it is; 30 (62.5%) of the identified research works propose or use an extension of BPMN 2.0. Some of those works make usage of BPMN only for the design of business process models (D), others propose entire architectures and frameworks based on BPMN and related tools so to enable the execution phase (E).

Not-BPMN. All approaches that do not use BPMN focus on the design phase except for [46] in which the authors propose Context-Adaptive Petri-Net (CAPN) a tool-supported formalism to construct Petri nets that are context-adaptive. In [49] the authors use CAPN for acquiring IoT-awareness in an industrial application. In [17] the authors use Cooperative Task Language (CoTaL), a subject-oriented and task-based approach for specifying activities in smart scenarios. In [16] the authors use Subject-Oriented Business Process Management (S-BPM), a modelling paradigm using a standard semantics of natural language with subjects, predicates, and objects to describe business processes of

Table 2. SLR extracted data (D = Design, E = Execution).

	Source	Modelling notation	Modelling tool	Notation usage	Application domain
Not-BPMN	[46]	CAPN	CPN Tools	E	Healthcare
	[49]	CAPN	Not specified	D	Industrial
	[17]	CoTaL	CoTaSE	D	Smart city
	[56]	S-BPM	Metasonic Build	D	Commercial
BPMN 2.0	[6]	BPMN 2.0	Oryx	D	Commercial
	[7]	BPMN 2.0	Oryx	D	Smart city
	[15]	BPMN 2.0	Not specified	D	Commercial
	[12]	BPMN 2.0	Eclipse Modeler	D	Smart city
	[48]	BPMN 2.0	Signavio	D	Commercial
	[29]	BPMN 2.0	Not specified	E	Environmental
	[14]	BPMN 2.0	jBPM	E	Environmental
	[30]	BPMN 2.0	Not specified	E	Smart city
	[18]	BPMN 2.0	bpmn.io	E	Smart city
	[31]	BPMN 2.0	Not specified	E	Commercial
	[45]	BPMN 2.0	Camunda	E	Industrial
	[37]	BPMN 2.0	Camunda	E	Industrial
	[57]	BPMN 2.0	Not specified	E	Smart city
	[41]	BPMN 2.0	Bonita	E	Healthcare
BPMN Extensions	[35]	IAPM	Not specified	D	Commercial
	[32]	IAPM	Activiti	D	Commercial
	[34]	IAPM	Signavio	D	Commercial
	[42]	IAPM	Signavio	D	General aspects
	[28]	IAPM	Not specified	D	Commercial
	[33]	IAPM	Not specified	D	General aspects
	[13]	IAPM	Not specified	D	General aspects
	[10]	IAPM	Not specified	D	Smart city
	[38]	IAPM	MagicDraw	D	Smart city
	[39]	I4PML	MagicDraw	D	Industrial
	[53]	IoT-BPO	Signavio Core C.	D	Commercial
	[52]	IoT-BPO	Signavio Core C.	D	Commercial
	[59]	uBPMN	Not specified	D	Industrial
	[58]	uBPMN	Not specified	D	Commercial
	[60]	uBPMN	Not specified	D	Commercial
	[20]	BPMN4CPS	Not specified	D	Healthcare
	[27]	BPMN-MDM	Not specified	D	Smart city
	[50]	Sperner et al.	Not specified	D	Not defined
	[40]	BPMN-E2	ARIS	D	Industrial
	[19]	Gao et al.	Not specified	D	Industrial
	[11]	Cheng et al.	jBPM	D	Environmental
	[26]	Kozel T.	Not specified	D	Commercial
	[43]	Sang et al.	Activiti	D	Healthcare
	[44]	Schonig S et al.	Not specified	D	Industrial
	[22]	Grefen P. et al.	Not specified	D	Industrial
	[8]	BPMN4WSN	Signavio Core C.	D	Smart city
	[55]	BPMN4WSN	Signavio Core C.	E	Smart city
	[51]	BPMN4WSN	Signavio Core C.	E	Smart city
	[36]	BPMN4WSN	Signavio Core C.	E	Smart city
	[1]	SPU	Not specified	E	Commercial

a smart environment; the Subject Behavior Diagram is generated for each subject involved in the business process to define its interactions with other subjects in the process.

BPMN. This category refers to all the research works that perceive BPMN, as it is, adequate in capturing process specifications of IoT scenarios; this means that they represent IoT aspects by means of the standard BPMN elements. In the following, we first report the research works that make use of BPMN mainly in the design phase of a model [6,7,12,15,48], then we refer to the ones also considering to model execution [14,18,29–31,37,41,45,57].

In [6] the authors use BPMN to capture IoT scenarios and propose to transform such models into code to be executed on a sensor network. In [15] the authors represent, by means of BPMN annotations attached to a pool, conditions that must always be valid within an IoT process. In [12] the authors propose a way to adapt an already available IoT platform to the needs of the BPM approach and analyse the difficulties that arise therein using BPMN in the design phase. In [48] the authors propose a framework to connect the IoT infrastructure to the context-aware BPM ecosystem using IoT-integrated ontologies and IoT-enhanced decision models, which enable the capabilities of IoT to make business processes modelled via BPMN and the involved decision making aware of the dynamic context. In [14] and [29] the authors directly focus on IoT and business processes, proposing first to use BPMN to model IoT scenarios, then to transform the models into an intermediate language, such as Callas Byte Code, to describe WSN systems, and finally to execute such code on the IoT devices. In [30] the authors use BPMN as a starting point to model IoT scenarios managed by BPM Systems, proposing an architecture for decentralised device-to-device business process execution over mobile nodes. In [18] the authors presented a contribution that allows to coordinate the devices used in an IoT application by means of a business process engine with the design of BPMN models for the process logic. In [31] the authors focus on monitoring the compliance of the execution of multi-party business processes. They exploit the IoT paradigm by instructing smart objects. The scenario is modelled in BPMN, then translated into a set of artifact-centric process models, rendered in another notation called Extended-GSM. In [45] the authors introduce an integrated approach for IoT-aware business process execution that exploits IoT for BPM with a particular focus on the management of IoT data. In [37] the authors use BPMN for and controlling the maintenance procedure in a scenario of an industrial cyber-physical production environment. In [57] the authors propose an architecture for a smart home service. Home business processes are analyzed and classified, and then a BPMN-based home business process method is presented. In [41] the authors propose a model based on BPM paradigm, and IT principles to model and enhance the process of a specific scenario.

BPMN Extensions. In this category are reported all the research works presenting extensions to the BPMN notation for better representing IoT aspects in a designed model. We can see that in some cases the same BPMN extension is used (i.e., IAPM [35], uBPMN [59], BPMN4WSN [8]) and in turn extended (see, e.g., [32,51,58]). Looking at the various extensions proposed to the BPMN notation,

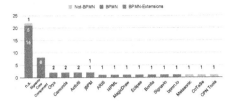

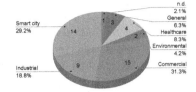

Fig. 4. Tools for modelling IoT-aware business processes.

Fig. 5. Application Domain of IoT-aware business processes.

we observed that none of them presents a completely new element, but they present characterizations of already available ones[2]. Several BPMN elements have been extended, such as Activities, Events, Data Objects, Pool/Lanes, and Gateways. IoT concepts like Sensing and Actuating are often represented using tasks: *Actuator task* is a physical task performed by an IoT device, while *Sensing task* indicates the retrieval of data from the physical world by a sensor. Data produced by IoT devices are represented using *Smart data objects* or *Stream Data objects*; they represent, either an input or an output, the data flow generally produced by IoT devices. Also, the *mobility* aspect, which characterizes some IoT devices, is represented using location markers added to Events or Pools/Lanes. Besides, it is worth noticing that only a few of the research works proposing a BPMN extension also provide approaches to execute the model designed with such a notation.

Answer to SRQ1 and SRQ2. For answering the research question *SRQ1 - Which are the available tools supporting the IoT-aware business process modelling?* - we can refer to column *Modelling Tool* of Table 2. For presentation purpose, we collected those data in the form of diagram reported in Fig. 4. We can see that 22 of the research works (45.8%) do not refer to any modelling tool that can be used to model IoT-aware business processes; 3 research works (6.25%) do not use BPMN and refer to dedicated tools; 9 research works (18.75%) make use of the BPMN notation without extending it but referring to standard modelling editors; 14 research works (29.2%) present a BPMN extension developed by adding new elements to the ones available in existing modelling editors. It results that for what concerns BPMN extensions for modelling IoT aspects, many of the proposed notations are limited to the conceptualization of an extension and do not provide any editor for actually being used. Usually, those that provide an editor refer to custom extensions of already available editors, which in many cases are not made available or are deprecated.

---

[2] For more details see the Graphical BPMN extension sheet available at http://pros. unicam.it/BP-meet-IoT-2020.

Column *Application Domain* in Table 2 permits to answer the research question *SRQ2* - *Which are the target application domains for the modelling of IoT-aware business processes?* For presentation purpose, we collected those data in a form of chart reported in Fig. 5. The application domains referred by the retrieved research works are: commercial (31.3%); smart city (29.2%); industrial (18.8%); healthcare (8.3%); general aspects (6.3%); environmental (4.2%). Only one of the research works (2.1%) does not refer to any application domain.

5   Related Work

While scouting the literature for retrieving scientific contributions concerning the modelling of IoT-aware business processes, we found some research works that, while being different, share the spirit of our work.

IBM provided a report [5] targeting the modelling of IoT-aware business processes. It gives, not in the form of a systematic literature review, an overview of some BPMN extensions used to incorporate IoT aspects in business process models. They describe extensions published in the period 2010–2018, which result in being a subset of those we retrieved from our study. In [9] the authors provide, not in the form of a systematic literature review, an analysis of existing BPMSs for IoT frameworks and identify the limitations, and their drawbacks based on a Mobile Cloud Computing perspective. They also provide a summary of some of the BPMN extensions used to incorporate IoT aspects in business process models. These extensions are a subset of those we retrieved from our study published in the period 2012–2016. The authors of [54] performed a systematic mapping study that investigates the modelling and automatic code generation initiatives for wireless sensor network applications based on the IEEE 802.15.4 standard. Even though this work presents a significant amount of retrieved research works, it differs from ours on the query, which is kept more general, also including terms referring to Model Driven Engineering. Besides, their contribution focuses on aspects linked to the technology used by the various approaches, the kind of supported middleware, the proposed service-oriented architecture. However, while limiting the notation comparison on the support for aspects linked to WSN (e.g. energy consumption), they miss some less specific notations, which we retrieved, and that target IoT in general. Moreover, the modelling notations that they identified have been published in the period 2005–2015, so not including more recent works. In [3], the authors provide a survey, not in the form of a systematic literature review, on domain-specific BPMN extensions. Their work is, therefore, more general and less precise on the topic of modelling IoT-aware business processes, resulting in a subset of the notations we identified. Their research only covers the period 2007–2014.

Table 3. IoT-aware extension language requirements comparison.

Source	RQ1	RQ2	RQ3	RQ4	RQ5	RQ6	RQ7	RQ8	RQ9	RQ10	RQ11
[46, 49]		✓		✓	Partly	✓	✓	Partly	✓	Partly	✓
[17]	Partly	✓	✓			✓	Partly		✓		✓
[56]	Partly	✓	✓	✓	Partly	✓	✓	✓	✓	Partly	✓
[6, 7, 12, 14, 15, 18, 29–31, 37, 41, 45, 48, 57]	Partly	Partly		Partly	Partly	✓		Partly	✓		✓
[10, 13, 28, 32–35, 38, 42]	✓	✓	✓	✓	Partly	✓	✓	Partly	✓	Partly	✓
[39]	✓		✓	✓		✓					✓
[52, 53]	Partly	✓				✓			✓		✓
[58–60]	✓		✓	✓		✓			✓		✓
[20]	✓		✓			✓					✓
[27]	✓		✓	✓		✓					✓
[50]	✓		✓			✓					✓
[40]		✓				✓					✓
[19]	✓			✓		✓					✓
[11]	✓		✓			✓					✓
[26]						✓	✓		✓		✓
[43]	✓		✓			✓			✓	✓	✓
[44]	✓		✓			✓			✓		✓
[22]		✓		✓		✓	✓				✓
[8, 36, 51, 55]	✓	✓	✓			✓					✓
[1]	✓	✓	✓			✓					✓

6   Concluding Remarks

In this paper, we performed an SLR on modelling notations for IoT-aware business processes using the Kitchenham guidelines. We organize the work in sections reflecting the followed SLR protocol: *planning* (Sect. 2), *conducting* (Sect. 3), and *reporting* (Sect. 4). We selected and analyzed 48 research works. The results confirmed the increasing relevance of the considered topic, witnessed by the increase in the number of the published research works on the subject till 2018.

Answering our research questions, we recognized that modelling notations for IoT-aware business processes result to be a hot topic from the BP-meet-IoT community. However, during our study, we observed the lack of a notation suitable to support all the IoT related requirements (reported in [35]) that are typically used by the community as a reference point for comparing IoT-aware business process notations. In this regard, while analyzing the various contributions, we synthesized an overview of the IoT requirements supported by the emerged notations. In Table 3 we present as columns the IoT requirements and as rows the identified notations (note that we group research works referring to the same notation). The comparison shows that almost all extensions integrate the *RQ1. Entity Based Concept* requirement. Other requirements, such as *RQ2. Distributed Execution, RQ3. Interactions, RQ4. Distributed Data, RQ6. Abstraction, RQ9. Flexibility - Event based,* and *RQ11. Real Time* are met by most of the proposed extensions. Other requirements, such as *RQ5. Scalability, RQ7. Availability - Mobility, RQ8. Fault Tolerance* and the *RQ10. Uncertainty of information* are requirements not met by existing extensions. This is because these requirements refer to data management, and most extensions avoid dealing with data. As a final consideration, we can see that none of the identified notations fully meets the IoT requirements.

Acknowledgement. The research has been partially supported by the MIUR projects PRIN "Fluidware" (A Novel Approach for Large-Scale IoT Systems, n. 2017KRC7KT) and "SEDUCE" (Designing Spatially Distributed Cyber-Physi-cal Systems under Uncertainty, n. 2017TWRCNB).

References

1. Appel, S., et al.: Modeling and execution of event stream processing in business processes. Inf. Syst. **46**, 140–156 (2014)
2. Asghari, P., Rahmani, A.M., Javadi, H.H.S.: Internet of things applications: a systematic review. Comput. Netw. **148**, 241–261 (2019)
3. Braun, R., Esswein, W.: Classification of domain-specific BPMN extensions. In: Frank, U., Loucopoulos, P., Pastor, Ó., Petrounias, I. (eds.) PoEM 2014. LNBIP, vol. 197, pp. 42–57. Springer, Heidelberg (2014). https://doi.org/10.1007/978-3-662-45501-2_4
4. Brereton, P., et al.: Lessons from applying the systematic literature review process within the software engineering domain. JSS **80**(4), 571–583 (2007)
5. Brouns, N., Tata, S., Ludwig, H., Asensio, E.S., Grefen, P.: Modeling IoT-aware business processes-a state of the art report. arXiv preprint arXiv:1811.00652 (2018)
6. Caracaş, A., Bernauer, A.: Compiling business process models for sensor networks. In: DCOSS, pp. 75–23. IEEE (2011)
7. Caracaş, A., Kramp, T.: On the expressiveness of BPMN for modeling wireless sensor networks applications. In: Dijkman, R., Hofstetter, J., Koehler, J. (eds.) BPMN 2011. LNBIP, vol. 95, pp. 16–30. Springer, Heidelberg (2011). https://doi.org/10.1007/978-3-642-25160-3_2
8. Casati, F., et al.: Towards business processes orchestrating the physical enterprise with wireless sensor networks. In: Software Engineering, pp. 1357–1360. IEEE (2012)
9. Chang, C., Srirama, S.N., Buyya, R.: Mobile cloud business process management system for the internet of things: a survey. ACM Comp. Surv. **49**(4), 1–42 (2016)
10. Chen, Y.T., Wang, M.S.: A study of extending BPMN to integrate IoT applications. In: Applied System Innovation, pp. 1797–1800. IEEE (2017)
11. Cheng, Y., Zhao, S., Cheng, B., Chen, X., Chen, J.: Modeling and deploying IoT-aware business process applications in sensor networks. Sensors **19**(1), 111 (2019)
12. Cherrier, S., Deshpande, V.: From BPM to IoT. In: Teniente, E., Weidlich, M. (eds.) BPM 2017. LNBIP, vol. 308, pp. 310–318. Springer, Cham (2018). https://doi.org/10.1007/978-3-319-74030-0_23
13. Chiu, H.H., Wang, M.S.: Extending event elements of business process model for internet of things. In: CIT/IUCC/DASC/PICom, pp. 783–788. IEEE (2015)
14. Domingos, D., Martins, F.: Using BPMN to model internet of things behavior within business process. Inf. Syst. and Proj. Manag. **5**(4), 39–51 (2017)
15. Ferreira, P., Martinho, R., Domingos, D.: Process invariants: an approach to model expected exceptions. Procedia Technol. **16**, 824–833 (2014)
16. Fleischmann, A., Schmidt, W., Stary, C., Obermeier, S., Börger, E.: The integrated S-BPM process model. Subject-Oriented Business Process Management, pp. 25–42. Springer, Heidelberg (2012). https://doi.org/10.1007/978-3-642-32392-8_3

17. Forbrig, P., Buchholz, G.: Subject-oriented specification of smart environments. In: Subject-oriented BPM, ACM (2017)
18. Friedow, C., Völker, M., Hewelt, M.: Integrating IoT devices into business processes. In: Matulevičius, R., Dijkman, R. (eds.) CAiSE 2018. LNBIP, vol. 316, pp. 265–277. Springer, Cham (2018). https://doi.org/10.1007/978-3-319-92898-2_22
19. Gao, F., Zaremba, M., Bhiri, S., Derguerch, W.: Extending BPMN 2.0 with sensor and smart device business functions. In: Enabling Technologies, pp. 297–302. IEEE (2011)
20. Graja, I., Kallel, S., Guermouche, N., Kacem, A.: BPMN4CPS: A BPMN extension for modeling cyberphysical systems. In: Enabling Technologies, pp. 152–157. IEEE (2016)
21. Greer, C., Burns, M., Wollman, D., Griffor, E.: Cyber-physical systems and internet of things. NIST Spec. Publ. **1900**, 202 (2019)
22. Grefen, P., et al.: Co-location specification for IoT-aware collaborative business processes. In: Cappiello, C., Ruiz, M. (eds.) CAISE Forum, LNBIP, vol. 350, pp. 120–132. Springer, Cham (2019) https://doi.org/10.1007/978-3-030-21297-1_11
23. Jalali, S., Wohlin, C.: Systematic literature studies: database searches vs. backward snowballing. In: ESEM, pp. 29–38. ACM-IEEE (2012)
24. Janiesch, C., Koschmider, A., et al.: The Internet-of-Things Meets Business Process Management: Mutual Benefits and Challenges. CoRR-Archive, 1709.03628 (2017)
25. Kitchenham, B., Charters, S.: Guidelines for performing Systematic Literature Reviews in Software Engineering. Technical Report, EBSE-2007-01 (2007)
26. Kozel, T.: BPMN mobilisation. In: WSEAS, p. 307–310. ACM (2010)
27. Lee, W.T., Ma, S.P.: Process modeling and analysis of service-oriented architecture-based wireless sensor network applications using multiple-domain matrix. J. Distrib. Sens. Netw. **12**(11), 667–675 (2016)
28. Martinho, R., Domingos, D.: Quality of information and access cost of IoT resources in BPMN processes. Procedia Technol. **16**, 737–744 (2014)
29. Martins, F., Domingos, D.: Modelling IoT behaviour within BPMN business processes. In: Procedia Computer Science, vol. 121, pp. 1014–1022. Elsevier (2017)
30. Mass, J., et al.: WiseWare: a device-to-device-based business process management system for industrial internet of things. In: IoT, pp. 269–275. IEEE (2017)
31. Meroni, G., Baresi, L., Montali, M., Plebani, P.: Multi-party business process compliance monitoring through IoT-enabled artifacts. Inf. Sys. **73**, 61–78 (2018)
32. Meyer, S.: Internet of Things Architecture IoT-A Project Deliverable D2.2-Concepts for Modelling IoT-Aware Processes. Technical Report (2012)
33. Meyer, S., Ruppen, A., Hilty, L.: The things of the internet of things in BPMN. In: Persson, A., Stirna, J. (eds.) CAiSE 2015. LNBIP, vol. 215, pp. 285–297. Springer, Cham (2015). https://doi.org/10.1007/978-3-319-19243-7_27
34. Meyer, S., Ruppen, A., Magerkurth, C.: Internet of things-aware process modeling: integrating IoT devices as business process resources. In: Salinesi, C., Norrie, M.C., Pastor, Ó. (eds.) CAiSE 2013. LNCS, vol. 7908, pp. 84–98. Springer, Heidelberg (2013). https://doi.org/10.1007/978-3-642-38709-8_6
35. Meyer, S., Sperner, K., Magerkurth, C., Pasquier, J.: Towards modeling real-world aware business processes. In: Workshop on Web of Things, ACM (2011)
36. Mottola, L., Picco, G.P., et al.: MakeSense: simplifying the integration of wireless sensor networks into business processes. IEEE TSE **45**(6), 576–596 (2019)
37. Panfilenko, D., et al.: BPMN for knowledge acquisition and anomaly handling in CPS for smart factories. In: ETFA, pp. 1–4. IEEE (2016)

38. Petrasch, R., Hentschke, R.: Towards an IoT-aware process modeling method. an example for a house surveillance system process model. In: MITiCON, pp. 168–172 (2015)
39. Petrasch, R., Hentschke, R.: Process modeling for industry 4.0 applications: towards an industry 4.0 process modeling language and method. In: JCSSE, pp. 1–5. IEEE (2016)
40. Ramos-Merino, M., et al.: BPMN-E2: a BPMN extension for an enhanced workflow description. Softw. Syst. Model. **18**(4), 2399–2419 (2019)
41. Ruiz-Fernández, D., et al.: Empowerment of patients with hypertension through BPM. IoT and remote sensing. Sensors **17**(10), 2273 (2017)
42. Ruppen, A., Meyer, S.: An approach for a mutual integration of the web of things with business processes. In: Barjis, J., Gupta, A., Meshkat, A. (eds.) EOMAS 2013. LNBIP, vol. 153, pp. 42–56. Springer, Heidelberg (2013). https://doi.org/10.1007/978-3-642-41638-5_3
43. Sang, K.S., Zhou, B.: BPMN Security Extensions for Healthcare Process. In: CIT/IUCC/DASC/PICom, pp. 2340–2345. IEEE (2015)
44. Schönig, S., Ackermann, L., Jablonski, S.: Internet of things meets BPM: a conceptual integration framework. In: SIMULTECH, pp. 307–314. SciTePress (2018)
45. Schönig, S., Ackermann, L., Jablonski, S., Ermer, A.: An integrated architecture for IoT-aware business process execution. In: Gulden, J., Reinhartz-Berger, I., Schmidt, R., Guerreiro, S., Guédria, W., Bera, P. (eds.) BPMDS/EMMSAD -2018. LNBIP, vol. 318, pp. 19–34. Springer, Cham (2018). https://doi.org/10.1007/978-3-319-91704-7_2
46. Serral, E., Smedt, J.D., Snoeck, M., Vanthienen, J.: Context-adaptive petri nets: supporting adaptation for the execution context. Expert Syst. Appl. **42**(23), 9307–9317 (2015)
47. Soffer, P., et al.: From event streams to process models and back: challenges and opportunities. Inf. Syst. **81**, 181–200 (2019)
48. Song, R.: Context-aware BPM using IoT-integrated context ontologies and IoT-enhanced decision models, pp. 541–550 (2019)
49. Song, R., et al.: Towards improving context interpretation in the IoT paradigm: A solution to integrate context information in process models. In: ICMSS, pp. 223–228. ACM (2018)
50. Sperner, K., Meyer, S., Magerkurth, C.: Introducing entity-based concepts to business process modeling. In: Dijkman, R., Hofstetter, J., Koehler, J. (eds.) BPMN 2011. LNBIP, vol. 95, pp. 166–171. Springer, Heidelberg (2011). https://doi.org/10.1007/978-3-642-25160-3_17
51. Sungur, C.T., Spiess, P., Oertel, N., Kopp, O.: Extending BPMN for wireless sensor networks. In: IEEE CBI, pp. 109–116. IEEE (2013)
52. Suri, K., Gaaloul, W., Cuccuru, A.: Configurable IoT-aware allocation in business processes. In: Ferreira, J.E., Spanoudakis, G., Ma, Y., Zhang, L.-J. (eds.) SCC 2018. LNCS, vol. 10969, pp. 119–136. Springer, Cham (2018). https://doi.org/10.1007/978-3-319-94376-3_8
53. Suri, K., Gaaloul, W., Cuccuru, A., Gerard, S.: Semantic framework for IoT aware business process development. In: WETICE, pp. 214–219. IEEE (2017)
54. Teixeira, S., et al.: Modeling and automatic code generation for wireless sensor network applications using model-driven or business process approaches: a systematic mapping study. JSS **132**, 50–71 (2017)
55. Tranquillini, S., et al.: Process-based design and integration of wireless sensor network applications. BPM. LNCS **7481**, 134–149 (2012)

56. Venkatakumar, H., Schmidt, W.: Subject-oriented specification of IoT scenarios. In: Subject-Oriented BPM, pp. 1–10. ACM (2019)
57. Xu, H., Xu, Y., Li, Q., Lv, C., Liu, Y.: Business process modeling and design of smart home service system. In: Service Sciences, pp. 12–17. IEEE (2012)
58. Yousfi, A., Bauer, C., Saidi, R., Dey, A.K.: UBPMN: a BPMN extension for modeling ubiquitous business processes. Inf. Soft. Tech. **74**, 55–68 (2016)
59. Yousfi, A., De Freitas, A., Dey, A.K., Saidi, R.: The use of ubiquitous computing for business process improvement. TSC **9**(4), 621–632 (2016)
60. Yousfi, A., Hewelt, M., Bauer, C., Weske, M.: Toward uBPMN-based patterns for modeling ubiquitous business processes. TII **14**(8), 3358–3367 (2018)

Workshop on Artificial Intelligence for Business Process Management (AI4BPM)

Workshop on Artificial Intelligence
for Business Process Management (AI4BPM)

The field of Artificial Intelligence (AI) continues to grow, with novel methodologies and techniques being applied across numerous areas. In the past few years, there has been a strong interest from both industry and academia in applying AI techniques in the area of Business Process Management (BPM). Indeed, the application of AI is impacting additional areas where process management perspectives become relevant, including industrial engineering, IoT, and emergency response. The use of AI in BPM has been discussed as the next disruptive technology that will touch almost all business process activities performed by humans. In some cases, AI will dramatically simplify human interaction with processes, while in other cases it will enable full automation of tasks that have traditionally required manual contributions. We believe that over time, AI may lead to entirely new paradigms for business process management in all of its aspects: modeling, analysis, automation, and monitoring. For example, instead of BPM models centered either on process or on case management, we anticipate models that are based fundamentally on goal achievement. Moreover, these models will fully enable continuous improvement and adaptation based on experiential learning with little to none human intervention after the learning phase has been completed.

The goal of this workshop is to establish a forum for researchers and professionals interested in understanding, envisioning and discussing the challenges and opportunities of moving from current, largely programmatic approaches for BPM, to emerging forms of AI-enabled BPM. The workshop attracted 14 international submissions on a large variety of topics including predictive and causal analysis, process optimization and management, process mining using AI and use of natural language processing in BPM. All submissions were reviewed by at least 3 committee members or their sub-reviewers and eventually 9 papers were accepted. We believe that the accepted papers provide a novel mix of conceptual and technical contributions that are of interest for the AI4BPM community.

Weinzierl et al. introduce XNAP, an LSTM-based approach for predicting next activities, while providing an explanation of the predictions. Agarwal et al. propose a new technique for encoding contextual state information for Predictive Process Monitoring tasks. Qafari and van der Aalst apply structural equation models in the context of root cause analysis to identify not only the feature(s) that caused a deviation in a process execution, but also the effects on the process outcome of an intervention on the feature(s). Zavatteri, Rizzi and Villa investigate the complexity of resource controllability in BPM, by proposing a resource-aware hierarchy of classes of business processes to manage resource assignments. Chakraborti et al. present D3BA, a tool for optimizing business processes using non-deterministic planning. Zebec and Štemberg propose a conceptualization of the AI adoption at the organizational level in the context of BPM, by taking a capability-based view. Park and van der Aalst present a framework for action-oriented process mining that supports the automated execution of actions to improve the performance of a running process. Gupta et al. propose to integrate unstructured data into event logs to discover process models from these enriched event logs. Chambers et al. introduce an algorithm for the automated extraction of processes from unstructured natural-language documents.

These contributions provide the reader with the latest advances in the AI4BPM research area.

October 2020

Chiara Di Francescomarino
Fabrizio Maria Maggi
Andrea Marrella
Arik Senderovich
Emilio Sulis

Organization

Program Committee

Han van der Aa	University of Mannheim, Germany
Annalisa Appice	University of Bari, Italy
Matteo Baldoni	University of Turin, Italy
Boualem Benatallah	University of New South Wales, Australia
Ralph Bergmann	University of Trier, Germany
Andrea Burattin	Technical University of Denmark, Denmark
Federico Chesani	University of Bologna, Italy
Marco Comuzzi	Ulsan National Institute of Science and Technology, Republic of Korea
Francesco Corcoglioniti	Free University of Bozen-Bolzano, Italy
Claudio Di Ciccio	Sapienza University of Rome, Italy
Joerg Evermann	Memorial University of Newfoundland, Canada
Stephan Fahrenkrog-Petersen	Humboldt University of Berlin, Germany
Peter Fettke	German Research Center for Artificial Intelligence, Germany
Francesco Folino	CNR, Italy
Avigdor Gal	Technion, Israel
Chiara Ghidini	Fondazione Bruno Kessler, Italy
Emna Hachicha Belghith	University of Namur, Belgium
Rick Hull	New York University, USA
Krzysztof Kluza	AGH University of Science and Technology, Poland
Francesco Leotta	Sapienza University of Rome, Italy
Massimo Mecella	Sapienza University of Rome, Italy
Paola Mello	University of Bologna, Italy
Roberto Micalizio	University of Turin, Italy
Marco Montali	Free University of Bozen-Bolzano, Italy
Hamid R. Motahari Nezhad	IBM Almaden Research Center, USA
Giulio Petrucci	Google, Switzerland
Artem Polyvyanyy	University of Melbourne, Australia
Luigi Pontieri	CNR, Italy
Manfred Reichert	University of Ulm, Germany
Andrey Rivkin	Free University of Bozen-Bolzano, Italy
Williams Rizzi	Fondazione Bruno Kessler, Italy
Tijs Slaats	University of Copenhagen, Denmark
Biplav Srivastava	IBM T.J. Watson Research Center, USA
Heiner Stuckenschmidt	University of Mannheim, Germany
Niek Tax	Booking.com, the Netherlands

Irene Teinemaa   Booking.com, the Netherlands
Daniele Theseider Dupré   University of Eastern Piedmont, Italy
Hagen Voelzer   IBM Zurich Research Lab, Switzerland
Matthias Weidlich   Humboldt University of Berlin, Germany

Artificial Intelligence Meets Business Process Management: Challenges, Opportunities, and Applications

Artem Polyvyanyy(ORCID)

The University of Melbourne, Parkville, VIC, 3010, Australia
artem.polyvyanyy@unimelb.edu.au

Abstract. In the past couple of centuries, humankind has achieved a significant improvement in the quality of life of the world's population, in large due to important advancements in the automation of wealth-generating activities. Business Process Management (BPM) studies concepts, methods, techniques, and tools that support and improve the way business processes are designed, performed, and analyzed in organizations, including workflow automation and control of business processes and decision-making practices. Artificial Intelligence (AI), in turn, strives to automate natural intelligence exhibited by humans, including the perception of the environment, taken decisions and actions, and learning and problem-solving. In this keynote, the discussion investigates how results in BPM inform and improve solutions to the problems addressed in AI, and vice versa. To exemplify potential synergies of the two fields, the keynote presents two concrete projects in the intersection of BPM and AI that Dr. Polyvyanyy works on together with his colleagues and Ph.D. students, namely applying the ideas from Process Mining, the subarea of BPM, to tackle the problems of Robotic Process Automation [1] and Goal Recognition [2] studied in AI. The screencast of the keynote is publicly available.[1]

References

1. Leno, V., Polyvyanyy, A., Dumas, M., La Rosa, M., Maggi, F.M.: Robotic process mining: Vision and challenges. Business & Information Systems Engineering (Mar 2020). https://doi.org/10.1007/s12599-020-00641-4
2. Polyvyanyy, A., Su, Z., Lipovetzky, N., Sardiña, S.: Goal recognition using off-the-shelf process mining techniques. In: Proceedings of the 19th International Conference on Autonomous Agents and Multiagent Systems, AAMAS'20, Auckland, New Zealand, May 9–13, 2020. pp. 1072–1080. International Foundation for Autonomous Agents and Multiagent Systems (2020). https://dl.acm.org/doi/abs/10.5555/3398761.3398886

---

[1] https://youtu.be/vIFTgrnj468.

XNAP: Making LSTM-Based Next Activity Predictions Explainable by Using LRP

Sven Weinzierl[1]([✉])[iD], Sandra Zilker[1][iD], Jens Brunk[2][iD], Kate Revoredo[3][iD], Martin Matzner[1][iD], and Jörg Becker[2][iD]

[1] Institute of Information Systems, Friedrich-Alexander-Universität Erlangen-Nürnberg, Fürther Straße 248, 90429 Nürnberg, Germany
{sven.weinzierl,sandra.zilker,martin.matzner}@fau.de
[2] European Research Center for Information Systems (ERCIS), University of Münster, Leonardo-Campus 3, 48149 Münster, Germany
{jens.brunk,becker}@ercis.uni-muenster.de
[3] Department of Information Systems and Operations, Vienna University of Economics and Business (WU), Welthandelsplatz 1, 1020 Vienna, Austria
kate.revoredo@wu.ac.at

Abstract. Predictive business process monitoring (PBPM) is a class of techniques designed to predict behaviour, such as next activities, in running traces. PBPM techniques aim to improve process performance by providing predictions to process analysts, supporting them in their decision making. However, the PBPM techniques' limited predictive quality was considered as the essential obstacle for establishing such techniques in practice. With the use of deep neural networks (DNNs), the techniques' predictive quality could be improved for tasks like the next activity prediction. While DNNs achieve a promising predictive quality, they still lack comprehensibility due to their hierarchical approach of learning representations. Nevertheless, process analysts need to comprehend the cause of a prediction to identify intervention mechanisms that might affect the decision making to secure process performance. In this paper, we propose XNAP, the first explainable, DNN-based PBPM technique for the next activity prediction. XNAP integrates a layer-wise relevance propagation method from the field of explainable artificial intelligence to make predictions of a long short-term memory DNN explainable by providing relevance values for activities. We show the benefit of our approach through two real-life event logs.

Keywords: Predictive business process monitoring · Explainable artificial intelligence · Layer-wise relevance propagation · Deep learning · Business process management · Process mining

1 Introduction

Predictive business process monitoring (PBPM) [14] emerged in the field of business process management (BPM) to improve the performance of operational

© Springer Nature Switzerland AG 2020
A. Del Río Ortega et al. (Eds.): BPM 2020 Workshops, LNBIP 397, pp. 129–141, 2020.
https://doi.org/10.1007/978-3-030-66498-5_10

business processes [5,20]. PBPM is a class of techniques designed to predict behaviour, such as next activities, in running traces. PBPM techniques aim to improve process performance by providing predictions to process analysts, supporting them in their decision making. Predictions may reveal inefficiencies, risks and mistakes in traces supporting process analysts on their decisions to mitigate the issues [7]. Typically, PBPM techniques use predictive models, that are extracted from historical event log data. Most of the current techniques apply "traditional" machine-learning (ML) algorithms to learn models, which produce predictions with a higher predictive quality [6]. The PBPM techniques' limited predictive quality was considered as the essential obstacle for establishing such techniques in practice [26]. Therefore, a plethora of works has proposed approaches to further increase predictive quality [22]. By using deep neural networks (DNNs), the techniques' predictive quality was improved for tasks like the next activity prediction [9].

In practice, a process analyst's choice to use a PBPM technique does not only depend on a PBPM technique's predictive quality. Márquez-Chamorro et al. [15] state that the explainability of a PBPM technique's predictions is also an important factor for using such a technique in practice. By providing an explanation of a prediction, the process analyst's confidence in a PBPM technique improves and the process analyst may adopt the PBPM technique [17]. However, DNNs learn multiple representations to find the intricate structure in data, and therefore the cause of a prediction is difficult to retrieve [13]. Due to the lack of explainability, a process analysts cannot identify intervention mechanisms that might affect the decision making to secure the process performance. To address this issue, explainable artificial intelligence (XAI) has developed as a sub-field of artificial intelligence. XAI is a class of ML techniques that aims to enable humans to understand, trust and manage the advanced artificial "decision-supporters" by producing more explainable models, while maintaining a high level of predictive quality [10]. For instance, in a loan application process, the prediction of the next activity "Decline application" (cf. (a) in Fig. 1) produced by a model trained with a DNN can be insufficient for a process analyst to decide if this is a normal behaviour or some intervention is required to avoid an unnecessary refusal of the application. In contrast, the prediction with explanation (cf. (b) in Fig. 1) informs the process analyst that some important details are missing for approving the application because the activity "Add details" has a high relevance on the prediction of the next activity "Decline application".

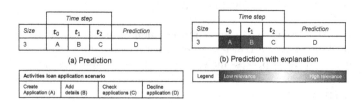

Fig. 1. Next activity prediction example without (a) and with explanation (b).

In this paper, we propose the explainable PBPM technique XNAP. XNAP integrates a layer-wise relevance propagation (LRP) method from XAI to make next activity predictions of a long short-term memory (LSTM) DNN explainable by providing relevance values for each activity in the course of a running trace. To the best of the authors' knowledge, this work proposes the first approach to make LSTM-based next activity predictions explainable.

The paper is structured as follows. Section 2 introduces the required background. In Sect. 3, we present related work on explainable PBPM and reveal the research gap. Section 4 introduces the design of XNAP. In Sect. 5, the benefits of XNAP are demonstrated based on two real-life event logs. In Sect. 6 we provide a summary and point to future research directions.

2 Background

2.1 Preliminaries[1]

Definition 1 (Vector, Matrix, Tensor). *A vector* $\mathbf{x} = (x_1, x_2, \ldots, x_n)$ *is an array of numbers, in which the ith number is identified by $\mathbf{x}_i$. If each number of vector $\mathbf{x}$ lies in $\mathbb{R}$ and the vector $\mathbf{x}$ contains n numbers, then the vector $\mathbf{x}$ lies in $\mathbb{R}^{1 \times n}$, and the vector $\mathbf{x}$'s dimension is $1 \times n$. A matrix $\mathbf{M} = \left(\mathbf{x}^{(1)}, \mathbf{x}^{(2)}, \ldots, \mathbf{x}^{(n)} \right)^T$ is a two-dimensional array of numbers, where $\mathbf{M} \in \mathbb{R}^{d \times n}$. A tensor T is an v-dimensional array of numbers. If $v = 3$, then T is a tensor of the third order with $\mathsf{T} = \left(\mathbf{M}^{(1)}, \mathbf{M}^{(2)}, \ldots, \mathbf{M}^{(d)} \right)$, where $\mathsf{T} \in \mathbb{R}^{d \times b \times u}$.*

Definition 2 (Event, Trace, Event Log). *An event is a tuple (c, a, t) where c is the case id, a is the activity (event type) and t is the timestamp. A trace is a non-empty sequence $\sigma = \langle e_1, \ldots, e_{|\sigma|} \rangle$ of events such that $\forall i, j \in \{1, \ldots, |\sigma|\}$ $e_i.c = e_j.c$. An event log L is a set $\{\sigma_1, \ldots, \sigma_{|L|}\}$ of traces. A trace can also be considered as a sequence of vectors, in which a vector contains all or a part of the information relating to an event, e.g. an event's activity. Formally, $\sigma = \langle \mathbf{x}^{(1)}, \mathbf{x}^{(2)}, \ldots, \mathbf{x}^{(t)} \rangle$, where $\mathbf{x}^{(i)} \in \mathbb{R}^{n \times 1}$ is a vector, and the superscript indicates the time-order upon which the events happened.*

Definition 3 (Prefix and Label). *Given a trace $\sigma = \langle e_1, \ldots, e_k, \ldots, e_{|\sigma|} \rangle$, a prefix of length k, that is a non-empty sequence, is defined as $f_p^{(k)}(\sigma) = \langle e_1, \ldots, e_k \rangle$, with $0 < k < |\sigma_c|$ and a label (i.e. next activity) for a prefix of length k is defined as $f_l^{(k)}(\sigma) = \langle e_{k+1} \rangle$. The above definition also holds for an input trace representing a sequence of vectors. For example, the tuple of all possible prefixes and the tuple of all possible labels for $\sigma = \langle \mathbf{x}^{(1)}, \mathbf{x}^{(2)}, \mathbf{x}^{(3)} \rangle$ are $\langle \langle \mathbf{x}^{(1)} \rangle, \langle \mathbf{x}^{(1)}, \mathbf{x}^{(2)} \rangle \rangle$ and $\langle \mathbf{x}^{(2)}, \mathbf{x}^{(3)} \rangle$.*

---

[1] Note definitions are inspired by the work of Taymouri et al. [23].

2.2   Layer-Wise Relevance Propagation for LSTMs

LRP is a technique to explain predictions of DNNs in terms of input variables [3]. For a given input sequence $\sigma = \langle \mathbf{x}^{(1)}, \mathbf{x}^{(2)}, \mathbf{x}^{(3)} \rangle$, a trained DNN model $\mathcal{M}_c$ and a calculated prediction $\mathbf{o} = \mathcal{M}_c(\sigma)$, LRP reverse-propagates the prediction $\mathbf{o}$ through the DNN model $\mathcal{M}_c$ to assign a relevance value to each input variable of σ [1]. A relevance value indicates to which extent an input variable contributes to the prediction. Note $\mathcal{M}_c$ is a DNN model, and c is a target class for which we want to perform LRP. In this paper, $\mathcal{M}_c$ is an LSTM model, i.e. a DNN model with an LSTM [11] layer as a hidden layer. The architecture of the "vanilla" LSTM (layer) is common in the PBPM literature for the task of predicting next activities [25]. For instance, an explanation of it can be found in the work of Evermann et al. [9].

To calculate the relevance values of the input variables, LRP performs two computational steps. First, it sets the relevance of an output layer neuron corresponding to the target class of interest c to the value $\mathbf{o} = \mathcal{M}_c(\sigma)$. It ignores the other output layer neurons and equivalently sets their relevance to zero. Second, it computes a relevance value for each intermediate lower-layer neuron depending on the neural connection type. A DNN's layer can be described by one or more neural connections. In turns, the LRP procedure can be described layer-by-layer for different types of layers included in a DNN. Depending on the type of a neural connection, LRP defines heuristic propagation rules for attributing the relevance to lower-layer neurons given the relevance values of the upper-layer neurons [3].

In case of recurrent neural network layers, such as LSTM [11] layers, there are two types of neural connections: *many-to-one weighted linear connections*, and *two-to-one multiplicative interactions* [2]. Therefore, we restrict the definition of the LRP procedure to these types of connections. For weighted connections, let $\mathbf{z}_j$ be an upper-layer neuron. Its value in the forward pass is computed as $\mathbf{z}_j = \sum_i \mathbf{z}_i \cdot \mathbf{w}_{ij} + b_j$, while $\mathbf{z}_i$ are the lower-layer neurons, and $\mathbf{w}_{ij}$ as well as $\mathbf{b}_j$ are the connection weights and biases. Given each relevance $\mathbf{R}_j$ of the upper-layer neurons $\mathbf{z}_j$, LRP computes the relevance $\mathbf{R}_i$ of the lower-layer neurons $\mathbf{z}_i$. Initially, $\mathbf{R}_j = \mathcal{M}_c(\sigma)$ is set. The relevance distribution onto lower-layer neurons comprises two steps. First, by computing relevance messages $\mathbf{R}_{i \leftarrow j}$ going from upper-layer neurons $\mathbf{z}_j$ to lower-layer neurons $\mathbf{z}_i$. The messages $\mathbf{R}_{i \leftarrow j}$ are computed as a fraction of the relevance $\mathbf{R}_j$ accordingly to the following rule:

$$\mathbf{R}_{i \leftarrow j} = \frac{\mathbf{z}_i \cdot \mathbf{w}_{ij} + \frac{\epsilon \cdot sign(\mathbf{z}_j) + \delta \cdot b_j}{N}}{\mathbf{z}_j + \epsilon \cdot sign(\mathbf{z}_j)} \cdot \mathbf{R}_j. \tag{1}$$

N is the total number of lower-layer neurons connected to $\mathbf{z}_j$, ϵ is a stabiliser (small positive number, e.g. 0.001) and $sign(\mathbf{z}_j) = (1_{\mathbf{z}_j \geq 0} - 1_{\mathbf{z}_j < 0})$ is the sign of $\mathbf{z}_j$. Second, by summing up incoming messages for each lower-layer neuron $\mathbf{z}_i$ to obtain relevance $\mathbf{R}_i$. $\mathbf{R}_i$ is computed as $\sum_j \mathbf{R}_{i \leftarrow j}$. If the multiplicative factor δ is set to 1.0, the total relevance of all neurons in the same layer is conserved. If it is set to 0.0, the total relevance is absorbed by the biases.

For two-to-one multiplicative interactions between lower-layer neurons, let z_j be an upper-layer neuron. Its value in the forward pass is computed as the multiplication of two lower-layer neuron values z_g and z_s, i.e. $z_j = z_g \cdot z_s$. In such multiplicative interactions, there is always one of two lower-layer neurons that represents a gate with a value range $[0, 1]$ as the output of a sigmoid activation function. This neuron is called gate z_g, whereas the remaining one is the source z_s. Given such a configuration, and denoting by R_j the relevance of the upper-layer neuron z_j, the relevance can be redistributed onto lower-layer neurons by: $R_g = 0$ and $R_s = R_j$. With this reallocation rule, the gate neuron already decides in the forward pass how much of the information contained in the source neuron should be retained to make the overall classification decision.

3   Related Work on Explainable PBPM

In the past, PBPM research has mainly focus on improving the predictive quality of PBPM approaches to foster the transfer of these approaches into practice. In contrast, the PBPM approaches' explainability was scarcely discussed although it can be equally important since missing explainability might limit the PBPM approaches' applicability [15]. In the context of ML, XAI has already been considered in different approaches [4]. However, PBPM research has just recently started to focus on XAI. Researchers differentiate between two types of explainability. First, ante-hoc explainability provides transparency on different levels of the model itself; thus they are referred to as transparent models. This can be the complete model, single components or learning algorithms. Second, post-hoc explainability can be provided in the form of visualisations after the model was trained since they are extracted from the trained model [8].

Concerning ante-hoc explainability in PBPM, multiple approaches have been proposed for different prediction tasks. For example, Maggi et al. [14] propose a decision-tree-based, Breuker et al. [5] a probabilistic-based, Rehse et al. [18] a rule-based and Senderovic et al. [21] a regression-based approach.

In terms of post-hoc explainability, research has focused on model-agnostic approaches. These are techniques that can be added to any model in order to extract information from the prediction procedure [4]. In contrast, model-specific explanations are methods designed for certain models since they examine the internal model structures and parameters [8]. Figure 2 depicts an overview of approaches for post-hoc explainability in PBPM. Verenich et al. [24] propose a two-step decomposition-based approach. Their goal is to predict the remaining time. First, they predict on an activity-level the remaining time. Next, these predictions are aggregated on a process-instance-level using flow analysis techniques. Sindhgatta et al. [22] provide both global and local explainability for XGBoost, this is for outcome and remaining time predictions. Global explanations are on a prediction-model-level. Therefore, the authors implemented permutation feature importance. On the contrary, Local explanations are on a trace-level, i.e. they describe the predictions regarding a trace. For this, the authors apply LIME [19]. This method perturbs the input, observes how predictions change and based on that, tries to provide explainability. Mehdiyev and Fettke [16] present an approach to make DNN-based process outcome predictions explainable. Thereby, they generate partial dependence plots (PDP) to provide causal explanations. Rehse et al. [18] create global and local explanations for outcome predictions. Based on a DL architecture with LSTM layers, they apply a connection weight approach to calculate the importance of features and therefore provide global explainability. For local explanations, the authors determine the contribution to the prediction outcome via learned rules and individual features.

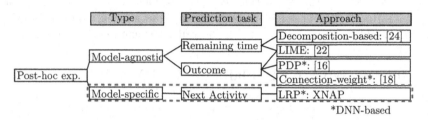

Fig. 2. Related work on XAI in PBPM.

In comparison to those approaches, LRP is not part of the training phase and presumes a learned model. LRP peaks into the model to calculate relevance backwards from the prediction to the input. Thus, through the use of LRP, we contribute by providing the first model-specific post-hoc explanations of LSTM-based next activity predictions.

4   XNAP: Explainable Next Activity Prediction

XNAP is composed of an offline and an online component. In the offline component, a predictive model is learned from a historical event log by applying a Bi-LSTM DNN. In the online component, the learned model is used for producing next activity predictions in running traces. Given the next activity predictions and the learned predictive model, LRP determines relevance values for each activity of running traces.

4.1   Offline Component: Learning a Bi-LSTM Model

The offline component receives as input an event log, pre-processes it, and outputs a Bi-LSTM model learned based on the pre-processed event log.

Pre-processing: The offline component's pre-processing step transforms an event log L into a data tensor X and a label matrix $\mathbf{Y}$ (i.e. next activities). The procedure comprises four steps. First, we transform an event log L into a matrix $\mathbf{S} \in \mathbb{R}^{|L| \times u}$. $|L|$ is the event log's size, whereas u is the number of an event tuple's elements. Note that we add an activity to the end of each sequence to predict their end. Second, we onehot-encode the string values of the activity attribute in $\mathbf{S}$ because a Bi-LSTM requires a numerical input for calculating forward and backward propagations. After this step, we get the matrix $\mathbf{S} \in \mathbb{R}^{|L| \times h}$, where h is the number of different activity values in the event log L. Third, we create prefixes and next activity labels. Thereby, a tuple of prefixes R is created from $\mathbf{S} \in \mathbb{R}^{|L| \times h}$ by applying the function f_p, whereas a tuple of labels K is created from $\mathbf{S} \in \mathbb{R}^{|L| \times h}$ through the function f_l. Lastly, we construct a third-order data tensor $\mathsf{X} \in \mathbb{R}^{|R| \times m \times h}$ based on the prefix tuple R as well as a label matrix $\mathbf{Y} \in \mathbb{R}^{|K| \times h}$ based on the label tuple K, where m is the longest trace in the event log L, i.e. $|max_\sigma(L)|$. The remaining space for a sequence $\sigma_c \in \mathsf{X}$ is padded with zeros, if $|\sigma_c| < |max_\sigma(L)|$.

Model Learning: XNAP learns a Bi-LSTM model $\mathcal{M}$ that maps the prefixes onto the next activity labels based on the data tensor X and label matrix $\mathbf{Y}$ from the previous step. We use the Bi-LSTM architecture, an extension of "vanilla" LSTMs since Bi-LSTMs are forward and backward LSTMs that can exploit control-flow information from two directions of sequences. XNAP's Bi-LSTM architecture comprises an input layer, a hidden layer, and an output layer. The *input layer* receives the data tensor and transfers it to the hidden layer. The *hidden layer* is a Bi-LSTM layer with a dimensionality of 100, i.e. the Bi-LSTM's cell internal elements have a size of 100. We assign the activation function *tanh* to the Bi-LSTM's cell output. To prevent overfitting, we perform a random dropout of 20% of input units along with their connections. The model connects the Bi-LSTM's cell output to the neurons of a dense *output layer*. Its number of neurons corresponds to the number of the next activity classes. For learning weights and biases of the Bi-LSTM architecture, we apply the *Nadam* optimisation algorithm with a *categorical cross-entropy loss* and default values for parameters. Note that the loss is calculated based on the Bi-LSTM's prediction and the next activity ground truth label stored in the label matrix $\mathbf{Y}$. Additionally, we set the batch size to 128. Following Keskar et al. [12], gradients are updated after each 128th trace of the data tensor X. Larger batch sizes tend to sharp minima and impair generalisation. The number of epochs (learning iterations) is set to 100, to ensure convergence of the loss function.

4.2   Online Component: Producing Predictions with Explanations

The online component receives as input a running trace, performs a pre-processing, creates a next activity prediction and concludes with the creation of a relevance value for each activity of the running trace regarding the prediction. The prediction is obtained by using the learned Bi-LSTM model from the offline component. Given the prediction, LRP determines the activity relevances by backwards passing the learned Bi-LSTM model.

Pre-processing: The online component's pre-processing step transforms a running trace σ_r into a data tensor and a label matrix, as already described in the offline component's pre-processing step. Note that we terminate the online phase if $|\sigma_r|$ is ≤ 1 since, for such traces, there is insufficient data to base prediction and relevance creation upon. Further, we assume that we have already observed all possible activities as well as the longest trace in the offline component. Thus, tensor X lies in $\mathbb{R}^{1 \times m \times h}$. In the offline component, next activity labels are not known and based on the data tensor X_r for a running trace σ_r a next activity is predicted.

Prediction Creation: Given the data tensor X_r from the previous step, the trained Bi-LSTM model $\mathcal{M}$ from the offline component returns a probability distribution $\mathbf{p}^{1 \times h}$, containing the probability values of all activities. We retrieve the prediction p from $\mathbf{p}$ through $argmax(\mathbf{p}[j])$, with $1 \leq j \leq h$.

Relevance Creation: Lastly, we provide explainability of the prediction p by applying LRP. For a next activity prediction p, LRP determines a relevance value for each activity in the course of a running trace σ_r towards it by decomposing the prediction, from the output layer to the input layer, backwards through the model. Note the prediction p was created in the previous step based on all activities of the running trace σ_r. In doing that, we apply the LRP approach proposed by Arras et al. [2] that is designed for LSTMs. As mentioned in Sect. 2, a layer of a DNN can be described by one or more neural connections. Depending on the layer's type, LRP defines rules for attributing the relevance to lower-layer neurons given the relevance values of the upper-layer neurons. After backwards passing the model by considering conversation rules of different layers, LRP returns a relevance value for each onehot-encoded input activity of the data tensor X. Finally, to visualise the relevance values, e.g. by a heatmap, positive relevance values are rescaled to the range $[0.5, 1.0]$ and negative ones to the range $[0.0, 0.5]$.

5   Results

5.1   Event Logs

We demonstrate the benefit of XNAP with two real-life event logs that are detailed in Table 1.

Table 1. Overview of used real-life event logs.

Event log	# Instances	# Instance variants	# Events	# Activities	# Events per instance*	# Activities per instance*
helpdesk	4,580	226	21,348	14	[2;15;5;4]	[2;9;4;4]
bpi2019	24,938	3,299	104,172	31	[1;167;4;4]	[1;11;4;4]

*[min; max; mean; median]

First, we use the *helpdesk* event log[2]. It contains data of a ticketing management process form a software company. Second, we make use of the *bpi2019* event log[3]. It was provided by a coatings and paint company and depicts an order handling process. Here, we only consider sequences of max. 250 events and extract a 10%-sample of the remaining sequences to lower computation effort.

---

[2] https://data.mendeley.com/datasets/39bp3vv62t/1.
[3] https://data.4tu.nl/repository/uuid:a7ce5c55-03a7-4583-b855-98b86e1a2b07.

5.2   Experimental Setup

LRP is a model-specific method that requires a trained model for calculating activity relevances to explain predictions. Therefore, we report the predictive quality of the trained models, and then demonstrate the activity relevances.

Predictive Quality: To improve model generalisation, we randomly shuffle the traces of each event log. For that, we perform a process-instance-based sampling to consider process-instance-affiliation of event log entries. This is important since LSTMs map sequences depending on the temporal order of their elements. Afterwards, for each event log, we perform a ten-fold cross-validation. Thereby, in every iteration, an event log's traces are split alternately into a 90%-training and 10%-test set. Additionally, we use 10% of the training set as a validation set. While we train the models with the remaining training set, we use the validation set to avoid overfitting by applying early stopping after ten epochs. Consequently, the model with the lowest validation loss is selected for testing. To measure predictive quality, we calculate the average weighted *Accuracy* (overall correctness of a model) and average weighted *F1-Score* (harmonic mean of *Precision* and *Recall*).

Explainability: To demonstrate the explainability of XNAP's LRP, we pick the Bi-LSTM model with the highest *F1-Score* value and randomly select two traces from all traces of each event log. One of these traces has a size of five; the other one has a size of eight. We use traces of different sizes to investigate our approach's robustness.

Technical Details: We conducted all experiments on a workstation with 12 CPU cores, 128 GB RAM and a single GPU NVIDIA Quadro RXT6000. We implemented the experiments in *Python* 3.6.8 with the DL library *Keras*[4] 2.2.4 and the *TensorFlow*[5] 1.14.1 backend. The source code can be found on Github[6].

5.3   Predictive Quality

The Bi-LSTM model of XNAP predicts the next most likely activities for the *helpdesk* event log with an average (Avg) *Accuracy* and *F1-Score* of 84% and 79.8% (cf. Table 2). For the *bpi2019* event log, the model achieves an Avg *Accuracy* and *F1-Score* of 75.5% and 72.7%. For each event log, the standard deviation (SD) of the *Accuracy* and *F1-Score* values is between 1.0% and 1.5%.

---

[4] https://keras.io.
[5] https://www.tensorflow.org.
[6] https://github.com/fau-is/xnap.

Table 2. Predictive quality of XNAP's Bi-LSTM model for the ten folds.

Event log	Metric	1	2	3	4	5	6	7	8	9	10	Avg	Sd
helpdesk	Accuracy	0.846	0.851	0.824	0.824	0.852	0.823	0.837	0.850	0.853	0.839	0.840	0.012
	F1-Score	0.807	0.811	0.779	0.780	0.813	0.777	0.794	0.810	**0.814**	0.798	0.798	0.015
bpi2019	Accuracy	0.758	0.759	0.748	0.762	0.754	0.758	0.753	0.734	0.748	0.772	0.755	0.010
	F1-Score	0.732	0.737	0.712	0.741	0.722	0.730	0.723	0.710	0.720	**0.742**	0.727	0.011

5.4   Explainability

We show the activity relevance values of XNAP's LRP on the example of two traces per event log (cf. Fig. 3). The time steps (columns) represent the activities that are used as input. For each trace, we predict the next activity for different prefix lengths (rows). We start with a minimum of three and make one next activity prediction until the maximum length of the trace is reached (five and eight in our examples). The data-given ground truth is listed in the last column. We use a heatmap to indicate the relevance of the input activities to the prediction of the same row. For example, in the traces (a) and (b), the activity "Resolve ticket" (C) has a high relevance on predicting the next activity "End (E)". With that, a process analyst knows that the trace will end since the ticket was resolved. Another example is in the traces (c) and (d), where the activity "Record Invoice Receipt (C)" has a high relevance on predicting the next activity "Clear Invoice (D)". Thus, a process analyst knows that the invoice can be cleared in the next step because the invoice receipt was recorded.

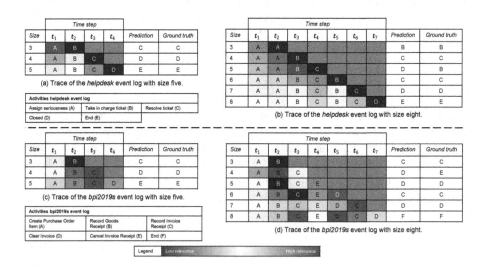

Fig. 3. Activity relevances of XNAP's LRP.

6   Conclusion

Given the fact that DNNs achieve a promising predictive quality at the expense of explainability and based on our identified research gap, we argue that there is a crucial need for making LSTM-based next activity predictions explainable. We introduced XNAP, an explainable PBPM technique, that integrates an LRP method from the field of XAI to make a BI-LSTM's next activity prediction explainable by providing a relevance value for each activity in the course of a running trace. We demonstrated the benefits of XNAP with two event logs. By analysing the results, we made three main observations. First, LRP is a model-specific XAI method; thus, the quality of the relevance scores depend strongly on the model's predictive quality. Second, XNAP performs better for traces with a smaller size and a higher number of different activities. Third, XNAP computes the relevance values of activities in very few seconds. In contrast, model-agnostic approaches, e.g. PDP [16], need more computation time.

In future work, we plan to validate our observations with further event logs. Additionally, we will conduct an empirical study to evaluate the usefulness of XNAP. We also plan on hosting a workshop with process analysts to better understand how a prediction's explainability contributes to the adoption of a PBPM system. Moreover, we plan to adapt the propagation rules of XNAP's LRP also to determine relevance values of context attributes. Another avenue for future research is to compare the explanation capability of a model-specific method like LRP to a model-agnostic method like LIME for, e.g. the DNN-based next activity prediction. Finally, XNAP's explanations, which are rather simple, might not capture an LSTM model's complexity. Therefore, future research should investigate new types of explanations that better represent this high complexity.

Acknowledgments. This project is funded by the German Federal Ministry of Education and Research (BMBF) within the framework programme *Software Campus* under the number 01IS17045. The fourth author received a grand from Österreichische Akademie der Wissenschaften.

References

1. Arras, L., et al.: Explaining and interpreting LSTMs. In: Samek, W., Montavon, G., Vedaldi, A., Hansen, L.K., Müller, K.-R. (eds.) Explainable AI: Interpreting, Explaining and Visualizing Deep Learning. LNCS (LNAI), vol. 11700, pp. 211–238. Springer, Cham (2019). https://doi.org/10.1007/978-3-030-28954-6_11
2. Arras, L., Montavon, G., Müller, K.R., Samek, W.: Explaining recurrent neural network predictions in sentiment analysis. In: Proceedings of the 8th Workshop on Computational Approaches to Subjectivity, Sentiment and Social Media Analysis, pp. 159–168. ACL (2017)
3. Bach, S., Binder, A., Montavon, G., Klauschen, F., Müller, K.R., Samek, W.: On pixel-wise explanations for non-linear classifier decisions by layer-wise relevance propagation. PLoS ONE **10**(7), e0130140 (2015)

4. Barredo Arrieta, A., et al.: Explainable artificial intelligence (XAI): concepts, taxonomies, opportunities and challenges toward responsible AI. Inf. Fusion **58**, 82–115 (2020)
5. Breuker, D., Matzner, M., Delfmann, P., Becker, J.: Comprehensible predictive models for business processes. MIS Q. **40**(4), 1009–1034 (2016)
6. Di Francescomarino, C., Ghidini, C., Maggi, F.M., Milani, F.: Predictive process monitoring methods: which one suits me best? In: Weske, M., Montali, M., Weber, I., vom Brocke, J. (eds.) BPM 2018. LNCS, vol. 11080, pp. 462–479. Springer, Cham (2018). https://doi.org/10.1007/978-3-319-98648-7_27
7. Di Francescomarino, C., Ghidini, C., Maggi, F.M., Petrucci, G., Yeshchenko, A.: An eye into the future: leveraging a-priori knowledge in predictive business process monitoring. In: Carmona, J., Engels, G., Kumar, A. (eds.) BPM 2017. LNCS, vol. 10445, pp. 252–268. Springer, Cham (2017). https://doi.org/10.1007/978-3-319-65000-5_15
8. Du, M., Liu, N., Hu, X.: Techniques for interpretable machine learning. Commun. ACM **63**(1), 68–77 (2019)
9. Evermann, J., Rehse, J.R., Fettke, P.: Predicting process behaviour using deep learning. Decis. Support Syst. **100**, 129–140 (2017)
10. Gunning, D.: Explainable artificial intelligence (XAI). Defense Adv. Res. Projects Agency **2**, 1–18 (2017)
11. Hochreiter, S., Schmidhuber, J.: Long short-term memory. Neural Comput. **9**(8), 1735–1780 (1997)
12. Keskar, N.S., Mudigere, D., Nocedal, J., Smelyanskiy, M., Tang, P.T.P.: On large-batch training for deep learning: generalization gap and sharp minima. In: Proceedings of the 5th International Conference on Learning Representations, pp. 1–16 (2017) openreview.net
13. LeCun, Y., Bengio, Y., Hinton, G.: Deep learning. Nature **521**(7553), 436 (2015)
14. Maggi, F.M., Di Francescomarino, C., Dumas, M., Ghidini, C.: Predictive monitoring of business processes. In: Jarke, M., Mylopoulos, J., Quix, C., Rolland, C., Manolopoulos, Y., Mouratidis, H., Horkoff, J. (eds.) CAiSE 2014. LNCS, vol. 8484, pp. 457–472. Springer, Cham (2014). https://doi.org/10.1007/978-3-319-07881-6_31
15. Márquez-Chamorro, A., Resinas, M., Ruiz-Cortás, A.: Predictive monitoring of business processes: a survey. Trans. Serv. Comput. **11**, 1–18 (2017)
16. Mehdiyev, N., Fettke, P.: Prescriptive process analytics with deep learning and explainable artificial intelligence. In: Proceedings of the 28th European Conference on Information Systems, AISeL (2020)
17. Nunes, I., Jannach, D.: A systematic review and taxonomy of explanations in decision support and recommender systems. User Model. User-Adap. Inter. **27**(3–5), 393–444 (2017)
18. Rehse, J.R., Mehdiyev, N., Fettke, P.: Towards explainable process predictions for industry 4.0 in the DFKI-smart-lego-factory. Künstliche Intelligenz **33**(2), 181–187 (2019)
19. Ribeiro, M.T., Singh, S., Guestrin, C.: "Why should I trust you?" Explaining the predictions of any classifier. In: Proceedings of the 22nd International Conference on Knowledge Discovery and Data Mining, pp. 1135–1144 (2016)
20. Schwegmann, B., Matzner, M., Janiesch, C.: preCEP: facilitating predictive event-driven process analytics. In: vom Brocke, J., Hekkala, R., Ram, S., Rossi, M. (eds.) DESRIST 2013. LNCS, vol. 7939, pp. 448–455. Springer, Heidelberg (2013). https://doi.org/10.1007/978-3-642-38827-9_36

21. Senderovich, A., Di Francescomarino, C., Ghidini, C., Jorbina, K., Maggi, F.M.: Intra and inter-case features in predictive process monitoring: a tale of two dimensions. In: Carmona, J., Engels, G., Kumar, A. (eds.) BPM 2017. LNCS, vol. 10445, pp. 306–323. Springer, Cham (2017). https://doi.org/10.1007/978-3-319-65000-5_18
22. Sindhgatta, R., Ouyang, C., Moreira, C., Liao, Y.: Interpreting predictive process monitoring benchmarks. arXiv:1912.10558 (2019)
23. Taymouri, F., La Rosa, M., Erfani, S., Bozorgi, Z.D., Verenich, I.: Predictive business process monitoring via generative adversarial nets: the case of next event prediction. arXiv:2003.11268 (2020)
24. Verenich, I., Dumas, M., La Rosa, M., Nguyen, H.: Predicting process performance: a white-box approach based on process models. J. Softw. Evol. Process **31**(6), e2170 (2019)
25. Weinzierl, S., et al.: An empirical comparison of deep-neural-network architectures for next activity prediction using context-enriched process event logs. arXiv:2005.01194 (2020)
26. Weinzierl, S., Revoredo, K.C., Matzner, M.: Predictive business process monitoring with context information from documents. In: Proceedings of the 27th European Conference on Information Systems, pp. 1–10. AISeL (2019)

Unsupervised Contextual State Representation for Improved Business Process Models

Prerna Agarwal[1]([✉]), Daivik Swarup[2], Sushruth Prasannakumar[3],
Sampath Dechu[1], and Monika Gupta[1]

[1] IBM Research AI, Bangalore, India
{preragar,sampath.dechu,mongup20}@in.ibm.com
[2] IIT Kharagpur, Kharagpur, India
daivikswarupov@gmail.com
[3] NMIT Bangalore, Bangalore, India
sushruthkonapur@gmail.com

Abstract. Predictive Business Process Monitoring tasks such as next activity prediction, next timestamp prediction, etc. are becoming crucial as new technologies are enabling intelligent automation of business processes. Recent works try to address this problem by using deep learning models that encode limited attribute information of past activities for a case independently w.r.t the other cases in execution. However, the predictions for a case can also depend on contextual information such as inter-case dependencies and domain-specific attributes, which is not considered in previous works. We propose a novel method of encoding the contextual state information i.e., encoding the state of on-going cases and multi-attribute domain-specific information along with intra-case information in an unsupervised manner. We train two widely used deep learning models i.e., LSTM and Transformer using the proposed representation, and compare their performance to show the improved results over the state-of-the-art models. We also investigate the influence of past activities and other on-going cases on prediction using self-attention, making the framework to provide interpretable predictions for a decision making business user.

Keywords: Contextual representation · Inter-case · Interpretability

1 Introduction

Predictive Business Process Monitoring tasks predict monitoring measures for the on-going cases based on the historical event logs [8,9]. These event logs contain sequence of activities enriched with multiple domain-specific case attributes. Business process systems have a finite quantity of resources, and the presence of other on-going cases can influence the resource allocation for other case. For instance, the number of loan applications processed by a resource can influence the time taken for the completion of another application. The progress of the application can also be dependent on the case attributes. If the *loan amount* is greater than $10k, it may get allocated to a different set of agents for detailed

© Springer Nature Switzerland AG 2020
A. Del Río Ortega et al. (Eds.): BPM 2020 Workshops, LNBIP 397, pp. 142–154, 2020.
https://doi.org/10.1007/978-3-030-66498-5_11

processing. Accurate prediction of the next timestamp and activity gives an approximate time taken for the completion of the current activity, which will help in appropriate assignment and allocation of resources for the next activity.

Recent techniques learn inputs to learning process system such as activities, traces, and logs [3] in the form of representation. However, these methods do not consider the context of the business process state i.e., the other on-going cases (inter-case information) and features from domain-specific attributes to capture their influence on the business process state while learning the representation. Therefore, there is a need for an unsupervised contextual representation method that unifies the features from other on-going cases with the domain information for every case state. This modeling should avoid handcrafting of such domain-specific features so that the method to generate representation can generalize across domains without the need for domain knowledge expert.

Deep learning (DL) models have shown high performance as compared to statistical and machine learning methods in the recent works such as Evermann et al. [4], Tax et al. [12], etc., which uses Long Short-Term Memory (LSTM) based models for these prediction tasks. Therefore, in this paper, we experiment with DL models with our proposed unsupervised representation. To overcome the long sequence dependency issue in LSTM [13], we build predictive models with the recently proposed Transformers model [1] for business process logs. This offers two benefits: (1) It alleviates the long-sequence dependency problem with the help of self-attention mechanism; (2) It enables the business user to interpret model's prediction.

Recent works have used attention weights to interpret DL model predictions in NLP and Image Processing domain [5,6,10], but not yet explored for business processes. In this paper, we attempt to employ a similar technique to enable the business users to interpret the underlying case dependencies influencing model's predictions using self-attention weights. It also validates the usefulness of the proposed representation in accurately learning the process features by the model.

To summarize, the main contributions of this paper are: (1) We propose a novel unsupervised method of learning the contextual business process state representation which encodes inter-case interactions and domain-specific attributes along with intra-case attributes in a domain agnostic manner (2) We compare the performance of the proposed representation with two widely used sequential models: LSTM and Transformers for the two tasks - next activity and next timestamp prediction. (3) We use self-attention mechanism over the two sequential models to interpret the model's predictions. We use publicly available datasets to show a comparison of performance between the method employing only intra-case features with our proposed representation and illustrate the use of self-attention which can be used to interpret learned model's weights by a business user to make certain decisions for business process design choices.

2   Related Work

State-of-the-art approaches such as Evermann et al. [4] embeds event sequence and include additional attributes such as resources to predict the next activity and remaining suffix but cannot handle numerical attributes and thus cannot predict timestamps. Similar inability is seen in Lin et al. [7] as well which uses encoders and decoders to transform the attributes of activities. Tax et al. [12] predict timestamp as well but use only one-hot-encoded variables which can deteriorate the performance with an increase in the number of activities. Manuel et al. [2] considers triplets of event type, role, and timestamp to be used by shared and specialized LSTM architectures for prediction with embedded dimensions. These methods encode mostly intra-case information to form the feature vector for different case (prefix) lengths. They do not handle concurrent activity executions as well. Senderovich et al. [11] proposed supervised methods to use hand-crafted inter-case features to improve the accuracy of these tasks which is not scalable for any other domain. DL models such as Transformers are not yet explored for business processes which can help in overcoming the long sequence dependency problem of LSTM. Some recent works in NLP and Image domain have used attention weights to interpret the predictions of DL models [1,15]. For reading comprehension task, [5] use attention weights to show where the model "looks" when answering a question. We use a similar technique in this paper to make a first attempt towards interpretability in business processes.

3   Basic Concepts

3.1   Problem Definition

For the next activity and its timestamp prediction task, the input data is an event log L consisting of a set of traces $T = \{t_1, t_2, ..., t_n\}$, where n is the total number of traces. Each activity e_j consists of a set of attributes $\{a_1, a_2, ..., a_k\}$ where k is the number of attributes. Each trace t_i consists of an ordered sequence of activities $t_i = \{(e_1, time_1), (e_2, time_2), ..., (e_{n_i}, time_{n_i})\}$, where n_i is the number of events in trace t_i and $time_i$ is the timestamp corresponding to each activity. The prediction task can be defined as: Given a partial trace $t = \{(e_1, time_1), (e_2, time_2), ..., (e_j, time_j)\}$, the output of the predictive model is the next activity and its timestamp $(e_{j+1}, time_{j+1})$. This predictive model can be used further to predict the remaining trace i.e., $\{(e_{j+2}, time_{j+2}), .., (e_{n_i}, time_{n_i})\}$.

3.2   Models: LSTM and Transformers

LSTM Networks. We use the proposed contextual representation to build a joint activity and timestamp prediction model using LSTM. The following equations shows how the cell states $\bar{c}_{ts}$ and c_{ts} at any timestamp ts are computed:

$$\bar{c}_{ts} = tanh(w_c.[x_{ts}, h_{ts-1}] + b_c) \qquad c_{ts} = i_{ts}.\bar{c}_{ts} + f_{ts}.c_{ts-1} \qquad (1)$$

where, w is the weight matrix, x is input vector and b is constant for different gates. Input gate i_{ts} and forget gate f_{ts} are computed as:

$$i_{ts} = \sigma(w_i.[x_{ts}, h_{ts-1}] + b_i) \qquad f_{ts} = \sigma(w_f.[x_{ts}, h_{ts-1}] + b_f) \qquad (2)$$

The output gate o_{ts} and next hidden state h_{ts} is computed as:

$$o_{ts} = \sigma(w_o.[x_{ts}, h_{ts-1}] + b_o) \qquad h_{ts} = o_{ts}.tanh(c_{ts}) \qquad (3)$$

For self-attention, we use the Bahdanau model (described in Sect. 4.4). The concatenation of the hidden and cell state of each previous event is used as keys instead of the feature vector of the on-going cases.

Transformer Networks. We use a variant of the Transformer model which has been proposed for machine translation [1]. We use multi-head self-attention over the previous activities of a case with feed-forward network (FFN). This kind of attention helps the model to access the previous timestamp information without facing the computational bottlenecks of RNNs. We use self-attention in each layer as used in [14]. Let $[x_1, x_2, x_3...x_{T_{max}}]$ be the outputs of the previous layer (input in first layer) with input sequence length T_{max}. To maintain causality, the attention for time ts is over $[x_1, x_2, ...x_{ts}]$. For the jth head:

$$q_{tsj} = FeedForward1_j(x_{ts}) \qquad k_{tsj_i} = FeedForward2_j(x_i) \qquad (4)$$

where q_{ts} is the current activity, k_{ts} is previous activities $1 \leq i \leq x_{ts-1}$ and $k_{tsj} = [k_{tsj_1}, k_{tsj_2}, .., k_{tsj_{x_{ts-1}}}]$. Attention weight at for j^{th} head is calculated as:

$$at_j = Softmax\left(\frac{k_{tsj}.q_{tsj}}{\sqrt{d_k}}\right) \qquad (5)$$

The multi-head context vector c_{ts} and the final output y_{ts} of the layer is given by:

$$c_{ts} = MultiHeadAttention([x_1, x_2..x_{ts}], x_{ts}) \qquad (6)$$

$$c_{ts} = FFN(Concat([at_1, at_2...at_{n_{heads}}])) \qquad (7)$$

$$y_{ts} = LayerNorm(y_{ts}^1 + FFN(y_1^1)) \qquad (8)$$

The output of the final layer is projected to prediction for the two tasks.

4   Proposed Unsupervised Contextual State Representation

The need for contextual state representation can be derived analogously to the need for word embeddings in Natural Language Processing (NLP). In NLP, it is important to capture the semantics of a word, taking care of the context it appears in. Similarly, it is important in event logs as well to capture the semantics of a case taking care of other on-going cases at that time. The dearth of labeled data makes the need for obtaining the contextual representation in an unsupervised manner necessary. Thus, we propose an unsupervised methodology to obtain the contextual state representation.

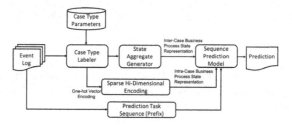

Fig. 1. Unsupervised contextual state representation framework

4.1 Framework of Proposed Representation

The proposed framework to obtain the contextual representation is shown in Fig. 1. It takes an event log L as input. For each event, we assume the availability of timestamp and activity name along with a set of attributes including domain-specific attributes. As shown in the figure, L is passed to *Case Type Labeler* which extracts features from different types of attributes such as categorical, numeric, etc. The attributes in the event log are generally available at 2 levels: (a) activity and (b) case level. Activity level attributes are present for each activity and can have different values for activities in a case whereas the case level attributes are the same for each activity in a case. For illustration, we take an example of one of the datasets used for the experiments i.e., BPI12W. This dataset contains an event log from a loan application system and consists of the following fields: {*Case ID, Activity, Complete Timestamp, REG_DATE, AMOUNT_REQ, org:resource, lifecycle:transition*}. Here, *Activity, Complete Timestamp, org:resource* and *lifecycle:transition* are activity level attributes, whereas, *AMOUNT_REQ* and *REG_DATE* are case level attributes. We broadly classify attributes into 3 types: (a) numeric, (b) categorical and (c) time. The attribute set A of an activity e at a timestamp ts is denoted as:

$$A_{ts,e} = \{a_{num}^i\}_{i=1}^{n_{num}} \cup \{a_{cat}^j\}_{j=1}^{n_{cat}} \cup \{a_{time}^k\}_{k=1}^{n_{time}} \tag{9}$$

where, n_{num} is the total number of numeric attributes, n_{cat} is the total number of categorical attributes, n_time is the total number of time attributes.

4.2 Case Type Labeler

It provides flexibility to the framework to accept only the required types of attributes. Case type parameters define the type of attributes the Case Type Labeler can accept. The Case Type Labeler then processes each type of attribute to extract features. It can handle both event and case level attributes. It also handles the presence of concurrent activities in the business process. Concurrent activities execute in parallel and hence, their representation is determined together. Considering these facts, each type of attribute is handled as follows:

1. **Categorical attribute:** Each categorical attribute is represented as a one-hot vector. The value at concurrent activities in vector will be set to 1.

2. **Numeric attribute:** The numeric attributes with continuous values are normalized between 0 and 1 using min-max normalization for each event e_i. For concurrent activities, the min and max value is used to bucket the continuous values into a fixed-size vector and the bucket in which the numeric values of the concurrent activities lies will be set to 1.
3. **Time attribute:** We represent a time attribute as the following vector inspired by [12]: $time_{attr} = \{time_m, w, time_e, time_s\}$. Here, $time_m$ is the time elapsed since midnight, w is the weekday in the timestamp, $time_e$ is the time elapsed since the last event i.e., the difference between the timestamp of the current event and the previous event, and $time_s$ is the time elapsed since the case started. For concurrent activities, each entry of $time_{attr}$ becomes the average of the value of all concurrent activities.

4.3 Sparse Hi-Dimensional Encoding

The representation of each prefix p consisting of set of activities $p = \{e_1, e_2, .., e_n\}$, is the concatenation of the feature vector of each event e_i. The feature vector v_{e_i} of event e_i having k attributes a_{ik}, is the concatenation of representation obtained for each a_{ik}. The following equation gives the sparse hi-dimensional encoding representation of intra-case state representation of a prefix v_p:

$$v_{e_i} = \cup_k FeatureVector(a_{ik}), \quad v_p = \cup_{i=1}^{i=n} v_{e_i} \tag{10}$$

4.4 Contextual Representation: State Aggregator Generator

State Aggregator Generator in Fig. 1 models the inter-case interactions to obtain aggregated contextual representation. We hypothesize that *the progress of the current case depends on the other on-going cases at that timestamp*. We define the duration of a case trace t_i in closed interval as:

$$duration(t_i) = [min_{j=1}^{n_i} f(e_{ij}, ts), max_{j=1}^{n_i} f(e_{ij}, ts)] \tag{11}$$

where, ts is the timestamp at which aggregation is done. For any time T ranging from $time_1$ to T_{max}, the set of on-going cases is defined as:

$$ongoing\_cases(T) = \{t_i \mid T \in duration(t_i)\} \tag{12}$$

We generate the vector for the last activity e in the prefix (as described in Sect. 4.2) of each on-going case that occurs on or before the timestamp of the current activity. Let the last activity in a case trace t_i before time T be

$$Last(t_i, T) = argmax_{e \in t_i, f(e,ts) <= T} f(e, ts) \tag{13}$$

The last activity is then used to form the context vector. Concretely, the set of vectors that forms the inter-case context for a particular activity e_i is given as:

$$Context(e_i) = \{FeatureVector(Last(t_k, f(e_i, ts)) \mid t_k \in ongoing\_cases(f(e_i, ts))\} \tag{14}$$

The context vectors for all on-going cases are then aggregated in the following described two different ways to model the inter-case interactions:

1. **Aggregation using Pooling:** The feature vector for each executing case is embedded using a FFN. The embeddings for an activity e_i with feature vector v are aggregated by performing a function of Mean/Sum/Max pooling.

$$E_v = FFN(v), \quad c = F_{aggregate}([E_v \; for \; v \in Context(e_i)]) \qquad (15)$$

2. **Aggregation using Attention:** We use Bahdanau attention [1] over the on-going cases. This attention is used because at time T it considers $T-1$ hidden states of the decoder to calculate alignment and context vector, unlike other attention. Figure 2a shows how attention is used to generate the aggregated Contextual State Vector Let h be the hidden state at activity e

$$s = Softmax(v \cdot h), v \in Context(e) \qquad c = \sum_{v_i \in Context(e)} (s_i \cdot v_i) \qquad (16)$$

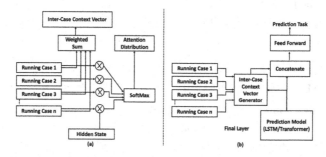

Fig. 2. (a) Contextual state vector with attention (b) Using contextual state vector in model

This gives the contextual inter-case business state representation. Figure 2b shows how the inter-case representation is used for a prediction task. The inter-case representation is used only at the final layer because the last LSTM cell determines the last activity of the partial trace and the timestamp of the last activity is considered as T in Eq. 14 to calculate the set of on-going cases.

Note: *Aggregation control the dimension of the contextual vector for any number of on-going cases by passing it to a FFN before using it for the prediction. This gives an advantage over other state-of-the-art models [7,12] where the dimension of feature vector increase exponentially with the number of events and deteriorates model's performance.*

4.5   Prediction Model

We train LSTM and Transformer to predict the next activity and next timestamp jointly. As shown in Fig. 3, we have n_{shared} shared layers followed by n_{act} layers for activity and n_{time} layers for timestamp prediction similar to Tax et al. [12]. The intra-case representation is fed as input to LSTM cells and the contextual inter-case state representation generated by any one method from Sect. 4.4 is concatenated to the hidden state and passed to a FFN for training.

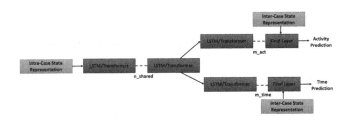

Fig. 3. Intra-case and Inter-case representation in Prediction Model

5   Experimental Details

In this section, we describe the datasets used to evaluate the proposed representation. We then describe the training details and hyper-parameters used.

5.1   Datasets

We experiment with widely used datasets (shown in Table 1) which are logs from different domains and vary in terms of dependencies and trace lengths.

1. **Helpdesk**: This dataset consists of ticket management system logs of an Italian software company[1].
2. **BPI12W**: It consists of loan application processing logs by a Dutch Financial institution[2]. We use subprocess W to compare with other baselines.
3. **BPI13**: It contains Volve IT incident and problem management[3]. We use the "complete" cases to compare with other baselines.
4. **BPI19**: This event data originates from MNC from the Netherlands for the purchase order handling process with different flows in data[4]. *None of the recent work has used this dataset for evaluation because of its very recent availability, hence we use it in this paper instead of BPI14, BPI15.*

---

[1] https://data.4tu.nl/repository/uuid:0c60edf1-6f83-4e75-9367-4c63b3e9d5bb.
[2] https://www.win.tue.nl/bpi/doku.php?id=2012:challenge.
[3] https://www.win.tue.nl/bpi/doku.php?id=2013:challenge.
[4] https://data.4tu.nl/repository/uuid:d06aff4b-79f0-45e6-8ec8-e19730c248f1.

Table 1. Statistics of experimental datasets

Dataset	#Traces	Avg. trace length	#Activities	#Categorical attributes	#Numeric attributes	Avg. duration
Helpdesk	4580	4.6	15	11	0	40.9 d
BPI12W	9658	17.6	7	3	1	8.8 d
BPI13	1487	4.47	7	10	0	179.2 d
BPI19	251734	6.33	42	12	1	36.75 d
Synthetic dataset 1	10000	3	3	0	0	8 h
Synthetic dataset 2	10000	3	3	0	0	8 h

Synthetic Dataset for Model Interpretability. We use the synthetic dataset for demonstration because it provides more control to induce task dependencies and hence, observe them in the output. We draw a simple BPMN with 3 tasks to generate synthetic logs (see Fig. 4). We ensure that all activities takes place on weekdays between 9:00 and 17:00 hrs. We create 2 datasets with 10K traces each for demonstration. The details of the two simulated datasets are:

1. *Dataset 1*: We assume that the processing time for each task is independent of the other tasks and distributed exponentially with different mean values.
2. *Dataset 2*: This is identical to Dataset 1 except that Task 3's processing time is drawn from two different exponential distributions depending on whether Task 1 or Task 2 was executed previously.

Table 2. Hyper-parameters search space

Hyperparameters	Values
# Shared layers	0,1,2
# Layers for prediction	1, 2, 3
# Heads in attention (for Transformers)	2,4,8
Dimension	32, 64, 128, 256

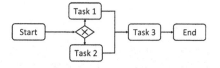

Fig. 4. Synthetic BPMN

We hypothesize that self-attention should attend only the current task in Dataset 1 because the time taken does not depend on previous tasks. In Dataset 2, we expect the self-attention to have a higher weight over task 1 or 2 because the current output is dependent on the previous one.

5.2 Training

The datasets are preprocessed to fill in the missing attribute values with mean values. The prefixes only of length 2 and above are used. We sort the traces in each dataset chronologically w.r.t start time of traces and divide it into Training, Validation and Test splits in the ratio 0.7 : 0.1 : 0.2. We observed that the optimal hyper-parameters for some tasks have shared layers, therefore, we perform joint training for both the prediction tasks. We use the sum of the negative log-likelihood of the next activity and the mean average error (MAE) of the next timestamp as a loss function L to train the parameters in all our experiments.

$$L(\theta) = -log(p(\hat{a}|e_0e_1e_2.., \theta)) + | \hat{ts} - ts(e_0e_1e_2.., \theta) | \tag{17}$$

Here, θ is a parameter, e_i's are previous activities, ts is the predicted timestamp, $\hat{t}$ and $\hat{a}$ are the ground truth next timestamp and next activity respectively. We performed grid search on search space shown in Table 2 to find the optimal hyper-parameters for all the experiments: (1,1,4,128) for Transformer and (1,1,128) for LSTM. We compare our results with: (1) Tax et al. [12]; (2) Evermann et al. [4]; (3) Shared cat and joint argmax configuration of Manuel et al. [2] for a fair comparison. Lin et al. [7] eliminate timestamp field, hence, not used as baseline.

6   Results and Discussion

We first discuss the results obtained from the proposed representation and then show the interpretability results obtained from self-attention using Transformers.

6.1   Results: Contextual State Representation

Table 3 shows the results obtained using different methods of generating intercase representation. For the next activity prediction, we observe a pronounced improvement of **15%** for LSTM and **15.3%** for Transformers w.r.t the baselines for Helpdesk and **27%** for LSTM and **35%** for Transformers for BPI13. Marginal improvement is seen for BPI12W (**0.57%** for LSTM and **1.4%** for Transformers). *The improvement observed on Helpdesk and BPI13 is large as compared to BPI12W because of a large number of available domain-specific attributes.*

Table 3. Performance of proposed approach w.r.t. baselines

Model variants	Activity accuracy				Timestamp MAE (days)			
	BPI12W	Helpdesk	BPI13	BPI19	BPI12W	Helpdesk	BPI13	BPI19
Evermann et al.	0.623	0.798	0.451	–	–	–	6.151	–
Tax et al.	0.76	0.7123	–	–	1.56	3.75	–	–
Manuel et al.	0.7855	0.5773	0.544	–	5.9	7.3	242.6	–
LSTM + Mean pool	0.762	0.786	0.68	0.73	1.63	3.788	2.83	117
LSTM + Sum pool	0.768	0.787	0.672	0.736	1.58	3.74	2.8	117.2
LSTM + Max pool	0.774	0.7874	0.687	0.74	1.33	3.69	2.72	117.4
LSTM + Attention	**0.79**	**0.949**	**0.694**	**0.763**	**1.3**	**3.64**	**2.7**	**110.4**
Transformer + Mean pool	0.762	0.786	0.695	0.804	1.676	3.788	2.75	111.8
Transformer + Sum pool	0.768	0.7857	0.71	0.807	1.58	3.74	2.4	112.3
Transformer + Max pool	0.779	0.791	0.715	0.812	1.58	3.71	2.81	110.5
Transformer + Attention	**0.797**	**0.951**	**0.734**	**0.863**	**1.26**	**3.617**	**2.21**	**108.9**

For the next timestamp prediction task, the improvement of **3%** for Transfomer and **2.6%** for LSTM is observed for BPI12W dataset w.r.t baseline. Similarly, an improvement of **3%** for LSTM and **3.5%** for Transformer is observed for Helpdesk and a major reduction in MAE of **3.45 d** for LSTM and **4 d** for Transformers is observed for BPI13. We discuss the observed improvement using attention over pooling in Sect. 6.2. We achieve **76.3%** accuracy for LSTM and **86.3%** for Transformers on BPI19 dataset. The MAE is large i.e., **110 d** for LSTM and **109 d** for Transformers. These results can act as the baseline to benchmark further advancements in this area.

LSTM v/s Transformers: Overall, the performance achieved by different variants of the Transformer model is on the higher end as compared to LSTM especially for BPI12W and BPI19 due to comparatively larger trace lengths where the performance of LSTM starts decreasing. For other datasets, the performance of both sequential models is similar because of shorter trace lengths.

A significant improvement in the performance of both the models w.r.t baselines shows that the proposed contextual state representation is able to model inter-case dependencies and the domain information appropriately.

6.2   Interpretability Using Self-attention

We derive insights that can be used by any business user who wants to use these predictions to make certain decisions. To validate the features learned by the model, we plot the probability distribution produced by the softmax in the self-attention layer. This provides an insight into the dependency of an activity on the previous activities. *We show the attention weights of a single head with two datasets i.e., Helpdesk and BPI12W for the ease of understanding.* Figure 5 shows the attention distribution for a case in the Helpdesk dataset (Left) and BPI12W dataset (Right). In Helpdesk case, we observe that the attention is mostly concentrated on the current activity (diagonal values in the plot) and there is less dependency on previous activities. But, when attention distribution is seen for on-going cases, it was observed that this case was dependent on the other on-going which shows that there are a lot of inter-case dependencies. Therefore, a significant improvement is seen in the performance of the models with attention as compared to aggregation methods. In BPI12W case, we observe that the attention is distributed across previous activities, depicting a dependency in prediction. The concentrated attention distribution for the on-going cases shows that there are fewer inter-case dependencies. Therefore, only marginal improvement is observed in the performance w.r.t the baselines. In Fig. 6, the distribution of attention for next timestamp prediction is shown on the simulated datasets. We observe that for the Simulated dataset 1 (Left), the attention is more focused on the current event. In the simulated dataset 2 (Right), we observe that the mean attention given to the event in time-step 2 when predicting the next time-step 3 is 0.2111 and 0.4458 respectively, which validates our intuition and qualitative observations of dependency.

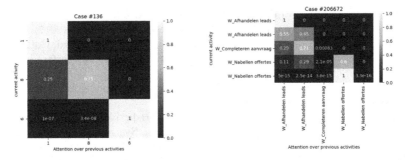

Fig. 5. Attention distribution at an activity in Helpdesk and BPI12W

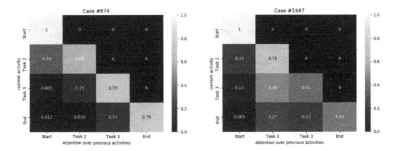

Fig. 6. Attention distribution at each activity for a case in Simulated datasets

7 Conclusion and Future Work

In this paper, we propose an unsupervised technique to obtain contextual state representation using intra-case, inter-case, and domain-specific attributes. We demonstrated that it outperforms the state-of-the-art LSTM based models. We also show the results of interpretability with self-attention to analyze the impact of other on-going cases, past events on the outcome of a case for decision making business users. This work only incorporates categorical, numeric and time attributes. It can be further extended to incorporate unstructured text information present in event logs as well which provides additional features and can lead to more improved models. Such kind of representation can be used to train sequential models for other tasks like trace clustering as well.

References

1. Bahdanau, D., Cho, K., Bengio, Y.: Neural machine translation by jointly learning to align and translate. CoRR (September 2014)
2. Camargo, M., Dumas, M., González-Rojas, O.: Learning accurate LSTM models of business processes. In: BPM, pp. 286–302 (2019)
3. De Koninck, P., vanden Broucke, S., De Weerdt, J.: act2vec, trace2vec, log2vec, and model2vec: representation learning for business processes. In: BPM, pp. 305–321 (2018)

4. Evermann, J., Rehse, J.R., Fettke, P.: A deep learning approach for predicting process behaviour at runtime. In: BPM Workshops, pp. 327–338 (2017)
5. Hermann, K.M., et al.: Teaching machines to read and comprehend. CoRR (2015)
6. Jo, J., Bengio, Y.: Measuring the tendency of CNNs to learn surface statistical regularities. CoRR (2017)
7. Lin, L., Wen, L., Wang, J.: Mm-pred: a deep predictive model for multi-attribute event sequence. In: SDM, pp. 118–126 (2018)
8. Metzger, A., et al.: Comparing and combining predictive business process monitoring techniques. IEEE Trans. Syst. Man Cybern. Syst. **45**(2), 276–290 (2015)
9. Márquez-Chamorro, A.E., Resinas, M., Ruiz-Cortés, A.: Predictive monitoring of business processes: a survey. IEEE Trans. Serv. Comput. **11**(6), 962–977 (2018)
10. Rauber, P.E., Fadel, S.G., Falcão, A.X., Telea, A.C.: Visualizing the hidden activity of artificial neural networks. IEEE Trans. Visual. Comput. Graphics **23**(1), 101–110 (2017)
11. Senderovich, A., Di Francescomarino, C., Ghidini, C., Jorbina, K., Maggi, F.M.: Intra and inter-case features in predictive process monitoring: A tale of two dimensions. In: BPM, pp. 306–323 (2017)
12. Tax, N., Verenich, I., La Rosa, M., Dumas, M.: Predictive business process monitoring with LSTM neural networks. In: AISE (2017)
13. Trinh, T.H., Dai, A.M., Luong, T., Le, Q.V.: Learning longer-term dependencies in RNNs with auxiliary losses. CoRR (2018)
14. Vaswani, A., et al.: Attention is all you need. In: CoRR (2017)
15. Xu, K., et al.: Show, attend and tell: Neural image caption generation with visual attention. CoRR (2015)

Root Cause Analysis in Process Mining Using Structural Equation Models

Mahnaz Sadat Qafari[(⊠)] and Wil van der Aalst

Rheinisch-Westfälische Technische Hochschule Aachen(RWTH), Aachen, Germany
{m.s.qafari,wvdaalst}@pads.rwth-aachen.de

Abstract. Process mining is a multi-purpose tool enabling organizations to monitor and improve their processes. Process mining assists organizations to enhance their performance indicators by helping them to find and amend the root causes of performance or compliance problems. This task usually involves gathering process data from the event log and then applying some data mining and machine learning techniques. However, using the results of such techniques for process enhancement does not always lead to any process improvements. This phenomenon is often caused by mixing up correlation and causation. In this paper, we present a solution to this problem by creating causal equation models for processes, which enables us to find not only the features that cause the problem but also the effect of an intervention on any of the features. We have implemented this method as a plugin ProM and we have evaluated it using two real and synthetic event logs. These experiments show the validity and effectiveness of the proposed method.

Keywords: Process mining · Root cause analysis · Causality inference

1 Introduction

One of the main purposes of using process mining is to enhance process performance indicators leading to reduced costs and response times and better quality. To enhance the performance of a process, we first need to identify friction points in the process. The second step is finding the root causes of each friction point and estimating the possible effect of changing each factor on the process. The final step is planning process enhancement actions and then re-engineering the process. While there are different techniques that help to find the friction points in processes, there is little work on root cause analysis. So, the focus of this paper is on the second step.

The task of finding the root cause of a problem in a given process is quite intricate. Each process involves many steps and in each step, many factors may be of influence. Also, the steps or the order of the steps that are taken for each case may vary. Another obstacle arises when using a classifier, which is basically designed for prediction and not for interventions, for finding the root cause of the problem. By judging the causal relationships among the features merely based on the findings of a classifier, we may fall into the trap of considering correlation as causation.

Consider a scenario where in an online shop, some of the package deliveries get delayed and there is a high correlation between the delayed orders and the resources that were responsible for them. We can infer causal relationships based on the observed

© Springer Nature Switzerland AG 2020
A. Del Río Ortega et al. (Eds.): BPM 2020 Workshops, LNBIP 397, pp. 155–167, 2020.
https://doi.org/10.1007/978-3-030-66498-5_12

correlations and declare these resources that were in charge of delivering those delayed items as the reason for delays. However, there may be a non-causal correlation between them. Changing the process based on an observed correlation may aggravate the problem (or create new problems). Two correlated events may have a confounder, i.e., a common unmeasured (hidden) cause. In this scenario, the delayed cases are related to the packages with a bigger size which are usually assigned to specific resources.

Two general frameworks for finding the causes of a problem and anticipating the effect of any intervention on the process are random experiments and the theory of causality [10,11]. Applying random experiments, i.e., randomly setting the values of those features that have a causal effect on the problem of interest and monitoring their effect, is usually too expensive (and sometimes unethical) or simply impossible. The other option is modeling the causal relationships between different features of the process using a *structural causal model* [10,11] and then studying the effect of changes imposed on each process feature on the process indicators.

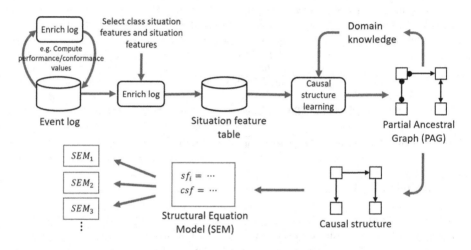

Fig. 1. The general structural causal equation discovery.

In this paper, we propose a framework based on the second approach for root cause analysis, which includes several steps. First, the event log is enriched by several process-related features derived from different data sources like the event log, the process model, and the conformance checking results. Then, depending on the identified problem and features that have a possible direct or indirect causal effect on it, a specific data table, which we call it *situation feature table*, is created. In the next step, a graph encoding the structure of causal relationships among the process features is provided by the customer. The other option for creating such a graph is using a causal structure learning algorithm, also called *search algorithm*, on the data. The resulting graph can be modified by adding domain knowledge as an input to the search algorithm or by modifying the discovered graph. Finally, the strength of each causal relationship and the effect of an intervention on any of the process features on the identified problem are estimated. The general overview of the proposed approach is shown in Fig. 1.

2   Motivating Example

Consider an IT company, that implements software for its customers, but does not maintain the implemented software after it has been released. The Petri-net model of this company is depicted in Fig. 2. Each trace, which is corresponding to the process of implementation of one project, has an associated attribute *priority*, indicating how urgent the software is for the customer. The manager of the company is concerned about the duration of the implementation phase of projects. By *implementation phase*, we mean the sub-model including two transitions "development" and "test" (marked with a blue rectangle in Fig. 2). For simplicity, we use the abbreviations mentioned in Table 1 in this paper.

The manager believes that *P*, *NT*, and *PBD* are the process features that might have a causal effect on *IPD*. The question is *which features have a causal effect on IPD and to what extent*. Also, the manager needs to know what is the effect of an intervention on any of the features on *IPD*. Mentioned questions are valid ones to be asked before planning for re-engineering and enhancing a process. Also, we consider *C* (the *complexity* of a project) as a feature that is not recorded in the event log and may have a causal effect on *IPD*. The answers to such questions are highly influenced by the structure of the causal relationship among the features. Some of the possible structures are depicted in Fig. 3. According to Fig. 3.a), the high correlation between *IPD* and *PBD* is not a causation. So, changing *PBD* does not have any effect on *IPD*. According to 3.b), one may conclude that all *P*, *PBD*, and *NT* play a role in the determination of the value of *IPD* and by changing each of these three features, one can influence the value of *IPD*. According to Fig. 3.c), the existence of a hidden feature in the model, captured by *C* (and depicted by the gray dashed oval in the model), has been considered that causally influences both *IPD* and *PBD*. So, their significant correlation relationship is due to having a common cause. If this is the case, then forcing *PB* activity to take a shorter or longer amount of time does not have any effect on the value of *IPD*.

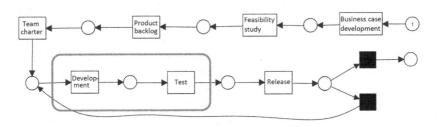

Fig. 2. The process of IT company described in Sect. 2.

Table 1. The list of abbreviations based on IT example, Sect. 2.

List of abbreviations			
FS feasibility study	*CID* case ID	*BC* business case development	*REL* release
PB product backlog	*DEV* development	*NT* number of employees in team	*TE* test
DD development duration	*TD* test duration	*PBD* product backlog duration	*P* priority
TC team charter	*C* complexity	*IPD* implementation phase duration	

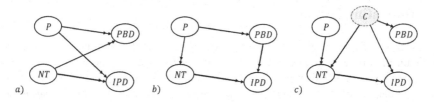

Fig. 3. Three possible causal structures.

It is worth noting that it is not possible to intervene on all the features. For example, the manager can assign more or fewer people to a project; but he cannot change the complexity of a project. So, it is possible to intervene on *NT*, but not on *C*. Just using common sense and domain knowledge we can judge whether a feature can be intervened.

The remainder of the paper is organized as follows. In Sect. 3, we briefly present some related work. In Sect. 4, an overview of the problem and the proposed approach is presented. The details and assumptions of the implemented plugin and the experimental results of applying it on synthetic and real event logs are presented in Sect. 5. Finally, in Sect. 6, we summarize our approach and its applications.

3   Related Work

The main approaches in process mining to find the root causes of a performance or compliance problem are classification [4,6], and rule mining [13]. Although the theory of causality has been studied deeply [11] and its methods have been successfully applied on a variety of domains (e.g. [7,17]), their application in the area of process mining is limited. However, there are some works in this field that use causality theory. For example, in [5] an approach based on time-series analysis has been proposed for discovering causal relationships between a range of business process characteristics and process performance indicators. The idea is to look at the performance indicators values as time-series, and investigating the causality relationship between them using the Granger causality test [3]. The problem with this approach is that the Granger test can only find predictive causality which might not be a true cause-and-effect relationship.

In [8], a methodology for structural causal model discovery in the area of process mining has been proposed. They propose discovering structural causal models using the event log and the BPMN model of a process. One of the assumptions in this work is that the BPMN model is an accurate model of a process, which is not always the case.

4   Approach

In the proposed method, we assume that we already know that a problem, such as a bottleneck or deviation, exists in the process. Several features may have a causal effect on the problem, which some of them may not exist in the event log. As the first step, the event log is enriched by adding some new derived features computed from the event log or possibly other sources. The derived features can be related to any of the process perspectives; the time perspective, the data flow-perspective, the control-flow perspective, the conformance perspective, or the resource/organization perspective of the process. The enriched event log is then used for creating the situation feature table. Finally, we discover the *Structural Equation Model (SEM)* of the situation features using the situation feature table. In the sequel, we explain the details of situation feature table creation and SEM inference. But first, we need to define an (enriched) event log.

An event log is a collection of traces where each trace itself is a collection of events. Let $\mathcal{U}_{act}$ be the universe of all *activity names*, $\mathcal{U}_{time}$ the universe of all *time stamps*, $\mathcal{U}_{att}$ the universe of all *attribute names*, and $\mathcal{U}_{val}$ the universe of all *values*. Also, consider $\mathcal{U}_{map} = \mathcal{U}_{att} \nrightarrow \mathcal{U}_{val}$ and $dom : \mathcal{U}_{att} \mapsto \mathbb{P}(\mathcal{U}_{val})$[1], the function that returns the set of all possible values of a given attribute name. We define an event log as follows:

Definition 1 (Event Log). *Each element of $\mathcal{U}_{act} \times \mathcal{U}_{time} \times \mathcal{U}_{map}$ is an event and $\mathcal{E}$ is the universe of all possible events. Let $\pi_{act}(e) = a$, $\pi_{time}(e) = t$, and $\pi_{map}(e) = m$ for a given event $e = (a, t, m) \in \mathcal{E}$. Each element of $\mathbb{P}(\mathcal{U}_{map} \times \mathcal{E}^+)$ is called an event log where $\mathcal{E}^+$ is the set of all non empty sequences of events such that for each $\langle e_1, \ldots, e_n \rangle \in \mathcal{E}^+$ we have $\forall_{1 \leq i < j \leq n} \pi_{time}(e_i) \leq \pi_{time}(e_j)$. The universe of all possible logs is denoted by $\mathcal{L}$ and each element (m, E) of an event log is called a* trace.

We assume that in a given log L each event has a unique time stamp. Also, given $E = \langle e_1, \ldots, e_n \rangle \in \mathcal{E}^+$, we define $tail(E) = e_n$, $prfx(E) = \{\langle e_1, \ldots, e_i \rangle | 1 \leq i \leq n\}$ which is the set of all the nonempty prefixes of E, and $set(E) = \{e_1, \ldots, e_n\}$.

4.1 Situation Feature Table Creation

An observed problem in the process might be related to either a trace or a specific activity. We assume that, given a problem, in each trace only that part of the data that has been recorded before the occurrence of the problem can have a causal effect on it. So the relevant part of a trace to a given problem is a prefix of it which we call a *situation*. More formally, a situation is an element of $\mathcal{U}_{map} \times \mathcal{E}^+$ and we use $\mathcal{U}_{sf}$ to denote the universe of all possible situations. Given an event log $L \in \mathcal{L}$, we define the set of all situations of L as $S_L = \bigcup_{(m,E) \in L} \{(m, E') | E' \in prfx(E)\}$. Considering $ActNames_L$ as the set of the activity names of all the events that appear in the traces of L, we define an *a-based situation subset* of L as $S_{L,a} = \{(m, E) \in S_L | \pi_{act}(tail(E)) = a\}$, where $a \in ActNames_L$ and *trace-based situation subset* of L as $S_{L,\perp} = L$.

There are two types of attributes in a given event log, attributes linked to the traces and attributes linked to the events. When extracting the data from event log, we need to distinguish these two levels of attributes. For that, we use *situation feature* function which is defined over the situations and is identified by an activity, a (possibly $a = \perp$), and an attribute, at. If $a = \perp$, $sf_{a,at}((m, E))$ returns the value of at in trace level (i.e. $m(at)$). However, if $a \neq \perp$, then $sf_{a,at}(s)$ returns the value of at in $e \in set(E)$ with the maximum time stamp for which $\pi_{act}(e) = a$. More formally:

---

[1] We define $\mathbb{P}(A)$ as the set of all non-empty subsets of set A.

Definition 2 (Situation Feature). *Let $L \in \mathcal{L}$, $a \in ActNames_L \cup \{\bot\}$, and at $\in \mathcal{U}_{att}$. We define a situation feature as $sf_{a,at} : \mathcal{U}_{sit} \not\to \mathcal{U}_{val}$. Given $(m, E) \in L$, we define,*

$$sf_{a,at}((m, E)) = \begin{cases} m(at) & a = \bot \\ \pi_{map}(\arg\max_{e \in \{e \in set(E) | \pi_{act}(e)=a\}} \pi_{time}(e))(at) & a \in ActNames_L \end{cases}.$$

We denote the universe of all possible situation features by $\mathcal{U}_{sf}$. With slight abuse of notation, we define $dom(sf_{a,at}) = dom(at)$. Also, for a given $n \in \mathbb{N}$, $SF \in \mathcal{U}_{sf}^n$ is a situation feature extraction plan of size n, where $\mathcal{U}_{sf}^n$ is defined as $\underbrace{\mathcal{U}_{sf} \times \cdots \times \mathcal{U}_{sf}}_{n \text{ times}}$.

We can interpret a situation feature extraction plan as the schema composed of those situation features that are relevant to the given problem in the process. In the sequel, we call the situation feature that captured the existence of the problem (or represents the quantity or quality of interest) in the process the *class situation feature* and denote it as *csf*. Moreover, in case of no ambiguity, we remove the subscripts of situation features to increase readability.

Now, we can concretely specify the problem in the process and the set of situation features that we need to investigate their causal effect on the occurrence of the problem.

Definition 3 (Causal Situation Specification). *A causal situation specification is a tuple $CSS = (SF, csf)$ in which $SF = (sf_1, \ldots, sf_n) \in \mathcal{U}_{sf}^n$ for some $n \in \mathbb{N}$, $csf \in \mathcal{U}_{sf}$, and $csf \notin set(SF)$ where with slight abuse of notation $set(SF) = \{sf_1, \ldots, sf_n\}$.*

Here, SF is the tuple that includes all of the situation features that we expect them to have a potential causal effect on the class situation feature. Here, we assume that there is no hidden common confounder exists any subset of situation features in $set(SF) \cup \{csf\}$. Using the causal situation specification, we can define a situation feature table as follows:

Definition 4 (Situation Feature Table). *Given an event log $L \in \mathcal{L}$ and a causal situation specification $CSS = (SF, csf)$, where $SF = (sf_1, \ldots, sf_n)$ and $csf = sf_{a,at}$, a situation feature table is a multi-set which is defined as:*

$$T_{CSS,L} = [(sf_1(s), \ldots, sf_n(s), csf(s)) | s \in S_{L,a} \wedge csf(s) \neq \bot].$$

We call $T_{CSS,L}$ a situation feature table of L.

4.2 SEM Inference

Given a log L, if we consider a trace or activity based situation subset of L as a sample and each situation feature as a random variable, then we can define a structural equation model for a given causal structure specification as follows[2]:

---

[2] Definition 5 and 8 are based on [11].

Definition 5 (Structural Equation Model (SEM)). *Given CSS* $= (\boldsymbol{SF}, csf)$, *a structural equation model is defined as a collection of assignments* $S = \{S_{sf_1}, \ldots, S_{sf_n}, S_{csf}\}$ *such that*

$$S_{sf} : sf = f_{sf}(\boldsymbol{PA}_{sf}, N_{sf}), \;\; sf \in set(\boldsymbol{SF}) \cup \{csf\},$$

where $\boldsymbol{PA}_{sf} \subseteq set(\boldsymbol{SF}) \cup \{csf\} \setminus \{sf\}$ *is called* parents *of sf and and* $N_{sf_1}, \ldots, N_{sf_n}, N_{csf}$ *are distributions of the noise variables, which we require to be jointly independent.*

Note that these equations are not normal equations but a way to determine how to generate the observational and the interventional distributions. The set of parents of a situation feature is the set of situation features that have a direct causal effect on it.

The structure of the causal relationships between the situation features in a SEM can be encoded as a directed acyclic directed $G = (V, \twoheadrightarrow)$ which is called a *causal structure*. In this graph, $V = set(\boldsymbol{SF}) \cup \{csf\}$, $\twoheadrightarrow = \{(sf_1, sf_2) \in V \times V | sf_1 \in \boldsymbol{PA}(sf_2)\}$.

The first step of inferring a SEM is discovering its causal structure and the second step is estimating a set of equations describing how each situation feature is influenced by its immediate causes. In the sequel, we describe these two steps.

Causal Structure Discovery. The causal structure can be determined by an expert who possesses the domain knowledge about the underlying process and the causal relationships between its features. But having access to such knowledge is quite rare. Hence, we support discovering the causal structure in a data driven manner.

Several search algorithms have been proposed in the literature (e.g., [2, 9, 15]). The input of a search algorithm is observational data in the form of a situation feature table (and possibly knowledge) and its output is a graphical object that represents a set of causal structures that cannot be distinguished by the algorithm. One of these graphical objects is *Partial Ancestral Graph (PAG)* introduced in [18].

A PAG is a graph whose vertex set is $V = set(\boldsymbol{SF}) \cup \{csf\}$ but has different edge types, including $\rightarrow, \leftrightarrow, \bullet\!\!\rightarrow, \bullet\!\!-\!\!\bullet$[3]. Each edge type has a specific meaning. Let $sf_1, sf_2 \in V$. $sf_1 \rightarrow sf_2$ indicates that sf_1 is a direct cause of sf_2, $sf_1 \leftrightarrow sf_2$ means that neither sf_1 nor sf_2 is an ancestor of the other one, even though they are probabalistically dependent (i.e., sf_1 and sf_2 are both caused by one or more hidden confounders), $sf_1 \bullet\!\!\rightarrow sf_2$ means sf_2 is not a direct cause of sf_1, and $sf_1 \bullet\!\!-\!\!\bullet sf_2$ indicates that there is a relationship between sf_1 and sf_2, but nothing is known about its direction. The formal definition of a PAG is as follows [18]:

Definition 6 (Partial Ancestral Graph (PAG)). *A PAG is a tuple* $(V, \rightarrow, \leftrightarrow, \bullet\!\!\rightarrow, \bullet\!\!-\!\!\bullet)$ *in which* $V = set(\boldsymbol{SF}) \cup \{csf\}$ *and* $\rightarrow, \leftrightarrow, \bullet\!\!\rightarrow, \bullet\!\!-\!\!\bullet \subseteq V \times V$ *such that* $\rightarrow$, $\leftrightarrow$, $\bullet\!\!\rightarrow$, *and* $\bullet\!\!-\!\!\bullet$ *are mutually disjoint. Moreover, there is at most one edge between every pair of situation features in* V.

The discovered PAG by the search algorithm represents a class of causal structures that satisfies the conditional independence relationships discovered in the situation table and ideally, includes the true causal structure of the causal situation specification. Now, it is needed to modify the discovered PAG to a compatible causal structure. As we

---

[3] We usually use $a \bullet b$ instead of $(a, b) \in \bullet$ for $\bullet \in \{\rightarrow, \leftrightarrow, \bullet\!\!\rightarrow, \bullet\!\!-\!\!\bullet\} \cup \{\twoheadrightarrow\}$.

assume no hidden common confounder exists, we expect that in the PAG, relation $\leftrightarrow$ be empty[4]. We can define the compatibility of a causal structure with a PAG as follows:

Definition 7 (Compatibility of a Causal Structure With a Given PAG). *Given a PAG* $(V, \rightarrow, \leftrightarrow, \bullet\!\!\rightarrow, \bullet\!\!-\!\!\bullet)$ *in which* $\leftrightarrow = \emptyset$, *we say a causal structure* $(U, \twoheadrightarrow)$ *is compatible with the given PAG if* $V = U$, $(sf_1 \rightarrow sf_2 \vee sf_1 \leftrightarrow sf_2) \implies sf_1 \twoheadrightarrow sf_2$, *and* $sf_1 \bullet\!\!-\!\!\bullet sf_2 \implies$ $(sf_1 \twoheadrightarrow sf_2 \oplus sf_2 \twoheadrightarrow sf_1)$, *where* $sf_1, sf_2 \in V$.

To transform the output PAG to a compatible causal structure, which represents the causal structure of the situation features in the situation feature specification, domain knowledge of the process and common sense can be used. These information can be use to directly modify the discovered PAG or by adding them to the search algorithm, as an input, in the form of *required directions* or *forbidden directions* denoted as D_{req} and D_{frb}, respectively. $D_{req}, D_{frb} \subseteq V \times V$ and $D_{req} \cap D_{frb} = \emptyset$. If $(sf_1, sf_2) \in D_{req}$ then $sf_1 \rightarrow sf_2$ or $sf_1 \leftrightarrow sf_2$ in the output PAG. However, if $(sf_1, sf_2) \in D_{frb}$, then in the discovered PAG it should not be the case that $sf_1 \rightarrow sf_2$.

Causal Strength Estimation. The final step of discovering the causal model is estimating the strength of each direct causal effect using the observed data. Suppose $\mathcal{G}$ is the causal structure of a causal situation specification $CSS = (SF, csf)$. As $\mathcal{G}$ is a directed acyclic graph, we can sort its nodes in a topological order γ. Now, we can statistically model each situation feature sf_2 as a function of the noise terms N_{sf_1} of the situation features sf_1 for which $\gamma(sf_1) \leq \gamma(sf_2)$, where $sf_1, sf_2 \in set(SF) \cup \{csf\}$. The result is $sf_2 = f_2((N_{sf_1})_{sf_1 : \gamma(sf_1) \leq \gamma(sf_2)})$ [11]. The set of these functions, for all $sf \in set(SF) \cup \{csf\}$, is the SEM of $CSS = (SF, csf)$.

Finally, we want to answer questions about the effect of an intervention on any of the situation features on the class situation feature. Here we focus on *atomic interventions* which are defined as follows:

Definition 8 (Atomic Intervention). *Given an SEM* S, *an intervention is obtained by replacing* $S_{sf} \in S \setminus \{S_{csf}\}$ *by* $sf = c$ *where* $c \in \mathbb{R}$.

Note that the corresponding causal structure of an SEM after intervention on sf is obtained from the original causal structure of M by removing all the incoming edges to sf [11]. When we intervene on a situation feature, we just replace the equation of that situation feature in the SEM and the others do not change as causal relationships are autonomous under interventions [11].

If in a given causal structure of a causal situation specification $CSS = (SF, csf)$, there is no directed path between $sf \in set(SF)$ and csf, they are independent and consequently, intervening on sf by forcing $sf = c$ has no effect on csf. Otherwise, we need to estimate the effect of that intervention on the class situation feature which is the function estimating (the distribution of) csf condition on $sf = c$, while controlling for situation features in PA_{sf}.

---

[4] If $\leftrightarrow \neq \emptyset$, the user can restart the procedure after adding some more situation features to the causal situation specification.

5   Experimental Results

To validate the proposed approach, we implemented it as a plugin in ProM [16], an open-source and extensible platform for process mining. The implemented plugin takes the event log, the Petri-net model of the process, and, the conformance checking results of replaying the given event log on the given model as input. The implemented plugin is available in the nightly-build of ProM under the name *causality inference in process mining*. In the following, we briefly mention some of the implementation details and the results of applying the plugin on both synthetic and real event logs.

5.1   Implementation Notes

As the search algorithm, we use the Greedy Fast Causal Inference (GFCI) algorithm [9]. GFCI is a hybrid search algorithm where its inputs are the situation feature table and possibly background knowledge and its output is a PAG with the highest score on the input data. In [9], it has been shown that if the assumptions mentioned in Sect. 4.2 hold, then under the large sample limit each edge in the PAG computed by GFCI is correct. Also, using empirical results on simulated data, it has been shown that GFCI has the highest accuracy among several other search algorithms [9]. In this plugin, we use the Tetrad [14] implementation of the GFCI algorithm. In the experiments, we use the following settings for the GFCI algorithm: cutoff for p-values = 0.05, maximum path length = -1, maximum degree = -1, and penalty discount = 2.

For estimating the effect of an intervention on the class situation feature, in the case of continuous data, we assume linear dependencies among the situation features and additive noise. We can represent the SEM graphically by considering the coefficient of sf_1 in S_{sf_2} as the weight of the edge from sf_1 to sf_2 in its corresponding causal structure. Thus, to estimate the magnitude of the effect of sf on the csf, it is enough to sum the multiplication of the weights of the edges for each directed path from sf to csf.

5.2   Synthetic Data

We have created the Petri-net model of the process described in Sect. 2 using CPN Tools [12] and generate an event log with 1000 traces. The log is enriched by adding IPD as a derived trace level attribute that indicates the duration of the sub-model including DEV and TE transitions in person-day. We generate the log with the following settings:

$$
\begin{aligned}
sf_{\perp,C} &= N_{sf_{\perp,C}} & N_{sf_{\perp,C}} &\sim Uniform(1,10) \\
sf_{BC,P} &= N_{sf_{BC,P}} & N_{sf_{BC,P}} &\sim Uniform(1,3) \\
sf_{PB,PBD} &= 10sf_{\perp,C} + N_{sf_{PB,PBD}} & N_{sf_{PB,PBD}} &\sim Uniform(-2,4) \\
sf_{TD,NT} &= 5sf_{\perp,C} + 3sf_{BC,P} + N_{sf_{TD,NT}} & N_{sf_{TD,NT}} &\sim Uniform(-1,2) \\
sf_{\perp,IPD} &= 50sf_{\perp,C} + 5sf_{TD,NT} + N_{sf_{\perp,IPD}} & N_{sf_{\perp,IPD}} &\sim Uniform(10,20)
\end{aligned}
$$

Then, we use the proposed approach for $CSS = ((sf_{\perp,P}, sf_{PB,PBD}, sf_{TC,TN}), sf_{\perp,IPD})$. The discovered causal structure is as depicted in Fig. 4.a. This causal structure does not say much about the direction of discovered potential causal relationships. Regarding the types of edges, we can guess that there might be another influential attribute that acts as a hidden common cause. If we consider $sf_{\perp,C}$ as one of the independent situation features, then the discovered causal structure is the one depicted in Fig. 4.b, which is more accurate. If the complexity of a project is not recorded in the event log, we can assume that the PB takes longer in more complex projects and compute it as the floor of the value of PBD divided by 10. Now, using domain knowledge, we can turn this PAG to the one depicted in Fig. 4.c. By doing the estimation, the SEM shown in Fig. 4.d is obtained. By comparing the estimated coefficients of situation features in Fig. 4.d (the weights of edges), and those in the mentioned equations, it is clear the estimated and real strengths of causal relationships are quite close which proves that the implemented plugin is capable of discovering the true SEM.

Now, if we want to see how each situation feature affects the class situation feature, we just need to click on its corresponding node in the causal structure to see the estimated interventional effect. Suppose c_1 is a constants. For example, we have $sf_{\perp,IPD} = 75.0004 \times sf_{\perp,C} + c_1 \times sf_{TD,NT} + noise$ which means that if we could enforce the complexity of the projects to be one unit more complex, then the implementation phase will take approximately 75 more person-days. The other estimated interventional effect is $sf_{\perp,IPD} = 0.0 \times sf_{PB,PBD}$ that means intervention on $sf_{PB,PBD}$ has no effect on $sf_{\perp,IPD}$.

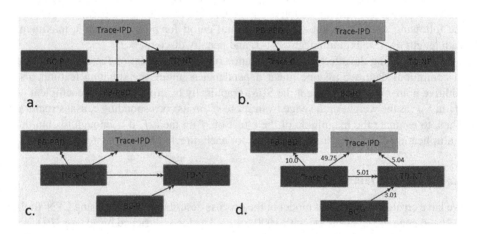

Fig. 4. a) The PAG of causal situation specification of Sect. 5.2, discovered by the implemented plugin. Using this PAG, we can guess that there are other influential situation features not recorded in the data. b) The resulting PAG after considering $sf_{\perp,C}$ as one of the influential situation features, c) The causal structure obtained by modifying the previous one using common sense and domain knowledge. d) The causal structure model generated by doing estimation on the strength of each discovered causal relationship.

5.3   Real Data

For the real data, we use *receipt phase of an environmental permit application process (WABO) CoSeLoG project* [1] event log (receipt log for short), which includes 1434 traces and 8577 activities. As a problem, we consider the delay in some of the cases where the threshold for the delay is set to 3% of the maximum duration of all traces.

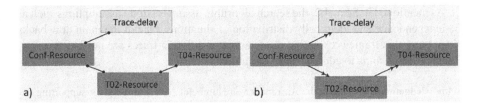

Fig. 5. a) The PAG of causal situation specification of receipt event log example in Sect. 5.3, discovered by the implemented plugin. b) The causal structure obtained by modifying the previous one using common sense and domain knowledge.

Note that the average trace duration in this event log is about 2% of the maximum duration of its traces. So, the class situation feature is $sf_{\perp,delay}$. Here, we want to investigate the effect of three situation features indicating the choice of recourse working on three activities "Confirmation of receipt" denoted as $Conf$, "T02 Check confirmation of receipt" denoted as $T02$, and "T04 Determine confirmation of receipt" denoted as $T04$, on $sf_{\perp,delay}$. Thus, the causal situation specification that we are interested in is $CSS = ((sf_{Conf,Resource}, sf_{T02,Resource}, sf_{T04,Resource}), sf_{\perp,delay})$. Note that using the Chi-square test, there is a high correlation between every pair of situation features.

The output of applying the implemented plugin on the created situation feature table for this problem is shown in Fig. 5.a). Using the temporal ordering of the activities (in this process, $Conf$ happens before $T02$, and $T02$ happens before $T04$ in all the traces) and common sense (the choice of the resource of an activity has no effect on the choice of the resource of another activity that happened before) we can infer that the true causal structure is the one shown in Fig. 5.b). This causal structure indicates that $sf_{Conf,Resource}$ has a causal effect on $sf_{\perp,delay}$ and there is no causal relationship between $sf_{T02,Resource}$ and $sf_{\perp,delay}$ and also between $sf_{T04,Resource}$ and $sf_{\perp,delay}$. Moreover, it indicates that the choice of $sf_{T02,Resource}$ is influenced by $sf_{Conf,Resource}$. By doing the estimation and clicking on each node of the graph we can see the interventional distributions of $sf_{\perp,delay}$ caused by intervention on the corresponding situation feature of that node. For example, we can see that the the probability of $sf_{\perp,delay} = delayed$ under an intervention which enforce $sf_{Conf,Resource} = Resource14$ is almost 0.159.

6   Conclusion

If an organization has performance and/or conformance problems, then it is vital to uncover the causes of these problems and the strength of their effect. It is also needed to investigate the effect of an intervention on the process. This information is essential to design and order the process enhancement and re-engineering steps. This is a very fundamental and influential step toward process enhancement. Process interventions based on correlations that are not causalities may lead to more problems. By using the framework proposed in this paper, the stakeholders can incorporate both domain knowledge and potential statistically supported causal effects to find the SEM of the features and indicators of the process. Using SEM in this framework enables us to estimate to what extent each feature contributes to the given problem using the observational data.

As mentioned in Sect. 4.2, the search algorithm assumes strong assumptions such as the independence and identically distribution of situations. One of the main drawbacks of applying this framework is that the features of different traces are not independent. Future research aims to address these limitations.

Acknowledgement. We thank the Alexander von Humboldt (AvH) Stiftung for supporting our research.

References

1. Buijs, J.: Receipt phase of an environmental permit application process ('wabo'), coselog project. Eindhoven University of Technology (2014)
2. Chickering, D.M.: Optimal structure identification with greedy search. J. Mach. Learn. Res. **3**, 507–554 (2002)
3. Granger, C.W.: Some recent development in a concept of causality. J. Econ. **39**(1–2), 199–211 (1988)
4. Gupta, N., Anand, K., Sureka, A.: Pariket: mining business process logs for root cause analysis of anomalous incidents. In: Chu, W., Kikuchi, S., Bhalla, S. (eds.) DNIS 2015. LNCS, vol. 8999, pp. 244–263. Springer, Cham (2015). https://doi.org/10.1007/978-3-319-16313-0_19
5. Hompes, B.F.A., Maaradji, A., La Rosa, M., Dumas, M., Buijs, J.C.A.M., van der Aalst, W.M.P.: Discovering causal factors explaining business process performance variation. In: Dubois, E., Pohl, K. (eds.) CAiSE 2017. LNCS, vol. 10253, pp. 177–192. Springer, Cham (2017). https://doi.org/10.1007/978-3-319-59536-8_12
6. de Leoni, M., van der Aalst, W.M., Dees, M.: A general process mining framework for correlating, predicting and clustering dynamic behavior based on event logs. Inf. Syst. **56**, 235–257 (2016). https://doi.org/10.1016/j.is.2015.07.003
7. Mothilal, R.K., Sharma, A., Tan, C.: Explaining machine learning classifiers through diverse counterfactual explanations. In: Proceedings of the 2020 Conference on Fairness, Accountability, and Transparency. pp. 607–617 (2020), http://arxiv.org/abs/1905.07697
8. Narendra, T., Agarwal, P., Gupta, M., Dechu, S.: Counterfactual reasoning for process optimization using structural causal models. In: Hildebrandt, T., van Dongen, B.F., Röglinger, M., Mendling, J. (eds.) BPM 2019. LNBIP, vol. 360, pp. 91–106. Springer, Cham (2019). https://doi.org/10.1007/978-3-030-26643-1_6
9. Ogarrio, J.M., Spirtes, P., Ramsey, J.: A hybrid causal search algorithm for latent variable models. In: Proceedings of Probabilistic Graphical Models - Eighth International Conference. pp. 368–379 (2016), http://proceedings.mlr.press/v52/ogarrio16.html
10. Pearl, J.: Causality. Cambridge University Press (2009)
11. Peters, J., Janzing, D., Schölkopf, B.: Elements of causal inference: foundations and learning algorithms. MIT press (2017)
12. Ratzer, A.V., et al.: CPN Tools for editing, simulating, and analysing coloured petri nets. In: van der Aalst, W.M.P., Best, E. (eds.) ICATPN 2003. LNCS, vol. 2679, pp. 450–462. Springer, Heidelberg (2003). https://doi.org/10.1007/3-540-44919-1_28
13. Fani Sani, M., van der Aalst, W., Bolt, A., García-Algarra, J.: Subgroup discovery in process mining. In: Abramowicz, W. (ed.) BIS 2017. LNBIP, vol. 288, pp. 237–252. Springer, Cham (2017). https://doi.org/10.1007/978-3-319-59336-4_17
14. Scheines, R., Spirtes, P., Glymour, C., Meek, C., Richardson, T.: The tetrad project: Constraint based aids to causal model specification. Multivar. Behav. Res. **33**(1), 65–117 (1998)

15. Spirtes, P., Glymour, C.N., Scheines, R., Heckerman, D.: Causation, prediction, and search. MIT press (2000)
16. Verbeek, H., Buijs, J., Van Dongen, B., van der Aalst, W.M.: Prom 6: The process mining toolkit. Proc. of BPM Demonstration Track **615**, 34–39 (2010)
17. Wang, Y., Liang, D., Charlin, L., Blei, D.M.: The deconfounded recommender: A causal inference approach to recommendation. CoRR abs/1808.06581 (2018), http://arxiv.org/abs/1808.06581
18. Zhang, J.: On the completeness of orientation rules for causal discovery in the presence of latent confounders and selection bias. Artif. Intell. **172**(16–17), 1873–1896 (2008). https://doi.org/10.1016/j.artint.2008.08.001

On the Complexity of Resource Controllability in Business Process Management

Matteo Zavatteri$^{(\boxtimes)}$, Romeo Rizzi, and Tiziano Villa

University of Verona, Verona, Italy
{matteo.zavatteri,romeo.rizzi,tiziano.villa}@univr.it

Abstract. Resource controllability of business processes (BPs) is the problem of executing a BP by assigning resources to tasks, while satisfying a set of constraints, according to the outcome of a few uncontrollable events that we only observe during execution. Recent research addressed resource controllability of acyclic BPs where the choices of the XOR paths to take were out of control. However, a formal model of BP to reason on resource controllability is still missing. Thus, the precise mathematical definitions of controllability problems, their semantics and complexity analysis, have remained unexplored. To bridge this gap, we propose a hierarchy of 8 classes of *Business Processes with Resources and Uncertainty (BPRUs)* to address controllable and uncontrollable resource assignments in combination with controllable and uncontrollable choices of the XOR paths to take. We define consistency of BPRs (i.e., BPRUs without uncertainty) and prove that deciding it is NP-complete. We define strong controllability of BPRUs and prove that deciding it is either NP-complete or Σ_2^p-complete depending on the class. We define weak and dynamic controllability of BPRUs and prove that deciding them is Π_2^p-complete and PSPACE-complete, respectively.

Keywords: BPRU · Resource controllability · Complexity analysis

1 Introduction and Related Work

Controllability studies systems that need to satisfy a set of constraints under a few sources of uncertainty that come from the external world. We need controllability because, after all, we always desire to achieve our goals regardless of what might go wrong. In BPM, controllability was investigated to deal with time and resources meaning that, depending on the observed uncontrollable events, we might decide to schedule the same tasks at different times or commit different resources for their execution. Weak controllability implies that, for each

This work was partially supported by MIUR, Project *Italian Outstanding Departments, 2018–2022* and by INdAM, GNCS 2020, Project *Strategic Reasoning and Automated Synthesis of Multi-Agent Systems*.

combination of uncontrollable events known in advance, there exists a way to operate on the controllable part satisfying all constraints. Strong controllability is the opposite case and says that there exists a unique way to operate on the controllable part satisfying all constraints no matter what will happen. Dynamic controllability implies the existence of a strategy reacting to what is going on. Each type of controllability is a two-player game between Controller and Nature.

Temporal controllability addresses BPs specifying tasks with uncontrollable but bounded durations in conjunction with uncontrollable XOR paths [4,5,9,15,19]. To achieve a correct scheduling of the activities, the BP is encoded into a temporal network. Simple Temporal Networks with Uncertainty (STNUs) [18] offer a framework to deal with uncontrollable durations. Conditional Temporal Problem (CTP) [17] and Conditional Simple Temporal Networks (CSTNs) [12] offer frameworks to deal with uncontrollable XOR paths. CSTNs with Uncertainty (CSTNUs) [11] offer a framework to handle both aspects simultaneously, whereas CSTNUs with Decisions (CSTNUDs) [21] offer a framework to also deal with controllable XOR paths in conjunction with all the previous. For temporal networks, the complexity of weak controllability ranges from coNP-completeness to Π_2-completeness, that of strong controllability from P to Σ_2-completeness, whereas that dynamic controllability from P to PSPACE-completeness [1].

Resource controllability addresses BPs under uncontrollable XOR paths. To achieve a correct resource allocation for executing tasks, an initial ad-hoc approach was provided in [19] where XOR paths were encoded into Constraint Networks [8] so that tasks belonging to common parts had the same resource assignments. This approach suffered from the problem that it was up to the designer to choose a fixed total order for all tasks so as to unfold the various XOR paths. It also pointed out that wrong total orders might lead to incompleteness of the approach, a conjecture that was later proved with the proposal of Conditional Constraint Networks with Uncertainty (CNCUs) [22] that handle ordering issues automatically. CNCUs do not offer a natural encoding to handle BPs with both controllable and uncontrollable XOR paths. Weak, strong and dynamic controllability of CNCUs are Π_2-complete, NP-complete, and PSPACE-complete [20]. Attempts to handle resource allocation in BPs are also provided in [3,10]. However, [3] is not a history-based allocation approach, whereas [10] focuses on finding time-optimal resource allocations w.r.t. availability constraints.

Supervisory Control [16] offers a framework for resource controllability once the workflow is encoded into a plant automaton and the constraints are encoded into specification automata. However, for acyclic BPs, this boils down to explode all resource constraints in a regular language that is finite but arbitrary big. A few other frameworks were proposed to handle time and resources together on top of temporal networks (e.g., [6,7]). However, dynamic controllability of those frameworks is reduced to controller synthesis for Timed Game Automata [14]. Also, weak and strong controllability of those frameworks were never addressed.

Despite all this, a natural formal model of BP suitable to reason on weak, strong and dynamic resource controllability is still missing. Consequently, the precise mathematical definitions of these three kinds of controllability, their semantics and complexity classification remain unexplored.

Contributions and Organization. Section 2 defines *Business Processes with Resources and Uncertainty (BPRUs)* to address *both* controllable and uncontrollable resource assignments *in combination* with *both* controllable and uncontrollable XOR path choices. BPRUs give rise to a hierarchy of 8 different classes of processes according to the combination of controllable and uncontrollable parts considered. Section 3 defines consistency of BPRs (without "U"), and weak, strong and dynamic controllability of BPRUs as two-player games. Section 4 proves that deciding consistency of BPRs is NP-complete, deciding strong controllability of BPRUs is either NP-complete or Σ_2^p-complete depending on the considered class, whereas deciding weak and dynamic controllability of BPRUs is Π_2^p-complete and PSPACE-complete, respectively. Section 5 sums up, discusses future work and current limitations for BPRUs involving loops.

2   Business Processes with Resources and Uncertainty

We start by defining structured BPRUs where each XOR split gateway is associated to a unique boolean variable whose truth value assignment is set upon the execution of the gateway. As usual, the components of the BP are represented by the nodes of a graph. Nodes can be labeled by conjunctions of literals over a given finite set of boolean variables abstracting conditionals. These labels can readily capture the nesting level which each node belongs to plus further conditions. In fact, nodes' labels are enough to drive the execution of the process itself. This way, the edges of the process (i.e., the precedence relation) can be left unlabeled, a convenient simplification w.r.t. [13]. We introduce notation little by little throughout the paper. Let B be finite set of boolean variables. A *label* ℓ is any consistent conjunction of literals drawn from B. A label ℓ entails another label ℓ' (in symbols, $\ell \Rightarrow \ell'$) iff ℓ contains ℓ'. We denote the empty label by ⊡.

Definition 1 (BP with Resources and Uncertainty). *A* BP with resources and uncertainty (BPRU) *is a tuple* $P = \langle N, E, B, X, L, R, C \rangle$, *where:*

1. $N := N_C \cup N_U$ *is a finite set of* nodes *partitioned in nodes with* controllable assignment *and nodes with* uncontrollable assignment, *respectively.*
2. $E \subseteq N \times N$ *is a set of directed edges regulating the precedence relation.*
3. $B := B_C \cup B_U$ *is a finite set of* boolean variables, *or* booleans, *partitioned in booleans with* controllable assignment *and booleans with* uncontrollable assignment, *respectively.*
4. $X : B \mapsto N$ *is a bijection associating a (XOR split) node $X(b)$ to each $b \in B$.*
5. *For each $n \in N$, $L(n)$ is the label of n saying when n must be executed, whereas $R(n)$ is a non-empty set of resources associated to n.*
4) *C is a finite set of constraints of the form (ℓ, F), where ℓ is a label and F a boolean formula over atoms of the form $n = r$ with $n \in N$ and $r \in R(n)$. For each $n = r$ appearing in F, we have that $\ell \Rightarrow L(n)$ (coherence), whereas for*

each $b, \neg b \in \ell$ it holds that $\ell \Rightarrow L(X(b))$ (honesty). Coherence and honesty properties are inherited from CSTNs [12]. A constraint (ℓ, F) is unconditional (i.e., must always be satisfied) iff $\ell = \boxdot$.

We assume that P is inductively generated according to the blocks in Fig. 1a-1e. This makes node labels honest by construction because of well-defined nesting levels. A BPRU where $N_U \cup B_U = \emptyset$ is a just BPR (without uncertainty).

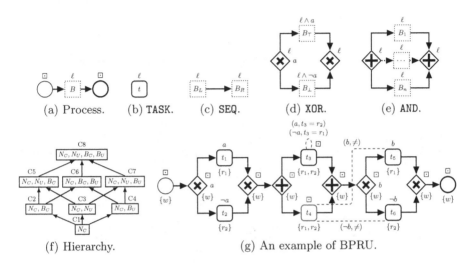

(a) Process.   (b) TASK.   (c) SEQ.   (d) XOR.   (e) AND.

(f) Hierarchy.   (g) An example of BPRU.

Fig. 1. Core process blocks for structured BPs where each component is labeled with a label resembling the nesting level in which the block appears (a)-(e). Hierarchy and example of BPRU (f)-(g).

In our figures, we decorate the BPMN [2] representation of a structured BP as follows. Node labels appear above them. Any XOR split gateway labeled on the right by b is the node $x_b = X(b)$ having j_b as its corresponding join. Any parallel split gateway is a node p in the specification having j_p as its corresponding join. Furthermore, the set of resources associated to nodes appear below them. If a node is uncontrollable (i.e., $n \in N_U$), we draw its border in red. Likewise, if a boolean b is uncontrollable (i.e., $b \in B_U$), we draw it red wherever it appears. We might sometimes draw constraints (ℓ, F) as labeled undirected dashed edges to give a graphical representation of them when they are clear from the context.

Definition 1 gives rise to a hierarchy of 8 classes of processes (C1-C8) each one differing from the others for the considered combination of non-empty N_C, N_U, B_C, and B_U sets. Since BPRs/BPRUs are structured, there always exists a start node s which is controllable. Therefore, each class in the hierarchy contains N_C. Figure 1f shows such a hierarchy. BPRs are processes of classes C1 and C2. BPRUs are processes of classes C3 to C8 as each of them specifies at least a source of uncertainty. Figure 1g provides an example of BPRU having the following specification. $P = \langle N, E, B, X, L, R, C \rangle$, where, by representing mappings m as sets of pairs $(x, m(x))$, we have $N := \{s, \, x_a, \, t_1, \, t_2, \, j_a,$

$p, t_3, t_4, j_p, x_b, t_5, t_6, j_b, e\}$, $E := \{(S, x_a), (x_a, t_1), (x_a, t_2), (t_1, j_a), (t_2, j_a),$
$(j_a, p), (p, t_3), (p, t_4), (t_3, j_p), (t_4, j_p), (j_p, x_b), (x_b, t_5), (x_b, t_6), (t_5, j_b), (t_6, j_b),$
$(j_b, e)\}$, $B := \{a, b\}$, $X := \{(a, x_a), (b, x_b)\}$, $L := \{(s, \square), (x_a, \square), (t_1, a), (t_2, \neg a),$
$(j_a, \square), (p, \square), (t_3, \square), (t_4, \square), (j_p, \square), (x_b, \square), (t_5, b), (t_6, \neg b), (j_b, \square), (e, \square)\}$,
$R := \{(s, \{w\}), (x_a, \{w\}), (t_1, \{r_1\}), (t_2, \{r_2\}), (j_a, \{w\}), (p, \{w\}), (t_3, \{r_1, r_2\}),$
$(t_4, \{r_1, r_2\}), (j_p, \{w\}), (x_b, \{w\}), (t_5, \{r_1\}), (t_6, \{r_2\}), (j_b, \{w\}), (e, \{w\})\}$ and
$C := \{(a, t_3 = r_2), (\neg a, t_3 = r_1), (b, t_4 = r_2 \wedge t_5 = r_1), (\neg b, t_4 = r_1 \wedge t_5 = r_2)\}$.
P is of class C8. However, turning either a or t_4 or both controllable in all
combinations allows us to obtain a process of any other class.

3   Decision Problems: Consistency and Controllability

We start by defining *consistency* of BPRs of class C2 as they employ no uncontrollable parts and contain BPRs of class C1. Then, we define weak, strong and
dynamic controllability of BPRUs of class C8 *as two-player games* between Controller and Nature so as to handle BPRUs of classes C3 to C7 as well. In general,
Controller assigns resources to controllable nodes for their executions and truth
values to the controllable booleans upon the execution of the corresponding XOR
split gateways. Nature does the same for uncontrollable nodes and uncontrollable booleans, respectively. Despite XOR paths discriminate which nodes (not)
to execute in a BP, in what follows we will always assume that all nodes are
executed and all booleans are assigned. Although this is superfluous, it is however not wrong because we will ignore irrelevant assignments when checking the
satisfaction of constraints. This way, we get rid of heavy XOR path reasoning
that would only uselessly complicate the achievement of our results.

Definition 2 (Scenario). *A scenario is a total mapping* $\sigma \colon B \mapsto \{0, 1\}$. *Given
any label* ℓ *and any scenario* σ, *the value of* ℓ *according to* σ *is* $\ell(\sigma) \in \{0, 1\}$.

Definition 3 (Resource assignment). *A resource assignment is a total mapping* ρ *assigning a resource* $\rho(n) \in R(n)$ *to each node* $n \in N$. *Given any boolean
formula* F *over atoms of the form* $n = r$ *with* $n \in N$ *and* $r \in R(n)$, *the value of
F according to* ρ *is* $F(\rho) \in \{0, 1\}$.

Definition 4 (Constraints satisfaction). *Given a scenario* σ *and a resource
assignment* ρ, *the set of constraints* C *is satisfied by* σ *and* ρ *(in symbols,
$C(\sigma, \rho) = 1$) iff for each* $(\ell, F) \in C$ *with* $\ell(\sigma) = 1$ *it holds that* $F(\rho) = 1$.

Definition 5 (Consistency of BPRs). *A BPR is* consistent *if there exists
a scenario* σ *and a resource assignment* ρ *such that* $C(\sigma, \rho) = 1$.

Consider Fig 1g and imagine that $a \in B_C$ and $t_4 \in N_C$. Then, the resulting
process is a consistent BPR of class C2. Indeed, if $\sigma := \{(a, 1), (b, 1)\}$ and $\rho :=
\{(s, w), (x_a, w), (t_1, r_1), (t_2, r_2), (j_a, w), (p, w), (t_3, r_2), (t_4, r_2), (j_p, w), (x_b, w),
(t_5, r_1), (t_6, r_2), (j_b, w), (e, w)\}$, then all constraints C are satisfied. Notice that
we are not discussing in which order to execute nodes. Indeed, once we have a
complete σ, we can project the process onto the unique XOR path coherent with

it and then compute an ordering for the survived nodes by a run of topological sort (at least one exists since the underlying process graph is acyclic).

For BPRUs, we move our analysis from consistency to controllability since we deal with uncontrollable parts. We define three main kinds of controllability for BPRUs: *weak*, *strong*, and *dynamic*. In weak and strong controllability one of the two players moves first and moves once by assigning all components under his/her control without any information about what the other one will do.

Game 1 (Weak controllability). *Let P be a BPRU.*

Phase 1 *For each $b \in B_U$, Nature sets $\sigma(b)$ to a truth value of her choice. For each $n \in N_U$, Nature sets $\rho(n)$ to a resource in $R(n)$ of her choice.*

Phase 2 *For each $b \in B_C$, Controller sets $\sigma(b)$ to a truth value of his choice. For each $n \in N_U$, Controller sets $\rho(n)$ to a resource in $R(n)$ of his choice.*

Controller wins if $C(\sigma, \rho) = 1$. Nature wins otherwise.

Definition 6. *A BPRU is weakly controllable (resp., uncontrollable) if Controller (resp., Nature) has a winning strategy for Game 1.*

The BPRU in Fig. 1g is weakly controllable. Again, since we have uncontrollable but fixed assignments here, we neglect ordering issues in the strategy:

Case 1. If Nature assigns true to a and r_1 to t_4, then Controller assigns false to b, r_1 to t_1, t_5, r_2 to t_2, t_3, t_6, and w to $s, x_a, j_a, p, j_p, x_b, j_b, e$.

Case 2. If Nature assigns true to a and r_2 to t_4, then Controller assigns true to b and makes the same resource assignment of Case 1.

Case 3. If Nature assigns false to a and r_1 to t_4, then Controller assigns false to b, r_1 to t_1, t_3, t_5, r_2 to t_2, t_6, and w to $s, x_a, j_a, p, j_p, x_b, j_b, e$.

Case 4. If Nature assigns false to a and r_2 to t_4, then Controller assigns true to b and makes the same resource assignment of Case 3.

In weak controllability Nature moves first and moves once. Therefore, all resource assignments to uncontrollable nodes and truth value assignments to uncontrollable booleans are known to Controller before he makes his move. Strong controllability models the dual situation where it is Controller to move first and once. Strong controllability is weak controllability with all quantifiers flipped.

Game 2 (Strong controllability). *Let P be a BPRU. The game is the same of that in Game 1 with the difference that Phase 2 goes before Phase 1.*

Definition 7. *A BPRU is strongly controllable (resp., uncontrollable) if Controller (resp., Nature) has a winning strategy for Game 2.*

The BPRU in Fig. 1g is not strongly controllable. Indeed, the controllable resource assignment to t_3 depends on the uncontrollable truth value of a. Likewise, the controllable truth value assignment to b depends on the uncontrollable resource assignment to t_4. Yet, a BPRU which is not strongly controllable might still be executable if Controller decides what truth values to assign to controllable booleans and what resources to assign to controllable nodes (and in which order) depending on how Nature is playing. *Dynamic controllability* addresses this case by means of a game proceeding in rounds until all nodes have been executed. Let $\pi(n) = \{n' \mid (n', n) \in E\}$ be the set of nodes preceding n in P.

Game 3 (Dynamic controllability). *Let P be a BPRU. Let $\overline{N} := N$.*

Phase 1 *If $\overline{N} = \emptyset$ the game is over. Otherwise, Controller retrieves the set of nodes ready for execution (i.e., those without predecessors in $\overline{N}$) by computing $N_R = \{n \mid n \in \overline{N}, \pi(n) \cap \overline{N} = \emptyset\}$. Controller chooses $n \in N_R$ and removes it from $\overline{N}$ by setting $\overline{N} := \overline{N} \setminus \{n\}$.*

Phase 2 *A player chooses a resource $r \in R(n)$ and sets $\rho(n) := r$. If $n \in N_C$ this player is Controller. If $n \in N_U$ this player is Nature.*

Phase 3 *takes place iff n is a XOR split gateway. Let $b \in B$ such that $X(b) = n$. A player set $\sigma(b)$ to a possible truth value of his/her choice. If $b \in B_C$ this player is Controller. If $b \in B_N$ this player is Nature.*

Controller wins if $C(\sigma, \rho) = 1$. Nature wins otherwise.

Definition 8. *A BPRU is dynamically controllable (resp., uncontrollable) if Controller (resp., Nature) has a winning strategy for Game 3.*

Figure 1g is dynamically controllable. The (relevant) part of the strategy is:

1. Controller chooses s and assigns w to it, then chooses x_a and assigns w to it. After that,
   (a) If Nature assigns true to a, then Controller chooses t_1 and assigns r_1 to it, then chooses j_a and assigns w to it, then chooses p and assigns w to it, then chooses t_3 and assigns r_2 to it.
   (b) If Nature assigns false to a, then Controller does the same of (1a) with the difference that he assigns r_1 to t_3.
2. Controller chooses t_4. After that,
   (a) If Nature assigns r_1 to t_4, then Controller chooses j_p and assigns w to it, then chooses x_b and assigns w to it, then assigns false to b, then chooses t_6 and assigns r_2 to it.
   (b) If Nature assigns r_2 to t_4, then Controller chooses j_p and assigns w to it, then chooses x_b and assigns w to it, then assigns true to b, then chooses t_5 and assigns r_1 to it.
3. Controller chooses j_p and assigns w to it, then chooses e and assigns w to it.

4   Complexity of Consistency and Controllability

We classify hardness for class C1 and membership for class C2 to classify the complexity of consistency of BPRs. We classify hardness for classes C3 and C4 and membership for classes C6 and C8 for the complexity of strong controllability of BPRUs. We classify hardness for classes C3 and C4 and membership for class

C8 to classify the complexity of weak and dynamic controllability of BPRUs. Before proceeding we discuss how to translate any quantified boolean formula (QBF) to a process of class C1 or C3 or C4 depending on the involved quantifiers. Let $\Phi := \mathcal{Q}_1 x_1, \ldots, \mathcal{Q}_n x_n \varphi(x_1, \ldots, x_n)$ be a QBF over n variables.

We can translate Φ to a process of class C1 or C3 with $n + 2$ nodes, $n + 1$ edges sequentially connecting them, 2 resources 0/1 associated to all nodes and 1 constraint $(\boxdot, \varphi(x_1, \ldots, x_n))$. The resulting process is of class C1 if all variables are existentially quantified or of class C3 if some variable is universally quantified. Let us call QBF-to-C1-C3(Φ) such a procedure.

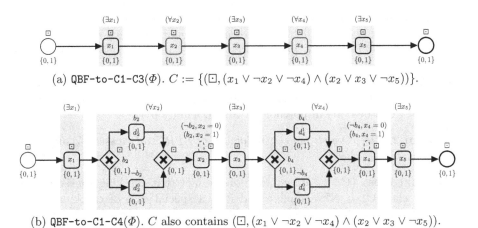

(a) QBF-to-C1-C3(Φ). $C := \{(\boxdot, (x_1 \vee \neg x_2 \vee \neg x_4) \wedge (x_2 \vee x_3 \vee \neg x_5))\}$.

(b) QBF-to-C1-C4(Φ). C also contains $(\boxdot, (x_1 \vee \neg x_2 \vee \neg x_4) \wedge (x_2 \vee x_3 \vee \neg x_5))$.

Fig. 2. $\Phi := \exists x_1 \forall x_2 \exists x_3 \forall x_4 \exists x_5 \varphi$, where $\varphi = (x_1 \vee \neg x_2 \vee \neg x_4) \wedge (x_2 \vee x_3 \vee \neg x_5)$.

Likewise, we can also translate Φ to a process of class C1 or C4 with $n_1 + 5n_2 + 2$ nodes, $n_1 + 6n_2 + 1$ edges and $2n_2 + 1$ constraints, where n_1 is the number of "$\exists x$" in Φ and n_2 is the number of "$\forall x$" in Φ ($n = n_1 + n_2$). Specifically, we encode each universally quantified variable $\forall x_i$ by means of the following gadget. We generate a sequence block containing a XOR block and a task. The XOR gateway is associated to an uncontrollable boolean b_i and contains two dummy tasks d_i^1 and d_i^0 that play no role on constraints (they are just there for the XOR block to exist). After the XOR block, we have a task x_i and a pair of constraints $(b_i, x_i = 1)$ and $(\neg b_i, x_i = 0)$ to *mirror* the uncontrollable truth value assignment to b_i to a resource assignment to x_i. After that we keep having the constraint $(\boxdot, \varphi(x_1, \ldots, x_n))$. The resulting process is either of class C1, if all variables are existentially quantified, or of class C4, if some variable is universally quantified. Let us call QBF-to-C1-C4(Φ) this second procedure.

It is clear that QBF-to-C1-C3 and QBF-to-C1-C4 run in polynomial time and that QBF-to-C1-C3(Φ) = QBF-to-C1-C4(Φ) if all variables are existentially quantified. This is what we hold on to for providing reductions from QBF. We provide intuitive examples in Fig. 2a and Fig. 2b.

Theorem 1. *Deciding consistency of BPRs is NP-complete.*

Proof. **Hardness (C1).** Let $\varphi(x_1, \ldots, x_n)$ be an instance of 3SAT. That is, a 3CNF formula over n variables and m clauses, where each clause is a disjunction of 3 literals over $x_1, \ldots, x_n$. Solving 3SAT is NP-complete and it is a special case of solving a QBF embedding the same $\varphi(x_1, \ldots, x_n)$ where all variables are existentially quantified. We claim that $\varphi(x_1, \ldots, x_n)$ is satisfiable iff $P :=$ QBF-to-C1-C3($\exists x_1, \ldots, \exists x_n \varphi(x_1, \ldots, x_n)$) = QBF-to-C1-C4($\exists x_1, \ldots, \exists x_n \varphi(x_1, \ldots, x_n)$) is consistent. Let t be any model of $\varphi(x_1, \ldots, x_n)$ and let $\rho :=$ $\{(x_i, t(x_i)) \mid 1 \leq i \leq n\} \cup \{(s, 0), (e, 0)\}$ be a resource assignment for P. Let $\sigma \colon B \to \{0, 1\}$ be an empty scenario (recall that $B = \emptyset$). Then, since tasks in P resemble variables in φ, ρ and t coincide on them and $(\Box, \varphi(x_1, \ldots, x_n)) \in C$, then $C(\sigma, \rho) = 1$. Likewise, let σ be the empty scenario and ρ a resource assignment such that $C(\sigma, \rho) = 1$. Let $t := \{(x_i, \rho(x_i)) \mid 1 \leq i \leq n\}$ be a truth value assignment for φ. Then, t satisfies $\varphi(x_1, \ldots, x_n)$ for the same reason. **Membership (C2).** A *yes certificate* is a pair scenario-resource assignment (σ, ρ) such that $C(\sigma, \rho) = 1$. We know that (σ, ρ) has size $|B| + |N|$ and that checking if $C(\sigma, \rho) = 1$ can be done in polynomial time (Definition 4). $\qquad\blacksquare$

Theorem 2. *Deciding strong controllability of BPRUs of classes C4 and C6 is NP-complete.*

Proof. **Hardness (C4).** Consequence of the fact that checking consistency of BPRs of class C1 is NP-complete and it is a special case of checking strong controllability of BPRUs of class C4 (when $B_U = \emptyset$). Therefore, strong controllability of class C4 (and thus class C6) is at least NP-hard. **Membership (C6).** A certificate of yes is a scenario σ_C over B_C and a resource assignment ρ which together amount to a size of $|B_C| + |N|$. Now, let $(\ell, F) \in C$ be any constraint such that ℓ is consistent with the assignments in σ_C (if any). Then, (ℓ, F) is satisfied for each possible scenario σ_N over B_N iff $F(\rho) = 1$, and since we have a finite number of constraints to check, the overall check is polynomial. $\qquad\blacksquare$

We now move to consider those classes of BPRUs where deciding controllability has complexity above NP $\cup$ coNP. The complexity classes Σ_i^p and Π_i^p are recursively defined as follows: $\Sigma_0^p = \Pi_0^p = P$, $\Sigma_1^p = \text{NP}$, $\Pi_1^p = \text{coNP}$, and for each $i \geq 0$, $\Sigma_{i+1}^p = \text{NP}^{\Sigma_i}$ and $\Pi_{i+1}^p = \text{coNP}^{\Sigma_i}$.

Theorem 3. *Deciding strong controllability of BPRUs of classes C3, C5, C7 and C8 is Σ_2^p-complete.*

Proof. **Hardness (C3).** Let $\Phi := \exists x_1, \ldots \exists x_k, \forall x_{k+1}, \ldots \forall x_n \varphi(x_1, \ldots, x_n)$ be a QBF with $\varphi(x_1, \ldots, x_n)$ in 3DNF. Solving such an instance of QBF is known to be Σ_2^p-complete. We claim that Φ is true iff $P :=$ QBF-to-C1-C3(Φ) is strongly controllable. The 0/1 resource assignments to the tasks of P match (in the same order) the true/false assignments to the homonymous boolean variables in the strategy to make Φ true resulting in a strategy to win Game 2 on P. And the true/false assignments to the boolean variables of Φ match (in the same order) the 0/1 resource assignments to the homonymous tasks in the strategy to win Game 2 on P resulting in a strategy to make Φ true. **Membership (C8).** A BPRU is strongly controllable if *there exists* a scenario σ_C over B_C and a resource assignment ρ_C over N_C such that *for every scenario* σ_N over B_N and *every resource assignment* ρ_N over N_N, $C(\sigma_C \cup \sigma_N, \rho_C \cup \rho_N) = 1$. Since $\sigma_C, \sigma_N, \rho_C$ and ρ_N admit a finite and compact encoding, this problem is in Σ_2^p by definition.

Theorem 4. *Deciding weak controllability of BPRUs is Π_2^p-complete.*

Proof. Since $\Pi_2^p = co\Sigma_2^p$, we provide a reduction from a QBF $\Phi := \forall x_1, \ldots \forall x_k, \exists x_{k+1}, \ldots \exists x_n \varphi(x_1, \ldots, x_n)$ with $\varphi(x_1, \ldots, x_n)$ in 3CNF. Solving such an instance of QBF is known to be Π_2^p-complete. Since neither BPRUs of class C3 contain BPRUs of class C4 nor vice versa we first show that any source of uncertainty (i.e., uncontrollable nodes or uncontrollable booleans) even considered in isolation makes the problem at least Π_2^p-hard. **Hardness (C3).** We claim that Φ is true iff $P := \mathtt{QBF\text{-}to\text{-}C1\text{-}C3}(\Phi)$ is weakly controllable. The reason is exactly the same of that in the proof of Theorem 3 (hardness). **Hardness (C4).** We claim that Φ is true iff $P := \mathtt{QBF\text{-}to\text{-}C1\text{-}C4}(\Phi)$ is weakly controllable. Existentially quantified variables are still treated the same way. For universally quantified variables in Φ we act in the following way. Let x_i a universally quantified variable in Φ. If x_i is assigned true/false in the strategy to make Φ true, then, the XOR gateway x_{b_i} is assigned any resource, $s(b_i)$ is set to the same truth value assignment to x_i in Φ, d_i^1, d_i^0 and j_{b_i} are assigned any resources and finally x_i is assigned 0 or 1 matching the truth value assignment of the homonymous variable in the strategy to make Φ true (in this order). This results in a strategy to win Game 1 on P. Again, $\varphi(x_1, \ldots, x_n)$ is satisfied by both Φ and P (and also self loops at tasks x_i in P since $\rho(x_i)$ is coherent with the truth value to b_i). We act similarly for the other direction by only mirroring the resource assignments 0/1 to x_i tasks in the strategy to win Game 1 on P into true/false assignments to the homonymous boolean variables in Φ. This results in a strategy to make Φ true. **Membership (C8).** A process is weakly controllable if *for every scenario* σ_N over B_N and *every resource assignment* ρ_N over N_N, *there exists* a scenario σ_C over B_C and a resource assignment ρ_C over N_C such that $C(\sigma_C \cup \sigma_N, \rho_C \cup \rho_N) = 1$. Since $\sigma_C, \sigma_N, \rho_C$ and ρ_N admit a finite and compact encoding, this problem is in Π_2^p by definition.

Theorem 5. *Deciding dynamic controllability of BPRUs is PSPACE-complete.*

Proof. **Hardness (C3,C4).** We consider a general QBF $\Phi := \exists x_1, \forall x_2, \ldots, \exists x_{n-1}, \forall x_n \varphi(x_1, \ldots, x_n)$ without bounds on the number of alternations of quantifiers. We claim that Φ is true iff $P := \mathtt{QBF\text{-}to\text{-}C1\text{-}C3}(\Phi)$ (C3) is dynamically controllable. Likewise, we claim that Φ is true iff $P := \mathtt{QBF\text{-}to\text{-}C1\text{-}C4}(\Phi)$ (C4) is dynamically controllable. The proofs are similar to those discussed above: the resource assignments in P match the truth assignments in Φ and vice versa *in the same order*. **Membership (C8).** Algorithm 1 is a polynomial space algorithm for deciding dynamic controllability of BPRUs: it explores an AND/OR search tree whose depth size is upper bounded by $\mathcal{O}(|N|)$.

Algorithm 1: BPRU-DC(P)

Input: A BPRU $P = \langle N, E, B, X, L, R, C \rangle$
Output: Yes, if P is dynamically controllable. No otherwise.

1   BPRU-DC (P)
2 | Let σ be an empty scenario, ρ an empty resource assignment, and $\overline{N} := N$.
3 | return Phase1($P, \sigma, \rho, \overline{N}$)

4   Phase1 ($P, \sigma, \rho, \overline{N}$)
5 | if $\overline{N} = \emptyset$ then return $C(\sigma, \rho)$;
6 | $N_R \leftarrow \{n \mid n \in \overline{N}, \pi(n) \cap \overline{N} = \emptyset\}$ $\triangleright \pi(n) = \{n' \mid (n', n) \in E\}$ (polytime step)
7 | for $n \in N_R$ do
8 | | if Phase2($P, \sigma, \rho, \overline{N} \setminus \{n\}, n$) then return Yes;
9 | return No

10   Phase2 ($P, \sigma, \rho, \overline{N}, n$)
11 | if $n \in N_C$ then return c_rassign($P, \sigma, \rho, \overline{N}, n$) ;
12 | return n_rassign($P, \sigma, \rho, \overline{N}, n$)

13   c_rassign ($P, \sigma, \rho, \overline{N}, n$)
14 | for $r \in R(n)$ do
15 | | if Phase3($P, \sigma, \rho \cup \{(n, r)\}, \overline{N}, n$) then return Yes;
16 | return No

17   n_rassign ($P, \sigma, \rho, \overline{N}, n$)
18 | for $r \in R(n)$ do
19 | | if $\neg$Phase3($P, \sigma, \rho \cup \{(n, r)\}, \overline{N}, n$) then return No;
20 | return Yes

21   Phase3 ($P, \sigma, \rho, \overline{N}, n$)
22 | if there does not exist $b \in B$ with $X(b) = n$ then return Phase1($P, \sigma, \rho, \overline{N}$) ;
23 | Let $b \in B$ such that $X(b) = n$.
24 | if $b \in B_C$ then return Phase1($P, \sigma \cup \{(b, 0)\}, \rho, \overline{N}$) $\vee$ Phase1($P, \sigma \cup \{(b, 1)\}, \rho, \overline{N}$);
25 | return Phase1($P, \sigma \cup \{(b, 0)\}, \rho, \overline{N}$) $\wedge$ Phase1($P, \sigma \cup \{(b, 1)\}, \rho, \overline{N}$)

---

5   Conclusions and Future Work

We defined business processes with resources and uncertainty (BPRUs). Resources might abstract users, machineries, autonomous agents, or a combination of them. BPRUs give rise to a hierarchy of 8 classes of BPs each one differing from the others for the considered combination of controllable and uncontrollable parts. BPRs of classes C1 and C2 specify no uncertainty, whereas BPRUs of classes C3 to C8 do. We defined consistency of BPRs and proved that deciding it is NP-complete. We defined strong controllability of BPRUs and proved that deciding it is NP-complete (classes C4 and C6) and Σ_2^p-complete (classes C3, C5, C7 and C8). We defined weak and dynamic controllability of BPRUs and proved that deciding them is Π_2^p-complete and PSPACE-complete, respectively.

As future work, we plan to design strategy synthesis algorithms for the corresponding search problems. We also plan to face the following limitation arising from loops containing constraints. Consider Fig. 3 in which all labels are empty. Regardless of the iterating condition, the problem is that t_1 is out of control with Nature that can assign different resources to it in different iterations making impossible to always satisfy the specified constraint.

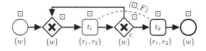

Fig. 3. Loop example. $F = (t_1 = r_1 \wedge t_2 = r_1) \vee (t_1 = r_2 \wedge t_2 = r_2)$.

References

1. Bhargava, N., Williams, B.C.: Complexity bounds for the controllability of temporal networks with conditions, disjunctions, and uncertainty. Artif. Intell. **271**, 1–17 (2019)
2. Business process modeling notation 2.0, http://www.omg.org/spec/BPMN/2.0/
3. Cabanillas, C., Resinas, M., del-Río-Ortega, A., Cortés, A.R.: Specification and automated design-time analysis of the business process human resource perspective. Inf. Syst. **52**, 55–82 (2015)
4. Combi, C., Gambini, M., Migliorini, S., Posenato, R.: Representing business processes through a temporal data-centric workflow modeling language: an application to the management of clinical pathways. IEEE Trans. Systems, Man, Cybernetics: Syst. **44**(9), 1182–1203 (2014)
5. Combi, C., Posenato, R.: Controllability in temporal conceptual workflow schemata. In: Dayal, U., Eder, J., Koehler, J., Reijers, H.A. (eds.) BPM 2009. LNCS, vol. 5701, pp. 64–79. Springer, Heidelberg (2009). https://doi.org/10.1007/978-3-642-03848-8_6
6. Combi, C., Posenato, R., Viganò, L., Zavatteri, M.: Access controlled temporal networks. In: ICAART 2017. pp. 118–131. ScitePress (2017)
7. Combi, C., Posenato, R., Viganò, L., Zavatteri, M.: Conditional simple temporal networks with uncertainty and resources. JAIR **64**, 931–985 (2019)
8. Dechter, R.: Constraint processing. Elsevier (2003)
9. Eder, J., Franceschetti, M., Köpke, J.: Controllability of orchestrations with temporal sla: Encoding temporal xor in cstnud. In: IIWAS. pp. 234–242. ACM (2018)
10. Havur, G., Cabanillas, C., Mendling, J., Polleres, A.: Automated resource allocation in business processes with answer set programming. In: Reichert, M., Reijers, H.A. (eds.) BPM 2015. LNBIP, vol. 256, pp. 191–203. Springer, Cham (2016). https://doi.org/10.1007/978-3-319-42887-1_16
11. Hunsberger, L., Posenato, R., Combi, C.: The dynamic controllability of conditional stns with uncertainty. In: PlanEx at ICAPS-2012. pp. 1–8 (2012)
12. Hunsberger, L., Posenato, R., Combi, C.: A sound-and-complete propagation-based algorithm for checking the dynamic consistency of conditional simple temporal networks. In: TIME 2015. pp. 4–18. IEEE CPS (2015)
13. Kiepuszewski, B., ter Hofstede, A.H.M., Bussler, C.J.: On structured workflow modelling. In: Wangler, B., Bergman, L. (eds.) CAiSE 2000. LNCS, vol. 1789, pp. 431–445. Springer, Heidelberg (2000). https://doi.org/10.1007/3-540-45140-4_29
14. Maler, O., Pnueli, A., Sifakis, J.: On the synthesis of discrete controllers for timed systems. In: Mayr, E.W., Puech, C. (eds.) STACS 1995. LNCS, vol. 900, pp. 229–242. Springer, Heidelberg (1995). https://doi.org/10.1007/3-540-59042-0_76
15. Posenato, R., Zerbato, F., Combi, C.: Managing Decision Tasks and Events in Time-Aware Business Process Models. In: Weske, M., Montali, M., Weber, I., vom Brocke, J. (eds.) BPM 2018. LNCS, vol. 11080, pp. 102–118. Springer, Cham (2018). https://doi.org/10.1007/978-3-319-98648-7_7

16. Ramadge, P., Wonham, W.: Supervisory control of a class of discrete event processes. SIAM J. Control Optim. **25**, 206–230 (1987)
17. Tsamardinos, I., Vidal, T., Pollack, M.E.: CTP: A new constraint-based formalism for conditional, temporal planning. Constraints **8**(4), 365–388 (2003)
18. Vidal, T., Fargier, H.: Handling contingency in temporal constraint networks: from consistency to controllabilities. J. Exp. Theor. Artif. Intell. **11**(1), 23–45 (1999)
19. Zavatteri, M., Combi, C., Posenato, R., Viganò, L.: Weak, strong and dynamic controllability of access-controlled workflows under conditional uncertainty. In: Carmona, J., Engels, G., Kumar, A. (eds.) BPM 2017. LNCS, vol. 10445, pp. 235–251. Springer, Cham (2017). https://doi.org/10.1007/978-3-319-65000-5_14
20. Zavatteri, M., Rizzi, R., Villa, T.: Complexity of weak, strong and dynamic controllability of CNCUs. In: OVERLAY. vol. 2509, pp. 83–88. CEUR-WS.org (2019)
21. Zavatteri, M., Viganò, L.: Conditional simple temporal networks with uncertainty and decisions. Theor. Comput. Sci. **797**, 77–101 (2019)
22. Zavatteri, M., Viganò, L.: Conditional uncertainty in constraint networks. In: van den Herik, J., Rocha, A.P. (eds.) ICAART 2018. LNCS (LNAI), vol. 11352, pp. 130–160. Springer, Cham (2019). https://doi.org/10.1007/978-3-030-05453-3_7

D3BA: A Tool for Optimizing Business Processes Using Non-deterministic Planning

Tathagata Chakraborti$^{(\boxtimes)}$, Shubham Agarwal, Yasaman Khazaeni, Yara Rizk, and Vatche Isahagian

IBM Research AI, Cambridge, MA, USA

Abstract. This paper builds on recent work in the declarative design of dialogue agents and proposes an exciting new tool – D3BA (Declarative Design for Digital Business Automation) – to optimize business processes using AI planning. The tool provides a powerful framework to build, optimize, and maintain complex business processes and optimize them by composing with services that automate one or more subtasks. We illustrate salient features of this composition technique, compare with other philosophies of composition, and highlight exciting opportunities for research in this emerging field of business process automation.

Keywords: Robotic process automation · Process composition · Conversational agents · Multi-agent orchestration · Automated planning

1 Introduction

A business process is a collection of tasks which in a specific sequence meet some business goal, such as producing a service or product for customers. People performing these tasks are referred to as case workers. In practice, the life cycle of a business process is riddled with repetitive tasks, severe bottlenecks and hot spots which impact the performance of the case worker, and quality of service. Recent advances in artificial intelligence can be leveraged to significantly revamp how we build and maintain business processes with the goal of improving the case worker experience. Indeed, the ability to inject manual business processes with artificial intelligence and sophisticated automation has received increased attention lately [11]. We want to determine where and when to deploy automation, to reduce the workload on the individual caseworkers and make such processes easier to author and maintain by stakeholders.

Our agent composition tool, D3BA, brings the notion of web service composition to the task of business process management [16]. The idea here is to change, evolve, and optimize a business process by injecting into it services that can perform specific computations, while providing the persona in charge of the process tools to manage it easily. The specific notion of optimization we focus on is that

© Springer Nature Switzerland AG 2020
A. Del Río Ortega et al. (Eds.): BPM 2020 Workshops, LNBIP 397, pp. 181–193, 2020.
https://doi.org/10.1007/978-3-030-66498-5_14

of maximizing automation in the composed process and thus reducing the load on individual case workers. Of course, if the automated components come with additional features, we can add those considerations (such as health, probability of success, cost, etc.) into the optimization criterion. Thus, D3BA provides two main features: (1) an interface to build, edit, visualize, and maintain complex processes concisely using a declarative specification that allows exponential scale-up from the representation to the realized process; and, (2) an interface to optimize the process by composing it with services that can automated one or more parts of it, while still maintaining the features from (1).

To achieve this, we build on a substrate of non-deterministic planning (a) to compute the composed process offline and allow the process manager to analyse and edit it; and (b) to plan for the inherent uncertain nature of execution with external services. The tool includes a built-in executor for the generated process once it is deployed. It optimizes part of a given business process or orchestrates the entire life-cycle of the process.

With regard to past work at the intersection of non-deterministic planning (c.f. Section 4) and declarative specification of processes, we would like to emphasize that our focus here is not on the declarative aspect as a means to define a process but as a means to facilitate a particular kind of optimization (through composition with automation). Also, it is not on non-determinism as a means to handle uncertainty (which comes for free) but as a means to facilitate the user interaction for this optimization process. There is indeed a lot of work [27,38] on how the declarative paradigm readily translates to business process applications that require the definition and composition of processes. That particular aspect of the problem is admittedly quite well explored but only acts as a means to an end here. The unique value proposition we are providing is the *business process + skills = optimized business process interaction patterns* with the human in the loop at design time [17]. The proposed interaction is a very specific flavor of human-in-the-loop composition that allows the business process manager persona to take in a new process or write a process in a declarative form, take in a catalog of skills, and author optimized processes offline. This is where the proposed framework built on non-deterministic planning comes into form in being able to surface different optimized compositions to the manager persona to edit, debug, visualize, and personalize further. The declarative form is necessary for this to happen (and account for the fact that skill catalogs are independent of the process).

2   System Overview: Process + Skills = Optimized Process

In AAAI 2019, we demonstrated D3WA [7] – a tool meant to reduce the effort and expertise required to design sophisticated *goal-directed* conversational agents e.g. for applications such as customer support. The state of the art [37] in the design of such agents requires the dialogue designer to either write down the entire dialogue tree by hand (e.g. Google Dialogue Flow or Watson Assistant), which quickly becomes cumbersome. Alternatively, they can train end-to-end systems, which provide no control over their emerging behavior [20] and are thus unusable in the enterprise scene. Instead, in [4,21], the authors proposed a paradigm shift in how such agents are built by conceiving a declarative way of specifying them. In this paper, we extend their framework for the purposes of the definition and composition of automated skills into business process specifications. A video demonstration can be viewed at http://ibm.biz/d3ba-video.

At the core of the declarative specification [21] is a set of variables that model the state of the world and actions that depend and operate on those variables to define an agent's capabilities to affect change to the world. In a conversational agent, such actions can be either speech actions that interact directly with the end user, or internal actions such as logical inferences or API calls. The latter is more relevant to our case.

Each action is defined by a set of NEEDs or statuses of variables which tells the agent when it can perform the action (these become actions preconditions available to the planner in the backend), and a set of OUTCOMES (recall the two outcomes in our API call example), one of which might occur at execution time. This specification, in terms of needs of an action and their corresponding outcomes, embodies the "can do" semantics that form the crux of the declarative specification of an agent (rather than "must do" semantics of the imperative setup).[1]

Each outcome produces a set of updates to the variables – these are compiled to the non-deterministic effects of the actions available to the planner in the backend. A non-deterministic planner [22,23] uses this specification to plan all possible outcomes offline and automatically generates the resulting dialogue tree (which would otherwise have had to be written manually). This results in exponential savings from the specification size to the complexity of the composed agent. More details on the D3WA specification can be found in [21] and a sample specification is available at: http://editor.planning.domains/#read_session=cH12680BRp. Next, we describe how we adopt this platform for business process management and optimization.

2.1   Composition Technique

The composition framework consists of skills and agents. Agents embody business processes and are composed of units of tasks. Skills, the building blocks

---

[1] Interestingly, as we discuss in more detail in [21], the declarative specification also allows for imperative patterns (e.g. "forced followups") to be brought in wherever deemed necessary by the process author – to enforce more control over the eventual process. The actual goal of the process itself can be quite open-ended: e.g. in the specification linked above in [21], the process ends when the user says that they are done, and this abstract goal can materialize in many ways. For most processes, the ability to model such abstract goals is imperative [1] – in general these manifest as conditions that can be planned with but whose exact values need to be sensed or "determined" at execution time [23] – either by interacting with the user, such as in the case of finding out if they are done, or by monitoring the system state.

of agents, are atomic functions that execute very specific tasks.[2] Our goal is to augment existing business processes with automated skills, so as to optimize the overall task.

Existing Process. The composition process starts with a business process written either in D3BA, or an existing process imported from outside, usually in the form of a finite state machine, a graph, a mind map, or a similar data structure. Though the D3BA tool currently does not support the latter yet, refer to [36] to see how this translation is done (note that this translation will lose the declarative specification's exponential savings).

The interface to specify a process remains almost identical to D3WA. This is not surprising since a key aspect of designing goal-oriented dialogue is the specification of the underlying process that must be maintained in conversation (e.g. in customer support). The only difference is that certain types of actions, namely API calls and logic actions, take precedence over speech actions. The latter may or may not occur in a process but still facilitates a conversational interface to the processes (as we will see later). These form a sufficient set of capabilities to represent any business process.

Skills. Skills are the unit of automation. They automate a task within a process: e.g. a skill that automatically retrieves the credit score of a customer may require as inputs the customer's name and account number and produce as output the credit score or an error if the retrieval fails. The skill specification interface thus follows a similar structure to that of a generic D3WA action with an even simpler abstraction to the outcome enumeration. It accepts as NEEDs and GOTs the inputs and outputs of a service. This specification, to be consumed alongside that of the rest of the business process, can be compiled internally into possible OUTCOMEs of invoking the service by considering a power set of the GOTs. The semantics of this compilation is that by invoking the service the planner is expecting to get back (n)one or more of the promised outputs.

Composed Process. The inputs to the composition step is thus a process and a set of skills and the output is a optimized process wherein the original process has been composed with skills wherever possible to maximize automation, as shown in Fig. 1. Once the skills have been compiled to the standard D3WA form the rest of the process remains same as in D3WA. This means we get all the rest of its features for free, including being able to visualize, debug, and iterate on the composed process once it has been computed. The reason declarative works well in this setting is two-fold: First, the sheer size of these composed processes, and the need to be able to be flexible with their management, makes it imperative that they are not written and maintained by hand. Furthermore, as we mentioned before, the source of skills and processes may be different. The declarative framework allows developers of either to develop without having to worry about how they relate to each other. The planner preforms an essential role in the background by providing a powerful tool to stitch them together.

---

[2] Similar to "skills" in Watson Assistant and Amazon Alexa.

This composition task is non-trivial since, given a complex process and a large set of skills, navigating the possible combinations and determining the optimal ones manually is infeasible. Also, the source of the process and the skills may be different: i.e. the developer may have no idea about the business processes that their skill is eventually going to be used in. Finally, the features of a skill, as well as the process itself, change over time, making it essential that the composition process is automated.

2.2 Features of Proposed Composition

Consider the travel application process again (Fig. 1). On the left, we see a part of the travel approval process dealing with acquiring information from the applicant. There are 3 back-and-forths to determine the name of the conference, title of the presentation, and estimated expenses, until this step of the process is complete.

Fallbacks. Fig. 1a (right) shows a process optimized with skills that estimate expenses given a conference and return the paper title given the employee information. In the optimized process, the added flows from to the automated skills are in blue, while handovers from automation to the original process are in red. Notice that the four original back-and-forths are still there, but two of them have been bypassed by the skills and are only resorted to as fallbacks, if they fail. This is noticeable in the dialogue in the inset, where the number of interactions with the caseworker is reduced.

Robustness. These fallbacks need not be restricted to the original process only. As shown in Fig. 1b, our approach figures out how to chain equivalent skills to maximize success of the automated components. The composition technique allows this optimized process to grow exponentially to increase the robustness of

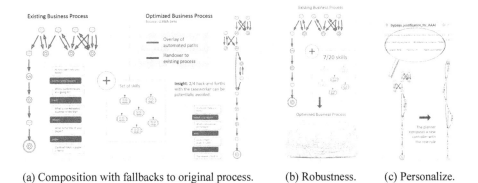

(a) Composition with fallbacks to original process.   (b) Robustness.   (c) Personalize.

Fig. 1. D3BA compositions illustrating (1a) fallback to equivalent services or (in the worst case) to the original manual process; (1b) automated filtering of relevant services + chaining of equivalent skills to increase robustness of the composition; and finally (1c) easy personalization and management of the composed process.

automation. This is an example of how quickly the composition task (let alone the orchestration – i.e. execution and monitoring – of the process once it is composed) can go out of hand for manual approaches even for the small process used here for illustrative purposes.

Relevancy. An additional feature in Fig. 1b is that the catalog is larger than the set of skills used in the optimized process. Our approach figures out which skills to use and which to ignore from the skill catalog. In addition to the complexity or chained composition demonstrated above, this too is often beyond the scope of manual composition.

Customization. Our approach also makes it easy to modify and personalize the composed process with new rules. In Fig. 1c, we add a rule to bypass the manager approval if the presentation is going to happen at a particular conference. This is automatically reflected again, in the newly composed process. This ability to make small edits and effect large changes in the process is a unique feature of declarative modeling.

3   Empirical Results

In order to better illustrate the properties of the processes composed in D3BA, let us consider three simplified business processes encountered in banking applications:

T_1 *Opening a new account:* This process requires the user to declare their name, age, and address to open a new bank account. Once the user asks to open a new account, the assistant asks the user for the required information and then makes an API call to get the application started. It follows up with the application number and sends the notification to the address once the application has been evaluated.

T_2 *Applying for a new credit card:* The application process for a new credit card is slightly more complicated. Once the user asks for a new credit card, the assistant determines if they already have an account with the bank or not. If they do not, it sets up an account first. In addition to the information to set up an account, this process also requires the credit score. Once the user has disclosed all the information, the assistant again puts in the application and provides the confirmation number, with an additional step of verification of the details provided.

T_3 *Applying for a loan:* The loan process is similar to the credit card application but requires an additional step of acquiring the loan amount.

We now explore the empirical properties of automated agent composition, especially in terms of the scale up achieved from the agent-skill orchestration framework.

Automated Agent Composition. Previously, we talked about generating agents on the fly from the task's end conditions or goals. In addition to this being an easier authoring construct for the bot developer, it has the advantage of bringing down the complexity of the compositions thereby still making them amenable for visualization and debugging before deployment. Table 1 illustrates this in terms of the relative sizes of the graphs composed for each task above.

Scale-up From Automated Composition. Table 2 illustrates the sizes (measured in edges of the composed graph, to compare against the complexity of manual build) of the composed processes and runtime with respect to increasing size of the skill catalog. Clearly, while composition is intractable without automation even for small catalog sizes, the planner is quite fast especially considering the compositions are offline. While the base process here is quite simple, for the ease of presentation, from the size of the composed graphs it should be quite clear that the tool is able to handle processes of much greater complexity. More empirical results can be found in [21].

Code Reuse in Composed Agents. One of the advantages of using the declarative form is reusing skills in the same (composed) process and across different processes. This allows the source of skills to be completely independent of the source of business processes, i.e. developers can contribute freely to the catalog without having to worry about where their skill would be used, while process authors can tap into this catalog to make robust compositions instead of manually writing and wiring together different possible automated paths. We measure this notion of "reuse" using three metrics, as follows:

m_1 Number of paths containing *only* automation, between any two manual nodes inherited from the original process, in the composed graph. This signifies all possible automation scripts needed to manually replicate the most robust composition.

m_2 Number of instances a fully automated path exists between any two manual nodes inherited from the original process in the composed graph: unlike m_1, m_2 does not double count different automated paths that are automating the same (sub-)process. This signifies the least number of scripts that one has to write to provide at least one of the automation options for all the ones provides by automated composition.

s Number of unique skills (out of S) from the catalog that got used in a composition.

Table 3 shows how the amount of reuse varies as we increase the number of skills in the catalog. For $S = 1$, we used a single skill that takes the account ID and returns the user information and the credit score. For $S = 5$, we had 4 additional skills that can achieve part of this, either return the user information only or just the credit score from several competing services. For $S = 10$, we had two skills with the same capabilities each, to measure the impact of robust composition. The gap between m_1 and s shows how intra-process reuse is facilitated by drastically reducing the number of scripts that needed to be written (even conservatively if one compares to m_2 instead) without the declarative approach. Similarly, all the skills in the catalog got used in all three processes, thereby demonstrating intra-process reuse as well.

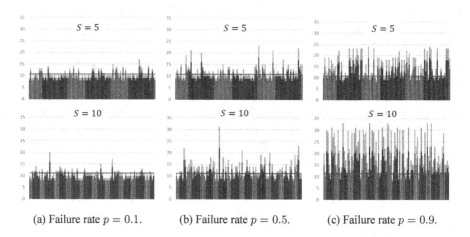

(a) Failure rate $p = 0.1$.   (b) Failure rate $p = 0.5$.   (c) Failure rate $p = 0.9$.

Fig. 2. Length of the manual (black) and the entire process (blue + black) during runtime with respect to varying failure probability of skills and size of the skill catalog. Each bar is one simulation out of 100. Average of manual only process is in red. (Color figure online)

Table 1. Size of composed agent when each task is considered separately

	T_1	T_2	T_3	$T_1 + T_2 + T_3$
Size of composition (#nodes)	8	27	32	56

Table 2. Average size of the composed process with respect to the number of skills.

Number of skills	Manual	1	5	15	20
Size of composition (#edges)	23	134	209	370	1039
Time to generate (secs)	-	0.02	0.24	3.10	19.48

Impact of Automation. So far, we have considered the offline properties of composition, its complexity with respect to the size of the original process and the size of the catalog. Now, we will simulate the actual runtime of these agents and measure the savings offered by the composition due to its automated components. We begin

	S = 1			S = 5			S = 10		
	m_1	m_2	s	m_1	m_2	s	m_1	m_2	s
T1	10	10	1	35	20	5	60	20	10
T2	8	8	1	34	24	5	55	18	10
T3	8	8	1	30	16	5	55	30	10

Table 3. Automated paths realized using composition.

from the start node, and at every node we pick an outcome at random and proceed along the process until we reach the goal. If we are at a skill with m

outcomes, we pick the failure outcome (where none of the GOTs are received) with probability p and the rest with probability $\frac{1-p}{m-1}$. For all other nodes, outcomes are picked with uniform probability. We measure two properties of the simulated process: T, the total length of the process during the simulation; and, t_m, the number of manual steps in the simulation.

Using T_3, the most complex of the three processes, and the skill catalog above, we note two key observations in Fig. 2. With low failure probabilities, a larger catalog with complementary skills does not help much, but the composition allows the simulations to remain below the average manual length without any automation, as desired. As the failure probability increases, the length of the simulation increases drastically as expected since a combination of one or more skills can complement failed skills. Interestingly, and rather counter-intuitively, the number of manual steps increases; the automated process, anticipating automation, takes the controller through steps that never materialize. Thus, while more accurate skills can support less accurate ones as fallbacks, they can also increase eventual number of steps if they do not work out. Thus, balancing skill-reuse with the actual skills' accuracy poses an interesting challenge.

4   Related Work

Business process management has been one of the most actively researched applications of planning technologies. A fantastic guide to existing work and challenges at the intersection of planning and business process management can be found in [16]. Salient problems in this area include specification and construction of business processes in the form of planning problems [29], robustification and adaptation to failures [12], validation, verification, and monitoring of processes [14], and prediction and mitigation of risk [36]. We address a somewhat overlooked promise of this union – on using planning to automate and optimize business processes. Specifically, we want to compose an existing process with "AI skills" or services to generate an optimized automated process. The motivation for this comes directly from an extensive body of work under the umbrella of "web service composition" [26].

Composing web services finds a ready ally [2, 8] in automated planning techniques which deals with composing actions in the service of constructing a course of action or a *plan*. There exists many planning techniques [30] with their own assumptions and features, and it follows that composition techniques built on top of them also demonstrate properties that can be traded off depending on the needs of the deployment. We refer to [30] for a great summary of work done in this area, while [33, 39] provides a very useful summary of many of the challenges involved. In the following, we will briefly survey existing composition techniques to uniquely place our tool among existing techniques for web service composition, and by extension the optimization of business processes using web services, for a reader choosing between competing technologies.

Automated planning offers a concise way to describe and maintaining processes – the exploration of web service composition in planning literature begun with classical (deterministic, fully observable) planning or compilations into it [9]. From the planning point of view, sophisticated interaction between services can be dealt with at the level of reasoning with capabilities of multi-agent systems [3]. A feature of web services is the uncertain nature of their execution [6]; this can also be modeled in classical form and dealt with at execution time under certain circumstances [28].

A natural way of handling uncertainty is to reason about it during planning, offline, by planning with a probability distribution [10] or set of possible outcomes [22]. Calling an API that returns employee information of a company, it may return the information or hit a 404 error. A non-deterministic planner will plan for both contingencies, instead of waiting till execution to reason about the outcome. A critical assumption is that nothing changes between planning and execution [34]. Since the planning is done offline, handling uncertainty in an open world is tricky. This is the paradigm adopted in this work to model the non-deterministic nature of automated skills.

Robust "model-lite" planning [13] maximizes success in the most number of possible models, i.e. the *robustness* [25] of a plan. In our running example, the planner would string together API calls that can potentially provide the same information to maximize chances of success (instead of contingent solutions like non-deterministic planners). Theoretically, every domain is deterministic if all the variables can be modelled. However, a complete model is rarely available, and incompleteness can manifest itself in a different form of uncertainty at planning time. In the API that returns employee information, imagine that it can only be pinged from inside the company firewall, but this constraint is not part of the planner's model. This uncertainty will be resolved at execution time (since the domain is deterministic, the outcome is the same) but the planner can account for this uncertainty at planning time.

Replanning offers a viable alternative to planning with uncertain outcomes upfront [15]. The replanning strategy varies largely based on assumptions about the underlying domain [41] – e.g. in robust planning, doing the same action multiple times will yield the same result since the underlying domain is deterministic. The replanning strategy also depends on the properties of the new plan to optimize – e.g. whether it is desired that the new process is as close to the older one as possible or preserves key properties or commitments [40]. Replanning significantly increases the runtime complexity, but is necessary since most models are not complete and will eventually require replanning even with the most carefully constructed plan dealing with possible contingencies. In business processes, replanning has been used in the past for the adaptation task [5].

Hierarchical Task Network-based planners such as SHOP2 [24] provide a powerful alternative framework for composition of services. It includes unique features like baking in considerations of quality of service into the action theory and sorting preconditions to effect this. This framework has seen continued interest for web service composition tasks [32] outside the scope of more traditional planning mechanisms. Action languages also provide an interesting alternative to address the adaptation task [18,19] and represent preferences using Golog-based templates [35].

5 Concluding Remarks

To conclude, we demonstrated how non-deterministic planning can provide a powerful way of constructing, maintaining, and optimizing business processes using AI microservices or skills. The larger effort towards the integration of this tooling into established business process automation suites, and the role of an agent in a multi-agent orchestrated assistant, can be read in [31]. Finally, an expanded version of this paper can be found at https://arxiv.org/abs/2001. 02619.

Acknowledgements. We thank IBM's digital business automation team and the research team including Scott Boag, Falk Pollok, Vinod Muthusamy, Sampath Dechu, Merve Unuvar, and Rania Khalaf for their support and ideas throughout the project. A special word of thanks to Christian Muise and the rest of the D3WA team [21] on whose work we built our D3BA extension. Finally, many thanks to Shirin Sohrabi and Michael Katz, also from IBM Research, who helped us navigate the fascinating world of planning, business processes, and web service composition.

References

1. Adamo, G., Borgo, S., Di Francescomarino, C., Ghidini, C., Guarino, N.: On the notion of goal in business process models. In: AI*IA (2018)
2. Araghi, S.S.: Customizing the Composition of Web Services and Beyond. Ph.D. thesis, University of Toronto (2012)
3. Au, T.C., Kuter, U., Nau, D.: Planning for interactions among autonomous agents. In: International Workshop on Programming Multi-Agent Systems (2008)
4. Botea, A., et al.: Generating Dialogue Agents via Automated Planning. Technical Report (2019)
5. Bucchiarone, A., Pistore, M., Raik, H., Kazhamiakin, R.: Adaptation of service-based business processes by context-aware replanning. In: SOCA (2011)
6. Carman, M., Serafini, L., Traverso, P.: Web service composition as planning. In: ICAPS Workshop on Planning for Web Services (2003)
7. Chakraborti, T., Muise, C., Agarwal, S., Lastras, L.: D3WA: an intelligent model acquisition interface for interactive specification of dialog agents. In: AAAI Demo (2019)
8. Dong, X., Halevy, A., Madhavan, J., Nemes, E., Zhang, J.: Similarity search for web services. In: VLDB (2004)
9. Hoffmann, J., Bertoli, P., Pistore, M.: Web Service Composition as Planning. AAAI, Revisited. In Between Background Theories and Initial State Uncertainty. In (2007)
10. Hoffmann, J., Brafman, R.I.: Conformant Planning via Heuristic Forward Search: A New Approach. AIJ (2006)
11. Hull, R., Nezhad, H.R.M.: Rethinking BPM in a cognitive world: transforming how we learn and perform business processes. In: BPM (2016)

12. Jarvis, P., Moore, J., Stader, J., Macintosh, A., Casson-du Mont, A., Chung, P.: Exploiting AI technologies to realise adaptive workflow systems. In: AAAI Workshop on Agent-Based Systems in the Business Context (1999)
13. Kambhampati, S.: Model-Lite planning for the web age masses: the challenges of planning with incomplete and evolving domain models. In: AAAI Senior Member Track (2007)
14. de Leoni, M., Lanciano, G., Marrella, A.: Aligning partially-ordered process-execution traces and models using automated planning. In: ICAPS (2018)
15. Little, I., Thiebaux, S.: Probabilistic planning vs replanning. In: ICAPS Workshop on IPC: Past, Present and Future (2007)
16. Marrella, A.: Automated planning for business process management. J. Data Sem. 8(2), 79–98 (2018). https://doi.org/10.1007/s13740-018-0096-0
17. Marrella, A., Lespérance, Y.: A planning approach to the automated synthesis of template-based process models. Serv. Orient. Comput. Appl. 11(4), 367–392 (2017). https://doi.org/10.1007/s11761-017-0215-z
18. Marrella, A., Mecella, M., Sardina, S.: Intelligent process adaptation in the smartpm system. TIST 8(2), 1–43 (2017)
19. Marrella, A., Mecella, M., Sardina, S.: Supporting adaptiveness of cyber-physical processes through action-based formalisms. AI Communications 31(1), 47–74 (2018)
20. Metz, R.: Microsoft's Neo-Nazi Sexbot was a Great Lesson for Makers of AI Assistants, MIT Tech. Review (2018)
21. Muise, C., et al.: Planning for Goal-Oriented Dialogue Systems. Technical Report (2020)
22. Muise, C., McIlraith, S., Beck, C.: Improved non-deterministic planning by exploiting state relevance. In: ICAPS (2012)
23. Muise, C., Vodolan, M., Agarwal, S., Bajgar, O., Lastras, L.: Executing contingent plans: challenges in deploying artificial agents. In: AAAI Fall Symposium (2019)
24. Nau, D.S., et al.: SHOP2: An HTN planning system. JAIR 20, 379–404 (2003)
25. Nguyen, T., Sreedharan, S., Kambhampati, S.: Robust planning with incomplete domain models. AIJ 245, 134–161 (2017)
26. Papazoglou, M.P., Georgakopoulos, D.: Service-oriented computing. Commun. ACM 46(10), 24–89 (2003)
27. Pesic, M., Van der Aalst, W.M.: A declarative approach for flexible business processes management. In: BPM (2006)
28. Pistore, M., Traverso, P., Bertoli, P.: Automated composition of web services by planning in asynchronous domains. In: ICAPS (2005)
29. R-moreno, M.D., Borrajo, D., Cesta, A., Oddi, A.: integrating planning and scheduling in workflow domains. Expert Syst. Appl. 33(2), 389–406 (2007)
30. Rao, J., Su, X.: A survey of automated web service composition methods. In: Cardoso, J., Sheth, A. (eds.) SWSWPC 2004. LNCS, vol. 3387, pp. 43–54. Springer, Heidelberg (2005). https://doi.org/10.1007/978-3-540-30581-1_5
31. Rizk, Y., et al.: A unified conversational assistant framework for business process automation. In: AAAI Worskhop on Intelligent Process Automation (IPA) (2020)
32. Sirin, E., Parsia, B., Wu, D., Hendler, J., Nau, D.: HTN Planning for Web Service Composition Using SHOP2. Science, Services and Agents on the World Wide Web, Web Semantics (2004)
33. Sohrabi, S.: Customizing the composition of actions, programs, and web services with user preferences. In: ISWC (2010)
34. Sohrabi, S., McIlraith, S.A.: Preference-Based web service composition: a middle ground between execution and search. In: ISWC (2010)

35. Sohrabi, S., Prokoshyna, N., McIlraith, S.: Web service composition via the customization of golog programs with user preferences. In: Conceptual Modeling (2009)
36. Sohrabi, S., Riabov, A.V., Katz, M., Udrea, O.: An AI planning solution to scenario generation for enterprise risk management. In: AAAI (2018)
37. Sreedhar, K.: What it Takes to Build Enterprise-Class Chatbots (2018)
38. Srivastava, B.: A decision-support framework for component reuse and maintenance in software project management. In: CSMR (2004)
39. Srivastava, B., Koehler, J.: Web service composition - current solutions and open problems. In: ICAPS Workshop on Planning for Web Services (2003)
40. Talamadupula, K., Smith, D.E., Cushing, W., Kambhampati, S.: A theory of intra-agent replanning. In: ICAPS Workshop on Distributed and Multi-Agent Planning (2013)
41. Yoon, S.W., Fern, A., Givan, R.: FF-Replan: a baseline for probabilistic planning. In: ICAPS (2007)

Conceptualizing a Capability-Based View of Artificial Intelligence Adoption in a BPM Context

Aleš Zebec[1]([⊠]) [iD] and Mojca Indihar Štemberger[2] [iD]

[1] School of Economics and Business, University of Ljubljana, Kardeljeva ploščad 17,
1000 Ljubljana, Slovenia
ales.zebec@student.uni-lj.si
[2] School of Economics and Business, Academic Unit for Business Informatics and Logistics,
University of Ljubljana, Kardeljeva ploščad 17, 1000 Ljubljana, Slovenia
mojca.stemberger@ef.uni-lj.si

Abstract. Advances in Artificial Intelligence (AI) technologies are creating new opportunities for organizations to improve their performance; however, as with other technologies, many of them have difficulties leveraging AI technologies and realizing performance gains. Research on the business value of information technology (IT) suggests that the adoption of AI should improve organizational performance, though indirectly, through improved business processes and other mediators, but research so far has not extensively empirically investigated the way AI creates business value. The paper proposes a capability-based view of AI adoption based on the conception that, with the adoption of AI, an organization develops AI-enabled capabilities – abilities to mobilize AI resources to effectively exploit, create, extend, or modify its resource base. This leads to higher organizational performance through cognitive process automation, innovation, and organizational learning. The first step in this research is to clarify the AI adoption construct. The goal of the paper is thus to provide a conceptual definition, and deeper insights into the components of the AI adoption construct at the organizational level.

Keywords: Artificial Intelligence · Adoption · Conceptualization · Business Process Management · Business value

1 Introduction

Artificial Intelligence is driving intelligent automation, augmentation, and innovation as well as transforming every aspect of society at individual, organizational, and societal levels [1]. Although it has no single accepted definition, we understand AI as *a simulation of human cognitive functions using intelligent agents*. Intelligent agents are capable of receiving percepts from the environment and performing actions [2]. AI is likely to be a general-purpose technology [3], characterized by pervasiveness, inherent potential for technical improvements, and innovational complementarities [4].

Although AI has been around since the 1960s, it has reemerged on the stage as a key technology that will likely play a central role in realizing performance and competitive

© Springer Nature Switzerland AG 2020
A. Del Río Ortega et al. (Eds.): BPM 2020 Workshops, LNBIP 397, pp. 194–205, 2020.
https://doi.org/10.1007/978-3-030-66498-5_15

value for organizations [5]. Vast amounts of data (Big Data), cloud computing, data management, programming frameworks, and AI services provide a readily available platform for adopting AI technology.

However, organizations have difficulties with leveraging AI technologies and realizing performance gains [6]. There is much research on the business value of IT, but AI is a distinct kind of technology because it can perform cognitive tasks usually performed by humans. This ability opens up a lot of possibilities and opportunities for innovation [5]. The key characteristics of the value proposition of AI are speed, scale, granularity, learning (accuracy), and AI-assisted decision-making [7, 8]. Furthermore, AI is highly dependent on data and domain knowledge, making it challenging to integrate and align with existing business processes [9]. The literature on this topic is scarce. There is a lack of empirical research on how AI adoption impacts the performance of business processes and organizations.

To address the gap, we draw from the concept of IT capability [10]. Researchers have shown that an organization's ability to effectively leverage its IT investments by developing a strong IT capability can lead to improved organizational and process performance. The concept has been adapted for technologies such as Business Analytics, Business Intelligence, and Big Data Analytics [11–13]. We posit that AI-specific ability to create intelligent agents capable of self-learning and decision-making can enable significant performance gains [14].

The motivation behind this study was to define a concept that would capture all components of AI adoption at an organizational level in the context of Business Process Management (BPM). The concept is an important and foundational element that will support and enhance efforts to measure the impact and value of AI technology. The conceptualization procedure included literature identification and nine in-depth semi-structured interviews to confirm and refine the definition. Interviews included managers, academics, and experts from financial services, insurance, government organization, AI technology providers, and the energy sector. The concept would then be operationalized as a measurement scale to conduct empirical research. In the paper, we describe AI adoption as *the implementation of AI resources (data, AI infrastructure, skills, competencies, etc.) in business processes*. The level (success) of adoption is measured by the development of AI-enabled capabilities (components of AI adoption), which represent *the ability to mobilize AI resources for specific business needs through the implementation of AI applications, tools, or technology.*

The remainder of the paper is structured as follows. In the next section, we present the adoption of AI in the BPM context. In Sect. 3, we describe the conceptualization procedure and the detailed results of the capability-based view of AI adoption and its dimensions. Finally, we summarize our study and discuss future work.

2 Adoption of AI and BPM Context

Our exploratory research (Table 1) shows that organizations can achieve significant performance gains when aligning AI adoption with business processes. Findings from empirical studies generally suggest that organizational processes mediate IT's impact on organizational performance [13, 15–17]. Thus, we adapted the integrative model of IT

Business Value [18] to analyze the impact of AI adoption at the process and organizational level. We study the adoption of AI in the setting of BPM, focusing on operational and dynamic capabilities developed to manage and improve business processes. Based on a literature review [5, 6, 8, 14, 19–21] and in-depth exploratory interviews on the topic of AI and BPM, we identified three key ways AI can generate business value: 1) Cognitive Business Process Automation, 2) Business Process Innovation, 3) Organizational learning and by extension improved Decision-Making Performance. These capabilities comprise the BPM context and process level, through which AI adoption impacts organizational performance.

Cognitive automation is recognized as one of the key ways AI adoption can produce business value [5, 22]. The aim of cognitive automation has often been to speed up information flow and to provide decision support [23]. However, automation systems lack many of the human cognitive skills now made possible by AI technologies.

Cockburn, Henderson, and Stern present AI as an enabler of innovation, with the ability to spawn complementary innovations [3]. AI adoption affects business process innovation mainly by modifying underlying process models through self-learning [24] and by organizational learning's ability to spark innovation [22]. Numerous cases show the impact of AI on innovation – for example, in reducing unconscious bias during the hiring process (Harver[1]), modeling a person's immune process system for drug development (CyroReason[2]), performing a visual search in e-commerce, personalizing clothing and accessories (Stitch Fix[3]), or optimizing agricultural performance (Agtech[4]).

Various authors present evidence of AI's ability to enable double-loop learning and promote organizational learning through human-computer collaboration [6, 25–27]. Aydiner et al. [19] report that technological infrastructure and related systems affect decision-making performance. We conclude that as AI includes cognitive decision-support capabilities, adoption of it will directly or indirectly (through organizational learning) affect decision-making performance, reducing the time and effort required to make a decision. Faster decision-making has a significant positive impact on business process performance [19].

This view incorporates both AI-enabled strategies of automation and augmentation. The following figure (Fig. 1) presents the contextualization of the IT Business Value Generation Process model [18], outlining the impact of AI adoption on business processes. This view identifies the areas where AI can be applied and supports the conceptualization of AI adoption.

---

[1] https://harver.com/.

[2] https://www.cytoreason.com/.

[3] https://www.stitchfix.com/.

[4] https://www.valuer.ai/blog/best-agtech-startups-in-europe#top_EU_startups.

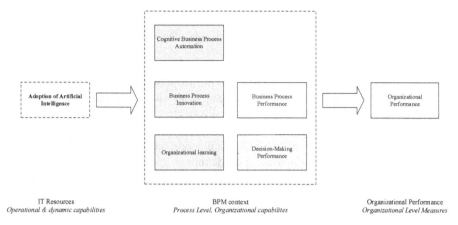

IT Resources
Operational & dynamic capabilities

BPM context
Process Level, Organizational capabilites

Organizational Performance
Organizational Level Measures

Fig. 1. AI adoption and business value framework

3 Conceptualization of Capability-Based View of AI Adoption

The conceptualization followed the recommended literature guidelines for the development of conceptual definitions [28, 29]. The definition of "adoption of artificial intelligence" was developed in three stages:

1. collecting possible attributes of the "adoption of artificial intelligence" construct by examining and assembling a set of definitions from the literature and in-depth semi-structured interviews,
2. compiling the key potential attributes and generating a preliminary definition of "adoption of artificial intelligence," and
3. refining the definition of "adoption of artificial intelligence."

3.1 Literature Identification

To identify any existing definitions, we conducted a review of information systems literature (SCOPUS and Web of Science). The literature review included papers on AI adoption constructs and models at the organizational level. We identified two empirical studies using diffusion of innovations (DOI) theory and the technology-organization-environment (TOE) framework [30, 31]. AI adoption constructs and measurement scales we found measured the adoption related to antecedents and determinants of readiness for adoption, the process of adoption, and adoption intention. Since we found no constructs assessing AI adoption level as an exogenous, component-based (unlike antecedents or determinants) variable related to the level of deployment, actual use, or utilization of specific applications and technologies, we deemed it necessary to develop a new construct.

To conduct a component-based conceptualization, we separately examined the concepts of "adoption" and "Artificial Intelligence." We followed the recommendations of Podsakoff et al. [28] and implemented several techniques to collect potential attributes and generate an illustrative set of definitions for the focal construct. The procedure included examining the definitions from dictionaries and antonyms of both concepts, a literature review, and in-depth semi-structured interviews with subject matters experts and practitioners.

We examined definitions of "adoption" in dictionaries and top MIS journals to extract common attributes centered on actual use after the adoption of the technology. The common attributes were *implementation, integration, deployment, use,* and *exploitation*. We discarded attributes centered on the adoption process: *investment decision, acceptance, selection, planning,* and *configuration*.

We examined definitions of "Artificial intelligence" in dictionaries and information systems literature. AI has no single accepted definition, and definitions are very general, focusing mostly on two key attributes: *learning* and *perception*. Some definitions emphasize computer capacity to mimic human intelligence; others are more precise, defining AI as the technologies enabling the simulation of human cognitive functions. The term *intelligent agent* is often presented as the central unifying theme [2]. The AI Group of Experts at the OECD [32] defined an intelligent agent or AI system as *a machine-based system that can, for a given set of human-defined objectives, make predictions, recommendations, or decisions influencing real or virtual environments. It does so by using machine and/or human-based inputs to 1) perceive real and/or virtual environments, 2) abstract such perceptions into models through analysis in an automated manner (e.g., with ML, or manually), and 3) use model inference to formulate options for information or action. AI systems are designed to operate with varying levels of autonomy.* To identify more specific characteristics and uncover conceptual themes of AI adoption, we further investigated AI types, AI features, AI technologies, and AI application domains.

We expect AI to have promising capabilities that enable or facilitate the transformation or redesign of business processes [1]. Thus we analyzed and extracted attributes from definitions of specific AI technologies, including *Biometrics, Collaborative Systems, Computer Vision, Deep Learning, Expert Systems, Generative Adversarial Networks, Image Analysis, Image Recognition, Knowledge Engineering, Knowledge Representation, Automated reasoning, Planning, Optimization, Verification, Logic Networks, Machine Learning, Natural Language Generation, Natural Language Processing, Natural Language Understanding, Neural Networks, Ontology Creation, Pattern Recognition, Robotic Process Automation (RPA), Robotics & Smart Robotics, Speech Recognition, Text Analysis, Video Analysis,* and *Virtual Agents*.

To organize the extracted attributes in conceptual themes, we used the lens of business capabilities or application domains rather than technological capabilities [5]: *Robotics and cognitive automation, Enhanced process automation, Cognitive insights, Cognitive engagement, Cognitive interaction,* and *Cognitive Decision Support*.

3.2 Exploratory Research

According to established guidelines [28, 29], the findings from the literature review were supplemented with interviews to extract additional definitions and attributes from experts. Organizations and experts were selected because of their work with AI or on AI-related projects. During all of the nine in-depth semi-structured interviews, we discussed the broader scope of AI adoption and the interviewees' experiences with AI implementation, deployment, and use. We aligned their views on the technology with the extracted conceptual themes (presented application domains), as these were the classifications experts and practitioners were most comfortable with. We present the excerpt of the results in Table 1.

Table 1. Excerpt of the main findings from the interviews.

Findings	Theme
Financial Services; 5900 employees; Chief Data Officer (CDO)	
AI adoption emerged in the analytics department. The organization adopted a new business strategy that relied on advanced techniques for pattern recognition to gain customer insights. The goals were to increase process performance and generate revenue based on AI capabilities. The new business strategy defined data as a critical resource. They appointed a CDO to manage a data management sector. The organization is dealing with automation in the scope of the Lean initiative, improving business processes by creating more value with fewer human resources. Are focused on processes where decisions are not deterministic (e.g., credit scoring). Integrated AI with their products (e.g., in personal finance management and in automating classification of revenue and expenses)	Business insights, Data management, Automated decision-making, Engagement, AI techniques
Insurance company; 5200 employees; head of the team responsible for developing DWH/BI/AI solutions	
Are integrating chatbots for customer support. Organization-wide deployment of advanced AI functionalities for decision support (e.g., a model for automated detection of insurance policy renewal). Are focused on automation and optimization of processes. They identified several opportunities for RPA. They are developing a central data repository to eliminate data silos and are gathering publicly available data from the environment. After data is available in the data warehouse, they search for opportunities with Business Intelligence and AI methods	Human–computer interaction, Decision support, AI technology, Data acquisition
Financial services; 1010 employees; head of Analytics Department	
Implemented and deployed a next-best-offer solution for salespeople. The organization is using machine learning and decision trees in marketing. They are mostly concerned with the propensity to buy and with churn management, and they see the most value in AI-enabled predictive analytics They are focused on ensuring high-quality data	Decision support, AI techniques, Business insights, Data management

3.3   Five-Dimensional Conceptualization

Based on the example of Sonenshein et al. [33], we organized the extracted attributes into a smaller set of themes and then aggregated them into dimensions. As indicated in Fig. 2, we organized the attributes into 17 related themes and five dimensions: *Data Acquisition and Preprocessing, Cognitive Insights, Cognitive Engagement, Cognitive Decision Assistance,* and *Cognitive Technologies.* The construct derived from the conceptual definition is a multidimensional, second-order construct, reflective – reflective type I [34].

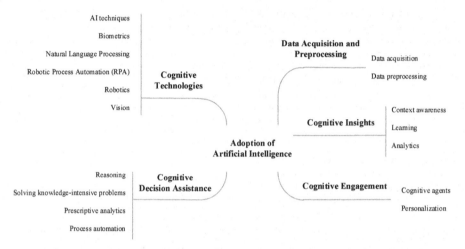

Fig. 2. Organizing attributes into common conceptual themes and dimensions.

In the last stage of the conceptual analysis, we refined the conceptual definition of the examined construct by discussing it with subject matter experts and peers to examine the definition and solicit questions about the concept. As a result, we modified the scope of the definition. Next, we present the resulting definition.

The Conceptual Definition of the Focal Construct: *"the implementation of AI resources in business processes."* We use the term AI resources for AI-related elements that must be brought together to ensure the successful deployment and use of AI technology. Key AI-related elements are scalable infrastructure, AI assets (data, trained models, etc.), AI skills, domain knowledge, expertise, capabilities, partnerships, AI talent, processes, privacy policies, etc. The definition of a construct must incorporate the "property" the concept characterizes and the "entity" to which that property relates [29]. We defined the property for "Adoption of Artificial Intelligence" as *"the organization's ability to develop a set of distinct AI-enabled capabilities (the ability to mobilize AI resources to exploit strategic assets and achieve innovative changes), through the implementation of AI applications, tools, or technology."* The property applies to the entity of an organization.

AI is primarily concerned with data, which is exploited, examined, renewed, or reconfigured through AI-enabled capabilities. We argue that AI, by itself, cannot be a capability. AI becomes a part of a capability when it is applied to a problem and given a goal. AI needs goals; otherwise, it cannot learn. This characteristic is not so different from one of humans; we cannot learn anything without a goal either. Without a goal, there is no frame of reference for evaluating performance and, thus, no way to improve. By conceptualizing "Adoption of Artificial Intelligence," we describe the level of the adoption through the development levels of five distinct and progressive AI-enabled capabilities (components of AI adoption). In a broader scope, these capabilities not only support business processes but become an integral part of an organization's ability to generate value from its data. Next, we present conceptualized dimensions.

Data Acquisition and Preprocessing: *"the organization's ability to extract data from structured and unstructured sources, new and legacy systems, and internal and external sources and to prepare it for analysis."* The three basic routines are data extraction, data preprocessing, and continuous ensuring of data quality. Their purpose is to deal with Big Data (ever-increasing volume, variety, and velocity of data) collected from internal and external sources. Preprocessing includes consolidation, organization, validation, cleaning, transformation, reduction, summarization, labeling, and loading into a data warehouse, data lake, NoSQL database, relational database (RDBMS), or other applications. High-quality data is an important business resource and asset and has tremendous impact on an entity's performance [35]. We propose to measure the dimension by assessing the successful deployment and use of data management applications and tools (e.g., Information propagation, Data warehousing/Data Lake, Data capturing system IoT/SCADA, Content Creation, Discovery, Creation, Computational Creativity, etc.).

Cognitive Insight: *"the organization's ability to use AI to detect patterns in data and interpret their meaning."* This dimension is conceptualized around themes of context awareness, learning, and analytics. AI enables a recognition of patterns or clusters of data that would be otherwise invisible to a human [36]. It can interpret events and contextualize recognized patterns to derive their true meanings. The learning aspect of AI allows for making predictions on the basis of past experiences [37] and, by the continuous learning process, for improving insight over time [5]. Cognitive analytics (Knowledge representation, Inference, Reasoning, Learning & adaptation, Hypothesis generation & validation, Domain cognitive models, and Machine learning, or Deep learning) offer better results in terms of speed, scale, accuracy, and granularity. We propose to measure the dimension by assessing the successful deployment and use of AI-analytics applications and tools (e.g., Predictive Sales, Churn Management, Fraud Detection, Risk Management, etc.).

Cognitive Engagement: *"the organization's ability to support AI-enhanced human–computer interaction and collaboration."* Engagement consists of several key elements, including understanding, perception of intention, and domain knowledge [38]. Understanding encompasses natural language processing, natural language understanding, automated speech recognition, and text-to-speech conversion. Leveraging contextual information about humans to develop human-like empathy and communication skills in human–computer interaction or collaboration applications involves perception of intention, tone, sentiment, emotional state, environmental conditions, and the strength and nature of a person's relationships [5]. All the elements are used for reasoning through the total of all structured and unstructured data to determine the optimal approach for engaging a person [39]. This ability enables automated interactions to reliably support customers' activities and prompt their engagement [40] in customer-facing business processes. Organizations are also increasingly using cognitive engagement to interact with employees (to support routine activities), augment information, improve knowledge acquisition/exploration/understanding, and to support the collaborative formulation of goals and decisions [5]. We propose to measure the dimension by assessing the successful deployment and use of AI-enabled applications and tools related to user engagement (e.g., Virtual assistants, Chatbots, Avatars, Recommendation systems, etc.).

Cognitive Decision Assistance: *"the organization's ability to use AI in decision-making processes."* AI technologies and techniques enable AI-assisted decision-making and render decision support more intelligent. Some common abilities descriptive of AI's capability are as follows: to speed up information flows, to provide predictive and adaptive decision support, to utilize automated reasoning to solve knowledge-intensive problems, to make sense out of ambiguous or contradictory messages in large data sets, to recognize the relative importance of different elements in a situation, to respond quickly and successfully to a new situation, and to apply knowledge to manipulate the environment [41]. We propose to measure the dimension by assessing the successful deployment and use of AI-assisted decision-making applications and tools (e.g., AI-enabled Decision Support System, Expert Systems, Fuzzy logic systems, Optimization, Knowledge Engineering, etc.).

Cognitive Technologies: *"the organization's ability to integrate AI technologies with other IT resources, services, and devices."* This dimension was isolated by the conceptualization process for cases where organizations do not deploy and use AI in a specific application domain as a particular application or a tool. *Cognitive Technologies* AI-enabled capability is the highest level of adoption, when AI is not merely used, but utilized (implying innovation or creative use beyond the intended use). AI technologies can radically transform data utilization and processing within existing value-creating processes. Their ability to learn and adapt continuously due to self-awareness and input from actors (with whom it interacts) and contexts (in which it is embedded) amplifies the resourcefulness of AI technologies [40]. The summative effects can be seen in the interactions between the knowledge of the AI-enabled device or service and the knowledge and actions of a human. We propose to measure the dimension by assessing the successful integration of AI technologies in other IT resources, services, and devices. The AI technologies most suitable for integration include Machine learning, Deep learning, and neural networks, Natural language processing, Genetic programming, Sensor networks, Augmented reality, Computer Vision, Speech recognition, and RPA [22].

We posit the presented dimensions impact all of the proposed ways (automation, innovation, organizational learning, decision-making) of business value generation, although to a different extent.

4   Discussion and Future Work

AI has the potential to change the workforce positively: automating repetitive tasks and freeing up workers to be more creative and productive with augmenting human capabilities. Although there is widespread consensus about the potential of AI [9], the technology and its impact on the workforce and by extension process performance is not fully understood. Organizations have difficulties leveraging AI technologies and extracting business value. A review of relevant studies and exploratory interviews by us confirmed that organizations are mostly at an early stage of AI adoption. Studies [9, 42] have reported that only 5% of organizations have extensively incorporated AI and that 20% have partially used it. Most are in the phase of evaluating a proof of concept and identifying business cases on which to apply the technology. There are no real "out of the box" solutions. AI is about data and domain knowledge. The results of our interviews show that organizations struggle with data management – especially data governance – and with the organization-wide deployment of AI. Thus, the study sought to identify the key AI-enabled capabilities to deploy and use the technology effectively. Building on these capabilities, organizations can realize opportunities that AI offers.

The objective of this study was to conceptualize AI adoption and capture specific dimensions that are potentially responsible for generating business value. Following Podsakoff's [28] guidelines, we developed dimensions based on extracted attributes and properties of the concept. Conceptualization was guided by the purpose of uncovering the details of the process of generating business value through AI. The goals were to assess adoption level, define AI adoption components as AI-enabled capabilities, and identify the most promising AI applications and technologies. The result is the refined definition of AI adoption as "the *successful deployment and use of AI resources in business processes.*" Five developed progressive levels of AI-enabled capabilities include 1) *Data Acquisition and Preprocessing*; 2) *Cognitive Insight*; 3) *Cognitive Engagement*; 4) *Cognitive Decision Assistance*; and 5) *Cognitive technologies.*

The next step of the research will be to operationalize the proposed dimensions to be able to assess the total level of AI adoption. Based on the proposed framework, we plan to develop a measurement model and conduct a comprehensive study using a questionnaire survey. The results will validate the proposed model and provide empirical evidence as well as a more detailed view of the business value generation process of AI technology.

Researchers can use the concept of AI adoption or its specific dimensions in different contexts to assess the level of adoption, uncover management strategies, and understand the impact and value of AI. (e.g., moving from organizational to the process level and assess the impact on specific process types, BPM goals, or specific organizational and environmental dimensions).

References

1. Bawack, R.E., Fosso Wamba, S., Carillo, K.: Artificial Intelligence in Practice: Implications for IS Research (2019)
2. Russel, S., Norvig, P.: Artificial intelligence: a modern approach. Pearson Education Limited (2016)

3. Cockburn, I.M., Henderson, R., Stern, S.: The impact of artificial intelligence on innovation. National bureau of economic research (2018)
4. Bresnahan, T.F., Trajtenberg, M.: General purpose technologies 'Engines of growth'? J. Econ. **65**, 83–108 (1995)
5. Davenport, T.H., Ronanki, R.: Artificial intelligence for the real world. Harvard Bus. Rev. **96**, 108–116 (2018)
6. Mishra, A.N., Pani, A.K.: Business value appropriation roadmap for artificial intelligence. VINE Journal of Information and Knowledge Management Systems (2020)
7. Agrawal, A., Gans, J., Goldfarb, A.: What to expect from artificial intelligence. MIT Sloan Management Review (2017)
8. Mikalef, P., Fjørtoft, S.O., Torvatn, H.Y.: Developing an artificial intelligence capability: a theoretical framework for business value. In: Abramowicz, W., Corchuelo, R. (eds.) BIS 2019. LNBIP, vol. 373, pp. 409–416. Springer, Cham (2019). https://doi.org/10.1007/978-3-030-36691-9_34
9. Chui, M.: Artificial intelligence the next digital frontier? McKinsey and Company Global Institute 47, (2017)
10. Santhanam, R., Hartono, E.: Issues in linking information technology capability to firm performance. MIS Q. **27**, 125–153 (2003)
11. Mikalef, P., Krogstie, J., Pappas, I.O., Pavlou, P.: Exploring the relationship between big data analytics capability and competitive performance: The mediating roles of dynamic and operational capabilities. Inf. Manage. **57**, 103169 (2020)
12. Shanks, G., Bekmamedova, N.: Achieving benefits with business analytics systems: an evolutionary process perspective. J. Decis. Syst. **21**, 231–244 (2012)
13. Krishnamoorthi, S., Mathew, S.K.: Business analytics and business value: A comparative case study. Inf. Manag. **55**, 643–666 (2018)
14. Wamba-Taguimdje, S.-L., Wamba, S.F., Kamdjoug, J.R.K., Wanko, C.E.T.: Influence of artificial intelligence (AI) on firm performance: the business value of AI-based transformation projects. Bus. Process Manag. J. (2020)
15. Marie Burvill, S., Jones-Evans, D., Rowlands, H.: Reconceptualising the principles of Penrose's (1959) theory and the resource based view of the firm: The generation of a new conceptual framework. J. Small Bus. Enterprise Develop. **25**, (2018)
16. Bhatt, G.D., Grover, V.: Types of information technology capabilities and their role in competitive advantage: An empirical study. J. Manag. Inf. Syst. **22**, 253–277 (2005)
17. Kim, G., Shin, B., Kim, K.K., Lee, H.G.: IT capabilities, process-oriented dynamic capabilities, and firm financial performance. J. Assoc. Inf. Syst. **12**, 1 (2011)
18. Melville, N., Kraemer, K., Gurbaxani, V.: Review: information technology and organizational performance: an integrative model of it business value. MIS Q. **28**, 283–322 (2004)
19. Aydiner, A.S., Tatoglu, E., Bayraktar, E., Zaim, S.: Information system capabilities and firm performance: Opening the black box through decision-making performance and business-process performance. Int. J. Inf. Manage. **47**, 168–182 (2019)
20. Liao, S.-H., Wu, C.-c.: System perspective of knowledge management, organizational learning, and organizational innovation. Expert Syst. Appl. **37**, 1096–1103 (2010)
21. Jiménez-Jiménez, D., Sanz-Valle, R.: Innovation, organizational learning, and performance. J. Bus. Res. **64**, 408–417 (2011)
22. Zasada, A.: How Cognitive Processes Make Us Smarter (2019)
23. Frohm, J.: Levels of Automation in production systems. Chalmers University of Technology Göteborg (2008)
24. Hull, R., Motahari Nezhad, H.R.: Rethinking BPM in a cognitive world: transforming how we learn and perform business processes. In: La Rosa, M., Loos, P., Pastor, O. (eds.) BPM 2016. LNCS, vol. 9850, pp. 3–19. Springer, Cham (2016). https://doi.org/10.1007/978-3-319-45348-4_1

25. Bohanec, M., Robnik-Šikonja, M., Borštnar, M.K.: Organizational learning supported by machine learning models coupled with general explanation methods: A Case of B2B sales forecasting. Organizacija **50**, 217–233 (2017)
26. Samek, W., Wiegand, T., Müller, K.-R.: Explainable artificial intelligence: Understanding, visualizing and interpreting deep learning models. arXiv preprint arXiv:1708.08296 (2017)
27. Banasiewicz, A.D.: Organizational Learning in the Age of Data (2019)
28. Podsakoff, P.M., MacKenzie, S.B., Podsakoff, N.P.: Recommendations for creating better concept definitions in the organizational, behavioral, and social sciences. Organ. Res. Methods **19**, 159–203 (2016)
29. MacKenzie, S.B., Podsakoff, P.M., Podsakoff, N.P.: Construct measurement and validation procedures in mis and behavioral research: integrating new and existing techniques. MIS Q. **35**, 293–334 (2011)
30. Alsheibani, S., Cheung, Y., Messom, C.: Artificial intelligence adoption: ai-readiness at firm-level. Artif. Intell. **6**, 26–2018 (2018)
31. Chen, H.: Success Factors Impacting Artificial Intelligence Adoption—Perspective From the Telecom Industry in China (2019)
32. OECD: Artificial Intelligence in Society (2019)
33. Sonenshein, S., DeCelles, K.A., Dutton, J.E.: It's not easy being green: The role of self-evaluations in explaining support of environmental issues. Acad. Manage. J. **57**, 7–37 (2014)
34. Jarvis, C.B., MacKenzie, S.B., Podsakoff, P.M.: A critical review of construct indicators and measurement model misspecification in marketing and consumer research. J. Cons. Res. **30**, 199–218 (2003)
35. Appelbaum, D., Kogan, A., Vasarhelyi, M., Yan, Z.: Impact of business analytics and enterprise systems on managerial accounting. Int. J. Account. Inf. Syst. **25**, 29–44 (2017)
36. Burgess, A.: AI capabilities framework. The Executive Guide to Artificial Intelligence, pp. 29–54. Springer, Cham (2018). https://doi.org/10.1007/978-3-319-63820-1_3
37. Bawack, R.E., Wamba, S.F.: Where Information Systems Research Meets Artificial Intelligence Practice: Towards the Development of an AI Capability Framework (2019)
38. Roeglinger, M., Seyfried, J., Stelzl, S., Muehlen, M.: Cognitive computing: what's in for business process management? an exploration of use case ideas. In: Teniente, E., Weidlich, M. (eds.) BPM 2017. LNBIP, vol. 308, pp. 419–428. Springer, Cham (2018). https://doi.org/10.1007/978-3-319-74030-0_32
39. Kelly, J.E.: Computing, cognition and the future of knowing. Whitepaper, IBM Reseach **2**, (2015)
40. Mele, C., Spena, T.R., Peschiera, S.: Value creation and cognitive technologies: opportunities and challenges. J. Creat. Value **4**, 182–195 (2018)
41. Phillips-Wren, G.: Ai tools in decision making support systems: a review. Int. J. Artif. Intell. Tools **21**(02), 1240005 (2012)
42. Sam Ransbotham, S.K., Ronny, F., Burt, L., David, K.: Winning With AI. MIT Sloan Management Review (2019)

A General Framework
for Action-Oriented Process Mining

Gyunam Park$^{(\boxtimes)}$ⓘ and Wil M.P. van der Aalstⓘ

Department of Computer Science, Process and Data Science Group (PADS),
RWTH Aachen University, Aachen, Germany
{gnpark,wvdaalst}@pads.rwth-aachen.de

Abstract. Process mining provides techniques to extract process-centric knowledge from event data available in information systems. These techniques have been successfully adopted to solve process-related problems in diverse industries. In recent years, the attention of the process mining discipline has shifted to supporting continuous process management and actual process improvement. To this end, techniques for operational support, including predictive process monitoring, have been actively studied to monitor and influence running cases. However, the conversion from insightful diagnostics to actual actions is still left to the user (i.e., the "action part" is missing and outside the scope of today's process mining tools). In this paper, we propose a general framework for *action-oriented process mining* that supports the continuous management of operational processes and the automated execution of actions to improve the process. As proof of concept, the framework is implemented in *ProM*.

Keywords: Action-oriented process mining · Continuous operational management · Insights turned into actions · Process improvement

1 Introduction

Process mining aims to discover, monitor, and improve business processes by extracting knowledge from event logs available in information systems [1]. Process mining techniques enable business managers to better understand their processes and gather insights to improve these processes. They are successfully deployed by a range of industries, including logistics, healthcare, and production [2].

Nowadays, attention in the process mining discipline is shifting to supporting continuous process management [2]. In order to manage operational processes properly, it is imperative to apply process mining techniques in a repetitive manner, rather than focusing on making a one-time report of process mining diagnostics. This repetitive application enables not only the identification of more relevant problems at stake, but also the continuous improvement of operational processes in a dynamically changing environment. The one-time report is

© Springer Nature Switzerland AG 2020
A. Del Río Ortega et al. (Eds.): BPM 2020 Workshops, LNBIP 397, pp. 206–218, 2020.
https://doi.org/10.1007/978-3-030-66498-5_16

likely to present less relevant problems in the current situation, failing to handle newly-introduced problems.

Online operational support techniques in process mining aim at enabling the continuous management of operational processes [1]. To that end, they continuously monitor and analyze cases that are still running, intended for controlling the problematic process instances [3,4]. These techniques have been effective in extracting practical diagnostics into performance and compliance problems [1]. However, they do not suggest how the diagnostics are exploited to achieve actual improvements in the operational processes.

For the actual process improvement, it is necessary to convert the insights from process mining diagnostics to management actions. For example, when a bottleneck emerges or is expected to occur, one should take actions, such as assigning more resources, alerting managers, and finding bypassing routes, to mitigate the risk caused by the problem. To fill the gap between the diagnostics and the improvement actions, in this paper, we propose a general framework for action-oriented process mining. This framework supports the continuous monitoring of operational processes and the automated execution of actions to improve the processes based on the monitoring results (i.e., diagnostics).

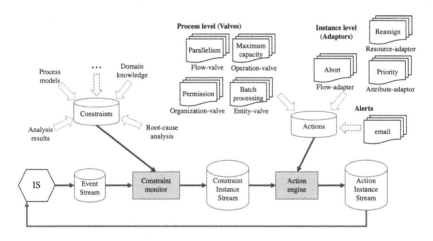

Fig. 1. Overview of the general framework for action-oriented process mining

Figure 1 shows an overview of the proposed framework. By analyzing the continuously updated event data (i.e., event stream), the *constraint monitor* evaluates a set of constraints that are defined with various diagnostics. As a result, it generates a constraint instance stream that is the description of monitoring results. By analyzing this constraint instance stream, the *action engine* assesses the necessity of actions and generates the actions ranging from process-level valves, instance-level adaptors, and alerts, as described in Fig. 1 with representative examples.

In order to advocate the effectiveness of the proposed framework on the continuous process management and the actual process improvement, it has been instantiated as a *ProM* plug-in. In addition, we have tested the implementation on an information system that supports a simulated order handling process. The details of implementation and the information system are publicly available via https://github.com/gyunamister/ActionOrientedProcessMining.

The remainder is organized as follows. We first present a motivating example in Sect. 2. Next, we explain the preliminaries and the general framework for action-oriented process mining in Sect. 3 and Sect. 4. Afterward, Sect. 5 and Sect. 6 present the implementation of the framework and experiments as a proof of concept. Section 6 discusses the related work, and Sect. 7 concludes the paper.

2 Motivating Example

Suppose we are operation managers in an e-commerce company like Amazon, responsible for an order handling process, where four main object types (i.e., *order, item, package,* and *route*) exist in the process as shown in Fig. 2a. Note that we do not assume a single case notion as in traditional process mining in the proposed framework. Instead, using the principles of object-centric process mining [5], we consider multiple object types and interacting processes. It is indispensable for acquiring precise diagnostics and deploying the framework at the enterprise level where multiple processes with different object types interact with each other.

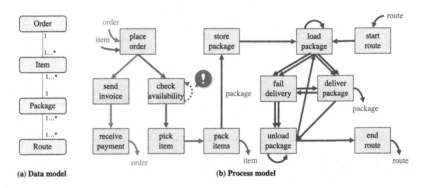

(a) **Data model**   (b) **Process model**

Fig. 2. Data model and the discovered process model of the order handling process. The discovered process model shows that *check availability* happens redundantly.

As operation managers, we analyze the event data using different process mining techniques. As an example, we discovered the process model shown in Fig. 2b where the arcs correspond to the specific object types. Interpreting the discovered process model, we observe that the activity, *check availability*, is redundantly repeated for some items, which should not happen according to a business rule and its negative effects on the overall operational performance. For the continuous management of this problem, we define a constraint $C1$ as follows:

– $C1$: there must be no more than one "check availability" for each item.

Afterward, we put it into the repository of constraints and let the *constraint monitor* evaluate if any item violates or is predicted to violate the constraint every morning (e.g., at 9 AM).

We consider that it is highly risky to have more than 10 items violating or predicted to violate $C1$ at any point in time, and in case this situation happens, it is most efficient to notify a case manager. Thus, we analyze the monitoring results every morning and take the following action to mitigate the risk:

– $A1$: If there exist more than 10 (possibly) violated items, send an e-mail to the case manager to warn for bad consequences.

This example shows how the insights from process discovery transform into mitigating actions (i.e., alerting a case manager). The proposed framework supports this process of insights turned into actions by continuously monitoring the violations and automatically generating proactive actions. In the following sections, we explain the major components of the framework, i.e., *constraint monitor* and *action engine*, using the above example as a running example.

3   Event Data

In this section, we introduce the basic concepts, including event streams, time windows, and time moments.

Real-life processes often have multiple candidate identifiers, as shown in Sect. 2. To enable precise analysis and enterprise-wide adoption of the proposed framework, we use a more realistic event data notion where multiple case notions (e.g., order, item, etc.) may coexist. Each event may refer to different objects from different object classes. Note that a conventional event log is a special case of this event data notion; hence one can use the proposed framework with the conventional event logs.

Definition 1 (Universes). *We define the following universes to be used in this paper:*

– $\mathcal{U}_{ei}$ *is the universe of event identifiers*
– $\mathcal{U}_{proc}$ *is the universe of process identifiers,*
– $\mathcal{U}_{act}$ *is the universe of activities,*
– $\mathcal{U}_{res}$ *is the universe of resources,*
– $\mathcal{U}_{time}$ *is the universe of timestamps,*
– $\mathcal{U}_{oc}$ *is the universe of object classes,*
– $\mathcal{U}_{oi}$ *is the universe of object identifiers,*

- $\mathcal{U}_{omap} = \mathcal{U}_{oc} \nrightarrow \mathcal{P}(\mathcal{U}_{oi})$ is the universe of object mappings where, for omap $\in$ $\mathcal{U}_{omap}$, we define omap(oc) $= \emptyset$ if oc $\notin$ dom(omap),
- $\mathcal{U}_{attr}$ be the universe of attribute names,
- $\mathcal{U}_{val}$ the universe of attribute values,
- $\mathcal{U}_{vmap} = \mathcal{U}_{attr} \nrightarrow \mathcal{U}_{val}$ is the universe of value mappings where, for vmap $\in$ $\mathcal{U}_{vmap}$, we define vmap(attr) $= \bot$ if attr $\notin$ dom(vmap).
- $\mathcal{U}_{event} = \mathcal{U}_{ei} \times \mathcal{U}_{proc} \times \mathcal{U}_{act} \times \mathcal{U}_{res} \times \mathcal{U}_{time} \times \mathcal{U}_{omap} \times \mathcal{U}_{vmap}$ is the universe of events.

We assume these universes are pairwise disjoint, e.g., $\mathcal{U}_{ei} \cap \mathcal{U}_{proc} = \emptyset$.

Each row in Table 1 shows an event of the order handling process introduced in Sect. 2.

Table 1. A fragment of event data where each line corresponds to an event

Event identifier	Process identifier	Activity name	Resource name	Timestamp	Objects involved				Attribute
					Order	Item	Package	Route	Type
...	...	...	...	...	...	...	...	...	...
746	OH	place order	Jane	01-01-2020 09:55	$\{o_7\}$	$\{i_8, i_9\}$	$\emptyset$	$\emptyset$	Gold
747	OH	check availability	Jansen	01-01-2020 10:15	$\{o_7\}$	$\{i_8\}$	$\emptyset$	$\emptyset$	
748	OH	pick item	Kevin	01-01-2020 11:55	$\{o_7\}$	$\{i_8\}$	$\emptyset$	$\emptyset$	
749	OH	check availability	Matthias	01-01-2020 17:55	$\{o_7\}$	$\{i_9\}$	$\emptyset$	$\emptyset$	
750	OH	check availability	Jansen	01-01-2020 19:05	$\{o_7\}$	$\{i_9\}$	$\emptyset$	$\emptyset$	
751	OH	pick item	Kevin	01-01-2020 19:55	$\{o_7\}$	$\{i_9\}$	$\emptyset$	$\emptyset$	
752	OH	place order	System	02-01-2020 09:15	$\{o_8\}$	$\{i_{10}\}$	$\emptyset$	$\emptyset$	Silver
753	OH	pack items	Robin	02-01-2020 15:05	$\{o_7\}$	$\{i_8, i_9\}$	$\{p_8\}$	$\emptyset$	
...	...	...	...	...	...	...	...	...	...

Definition 2 (Event Projection). Given an event $e = (ei, proc, act, res,$ $time, omap, vmap) \in \mathcal{U}_{event}$, $\pi_{ei}(e) = ei, \pi_{proc}(e) = proc, \pi_{act}(e) = act, \pi_{res}(e) = res, \pi_{time}(e) = time, \pi_{omap}(e) = omap,$ and $\pi_{vmap}(e) = vmap$.

Let e_{746} be the first event depicted in Table 1. $\pi_{ei}(e_{746}) = 746,$ $\pi_{proc}(e_{746}) = OH, \pi_{act}(e_{746}) = place\ order, \pi_{res}(e_{746}) = Jane, \pi_{time}(e_{746}) =$ 01-01-2020 09:55, $\pi_{omap}(e_{746})(Order) = \{o_7\}, \pi_{omap}(e_{746})(Item) = \{i_8, i_9\},$ and $\pi_{vmap}(e_{746})(Type) = Gold.$
We adopt the notion of online event stream-based process mining, in which the data are assumed to be an infinite collection of unique events. An event stream is a collection of unique events that are ordered by time.

Definition 3 (Event Stream). An event stream S is a (possibly infinite) set of events, i.e., $S \subseteq \mathcal{U}_{event}$ such that $\forall_{e_1, e_2 \in S} \pi_{ei}(e_1) = \pi_{ei}(e_2) \implies e_1 = e_2$. We let $\mathcal{U}_{stream}$ denote the set of all possible event streams.

A time window indicates the range of time to be analyzed.

Definition 4 (Time Window). *A time window $tw = (t_s, t_e) \in \mathcal{U}_{time} \times \mathcal{U}_{time}$ is a pair of timestamps such that $t_s \leq t_e$. Given a time window $tw = (t_s, t_e)$, $\pi_{start}(tw) = t_s$ and $\pi_{end}(tw) = t_e$. $\mathcal{U}_{tw}$ is the set of all possible time windows.*

A time moment represents the time when we start analyzing processes and the time window that the analysis addresses.

Definition 5 (Time Moment). *A time moment $tm = (t, tw) \in \mathcal{U}_{time} \times \mathcal{U}_{tw}$ is a pair of a timestamp t and a time window tw such that $t \geq \pi_{end}(tw)$. Given $tm = (t, tw)$, we indicate $\pi_t(tm) = t$ and $\pi_{tw}(tm) = tw$. $\mathcal{U}_{tm}$ is the set of all possible time moments.*

4 A General Framework for Action-Oriented Process Mining

The proposed framework is mainly composed of two components. Firstly, the *constraint monitor* converts an event stream into a *constraint instance stream* where each *constraint instance* describes the (non) violation of a constraint. Second, the *action engine* transforms the constraint instance stream into an *action instance stream* where each *action instance* depicts a *transaction* to be executed by the information system to mitigate the risks caused by the violations.

4.1 Constraint Monitor

Each (non) violation of a constraint has a context where it occurs. For instance, *C1* in Sect. 2 could be violated by item i_9, which is a part of an order by a Gold customer in the process *OH*, when processed by *Joe* for the activity *check availability*.

Definition 6 (Context). *A context $ctx \in \mathcal{P}(\mathcal{U}_{proc}) \times \mathcal{P}(\mathcal{U}_{act}) \times \mathcal{P}(\mathcal{U}_{res}) \times \mathcal{U}_{omap} \times \mathcal{U}_{vmap}$ is a tuple of a set of process identifiers Proc, a set of activities Act, a set of resources Res, an object mapping omap, and a value mapping vmap. $\mathcal{U}_{ctx}$ is the set of all possible contexts.*

The above context is denoted as $ctx_1 = (\{OH\}, \{check\ availability\}, \{Joe\}, omap_1, vmap_1)$, where $omap_1(Item) = \{i_9\}$ and $vmap_1(Type) = $ Gold.

Given an event stream and a time window, the constraint formula evaluates if violations happen in a specific context by analyzing the events in the event stream, which are relevant to the time window.

Definition 7 (Constraint Formula). *We define $\mathcal{U}_{outc} = \{OK, NOK\}$ to be the universe of outcomes. $cf \in (\mathcal{U}_{stream} \times \mathcal{U}_{tw}) \to \mathcal{P}(\mathcal{U}_{ctx} \times \mathcal{U}_{outc})$ is a constraint formula. $\mathcal{U}_{cf}$ is the set of all possible constraint formulas.*

Suppose cf_1 is instantiated to evaluate the constraint described in *C1* of the motivating example. Given the event stream S that contains events listed in Table 1 and time window tw_1 = (01-01-2020 09:00, 02-01-2020 09:00), it evaluates if any item in the time window experience more than one *check availability*. Since there are two *check availability* for item i_9, $(ctx_1, NOK) \in cf_1(S, tw_1)$.

In this paper, we do not assume specific approaches to instantiate the constraint formula. Several approaches are proposed in the field of process mining, including conformance checking techniques [6] and rule-driven approaches based on Petri-net patterns [7] and Linear Temporal Logic [8] (see also Sect. 7)

A constraint consists of a constraint formula and a set of time moments, where the former explains what to monitor, and the latter specifies when to monitor.

Definition 8 (Constraint). *A constraint* $c = (cf, TM) \in \mathcal{U}_{cf} \times \mathcal{P}(\mathcal{U}_{tm})$ *is a pair of a constraint formula* cf *and a set of time moments* TM. $\mathcal{U}_c$ *is the set of all possible constraints.*

Suppose c_1 = (cf_1, TM_1) where (02-01-2020 09:00, (01-01-2020 09:00, 02-01-2020 09:00)) $\in TM_1$. For instance, we evaluate cf_1 at 02-01-2020 09:00 with the events related to time window (01-01-2020 09:00,02-01-2020 09:00).

A constraint instance specifies when and whether a violation happens in a certain context by a constraint formula.

Definition 9 (Constraint Instance). *A constraint instance* $ci \in \mathcal{U}_{cf} \times \mathcal{U}_{ctx} \times \mathcal{U}_{time} \times \mathcal{U}_{outc}$ *is a tuple of a constraint formula* cf, *a context* ctx, *a timestamp time, and an outcome outc.* $\mathcal{U}_{ci}$ *is the set of all possible constraint instances.*

For instance, a constraint instance ci_1 = $(cf_1, ctx_1, 02\text{-}01\text{-}2020~09\text{:}00, NOK)$ denotes that cf_1 is violated at 02-01-2020 09:00. in context ctx_1.

A constraint instance stream is a collection of unique constraint instances.

Definition 10 (Constraint Instance Stream). *A constraint instance stream* CIS *is a (possibly infinite) set of constraint instances, i.e.,* $CIS \subseteq \mathcal{U}_{ci}$. $\mathcal{U}_{CIS}$ *is the set of all possible constraint instance streams.*

Given an event stream, a constraint monitor evaluates a set of constraints and generates a constraint instance stream.

Definition 11 (Constraint Monitor). *Let* $C \subseteq \mathcal{U}_c$ *be a set of constraints to be used for monitoring.* $cm_C \in \mathcal{U}_{stream} \to \mathcal{U}_{CIS}$ *is the constraint monitor such that, for any* $S \in \mathcal{U}_{stream}$, $cm_C(S) = \{(cf, ctx, time, outc) \in \mathcal{U}_{ci} | \exists_{TM,tm} (cf, TM) \in C \wedge tm \in TM \wedge time = \pi_t(tm) \wedge (ctx, outc) \in cf(S, \pi_{tw}(tm))\}$.

Note that the definition of a constraint monitor is abstracted in a way that we are able to analyze future events. In reality, it analyzes only the historical events from an event stream and outputs the constraint instance stream relevant to them.

4.2   Action Engine

The action engine aims at producing an action instance stream describing transactions that source information systems need to execute to mitigate the risk incurred by the constraint violations.

Definition 12 (Transaction). *Let $\mathcal{U}_{op}$ be the universe of operations that are executed by information systems (e.g.., send emails). A transaction $tr = (op, vmap) \in \mathcal{U}_{op} \times \mathcal{U}_{vmap}$ is a pair of an operation op and a parameter mapping vmap. $\mathcal{U}_{tr} \subseteq \mathcal{U}_{op} \times \mathcal{U}_{vmap}$ denotes the set of all possible transactions.*

For instance, the action description *A1* in Sect. 2 represents a transaction, $tr_1 = (send\text{-}an\text{-}email, vmap')$ where *vmap'(recipient)*= *"case manager"* and *vmap' (message)* = *"Frequent violations of C1"*.

Given a constraint instance stream and a time window, the action formula produces required transactions by analyzing the constraint instances in the constraint instance stream, which are relevant to the time window.

Definition 13 (Action Formula). *An action formula $af \in (\mathcal{U}_{CIS} \times \mathcal{U}_{tw}) \rightarrow \mathcal{P}(\mathcal{U}_{tr})$ is a function that maps a constraint instance stream and time window to a set of transactions. $\mathcal{U}_{af}$ is the set of all possible action formulas.*

Assume af_1 to assess the condition that is specified by the action description *A1* in Sect. 2, and to produce the corresponding transaction. Given constraint instance stream CIS and time window tw_1 = (01-01-2020 09:00, 02-01-2020 09:00), it assesses if there exist more than 10 constraint instances whose outcomes are "NOK" in the time window. If so, $tr_1 = (send\text{-}an\text{-}email, vmap') \in af(CIS, tw_1)$.

An action consists of an action formula and a set of time moments. The action formula specifies which transactions to generate in which conditions, and the set of time moments indicates when to assess the conditions and to generate transactions.

Definition 14 (Action). *An action $a = (af, TM) \in \mathcal{U}_{af} \times \mathcal{P}(\mathcal{U}_{tm})$ is a pair of an action formula af and a set of time moments TM. $\mathcal{U}_a$ denotes the set of all possible actions.*

Suppose a_1 = (af_1, TM_1) where (02-01-2020 09:00, (01-01-2020 09:00, 02-01-2020 09:00)) $\in TM_1$. We implement af_1 at 02-01-2020 09:00 with the constraint instances related to time window (01-01-2020 09:00,02-01-2020 09:00).

An action instance indicates when and which transaction is required.

Definition 15 (Action Instance). *An action instance $ai = (af, tr, time) \in \mathcal{U}_{af} \times \mathcal{U}_{tr} \times \mathcal{U}_{time}$ is a tuple of an action formula af, a transaction tr, and a timestamp time. $\mathcal{U}_{ai}$ is the set of all possible action instances.*

For instance, an action instance $ai_1 = (af_1, tr_1, $ 02-01-2020 09:00) denotes that the transaction tr_1 needs to be executed at 02-01-2020 09:00 according to af_1.

An action instance stream is a collection of unique action instances.

Definition 16 (Action Instance Stream). *An action instance stream AIS is a (possibly infinite) set of action instances, i.e., $AIS \subseteq \mathcal{U}_{ai}$. $\mathcal{U}_{AIS}$ is the set of all possible action instance streams.*

Given a constraint instance stream, an action engine continuously assesses the necessity of transactions by analyzing the action formulas in predefined actions at their appointed times.

Definition 17 (Action Engine). *Let $A \subseteq \mathcal{U}_a$ be a set of actions used by the action engine. $ae_A \in \mathcal{U}_{CIS} \to \mathcal{U}_{AIS}$ is the action engine such that, for any $CIS \in \mathcal{U}_{CIS}$, $ae_A(CIS) = \{(af, tr, time) \in \mathcal{U}_{ai} | \exists_{TM,tm} (af, TM) \in A \wedge tm \in TM \wedge time = \pi_t(tm) \wedge tr \in af(CIS, \pi_{tw}(tm))\}$.*

We abstract that the action engine is able to assess future constraint instances. In fact, it analyzes the historical constraint instance stream and produces transactions which mitigate risks caused by the past constraint violations.

5   Implementation

The general framework discussed above is implemented as a plug-in of *ProM*[1], an open-source framework for the implementation of process mining tools in a standardized environment. Our new plug-in is available in a new package named *ActionOrientedProcessMining*. The main input objects of our plug-in are an event stream, a constraint formula definition, and an action formula definition, whereas the output is an action instance stream.

The input event stream is in an XML-based *Object-Centric Log (OCL)* format storing events along with their related objects, while the constraint formula and action formula are defined by *Constraint Formula Language (CFL)* and *Action Formula Language (AFL)*, respectively. The schema of *OCL* and the syntax of *CFL* and *AFL* are explained in the tool manual[2] with examples.

The output action instance stream is in an XML-based *Action Instance Stream (AIS)* format (See footnote 2) storing action instances describing the transactions that need to be applied by source information systems. A dedicated gateway implemented in the source system parses the resulting *AIS* file and translates it into the system-readable transactions.

6   Proof-of-concept

Based on the implementation, we conducted experiments with the artificial information system supporting a simulated process to evaluate the feasibility of the framework. Specifically, we are interested in answering the following research questions.

– RQ1: Does the constraint monitor effectively detect violations?
– RQ2: Does the action engine effectively generate corresponding transactions?
– RQ3: Does the application of the transactions improve operational processes?

---

[1] http://www.promtools.org.

[2] https://github.com/gyunamister/ActionOrientedProcessMining.

Experimental Design. The information system used for the evaluation supports the order handling process described in Sect. 2. There are 16 available resources in total at any point in time, and each of them is responsible for multiple activities in the process. Orders are randomly placed and queued for the resource allocation after each activity. The resource is allocated according to the *First-in First-out* rule.

To answer RQ1-RQ3, we carried out the following steps repeatedly:

1. The information system generates events and updates an event stream.
2. The constraint monitor evaluates the constraint formula, which formulates "An order must be delivered in 72 h", every 24 h.
3. The action engine assesses the following action formulas every 24 h:
   - "If there is a violated order in the last 24 h, set a higher priority for it" (*AF1*), and
   - "If there is a violation that lasted longer than 24 h, send a notification to a case manager". (*AF2*)
4. The dedicated gateway for the information system translates the action instance stream into transactions that are executed by the information system.

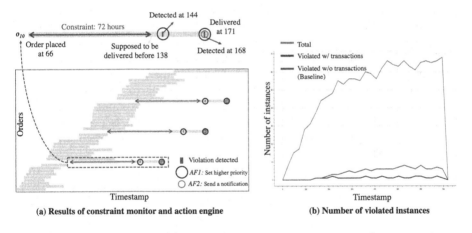

(a) Results of constraint monitor and action engine   (b) Number of violated instances

Fig. 3. Experimental results: (a) The results of constraint monitor and action engine on 40 selected orders. (b) Number of violated instances for 30 days. (Color figure online)

Experimental Results. Figure 3a reports the results related to RQ1 and RQ2. The figure shows the history of 40 orders by time, where the gray box indicates the delivery time (i.e. from order placement to delivery) and the green arrow denotes allowable delivery time (i.e., 72 h). The red box represents the time when the violation happens. As shown in Fig. 3a, every order whose delivery time is outside the green arrow is detected by the constraint monitor every 24 h. Moreover, *AF1* is generated for every violation and *AF2* is generated if the violation lasts longer.

Figure 3b reports an experimental result related to RQ3. The figure shows the number of violated instances for 30 days. The yellow line indicates the total number of instances, while the red/green lines represent the number of violated instances with/without applying mitigating transactions. The number of violated instances decreases when the transactions are applied.

7   Related Work

A constraint formula is a core component of the proposed framework, enabling the constraint monitor. The process mining discipline provides many techniques to be deployed to instantiate it. Conformance checking [6] can be deployed to find discrepancies between the modeled and the observed behavior of running instances. More tailor-made rules can be formalized into Petri-net patterns [7] or Linear Temporal Logic (LTL) [8] to evaluate whether process executions comply with them.

In addition, deviation detection techniques detect deviations in process executions with the user-defined compliance rules not being given. Instead, the model-based [9] and clustering-based [10] approaches learn the rules by analyzing event data and evaluate the violation of them in the execution of processes.

Furthermore, a constraint formula can be extended by more forward-looking techniques that are able to predict what will happen to individual cases and where bottlenecks are likely to develop [3,4]. The resulting predictions can be incorporated into the compliance rules [7,8] to evaluate future violations.

A commercial process mining tool, *Celonis Action Engine* [11], is a representative effort to turn analysis results into actions. It generates signals by analyzing the event data, and executes the actions corresponding to these signals to the source system. However, it does not support processing streaming data, which limits the continuous process management, analyzing signals (i.e., monitoring results), which inhibits the generation of relevant actions, and executing actions at process levels.

Several methods have been developed to generate proactive actions from the process mining diagnostics. In [12], the resource allocation is proactively optimized with the risk predictions of running instances. A prescriptive alarm system [13] generates alarms for the process instances that are predicted to be problematic with the aid of a cost model to capture the trade-off between different interventions. These approaches focus on improving the process by dealing with specific problems, mostly at the instance level. Instead, our proposed framework supports the management of comprehensive process-related problems and the execution of actions at both instance and process levels.

Robotic Process Automation (RPA) also aims at improving operational processes by automating repetitive tasks performed by humans by mimicking the execution on the user interface (UI) level [14]. While having the shared goal, the proposed framework has more emphasis on effectively managing the process in a continuous manner by identifying the problems based on diagnostics and executing proactive actions. Automating the problematic part of process executions with RPA techniques is one of those effective management actions.

8   Conclusion

In this paper, we proposed the general framework for action-oriented process mining, which continuously transforms process diagnostics into proactive actions for the process improvement. It is mainly composed of two parts: the constraint monitor and the action engine. The constraint monitor supports continuous monitoring of constraints, and the action engine generates the necessary transactions to mitigate the risks caused by the constraint violations.

The framework is instantiated in a *ProM* plug-in and tested on an information system. In fact, this paper is the starting point for a new branch of research in process mining. As future works, we plan to develop a concrete technique to support the efficient analysis of constraint instance streams by incorporating the concept of data cube. We also plan to validate the effectiveness of the proposed framework in real-life processes supported by information systems like SAP, Salesforce, Microsoft Dynamics, etc. Another important direction of future work is to provide a comprehensive taxonomy of constraints and actions to support the elicitation of relevant constraints and actions.

Acknowledgements. We thank the Alexander von Humboldt (AvH) Stiftung for supporting our research.

References

1. Aalst, W.: Data science in action. Process Mining, pp. 3–23. Springer, Heidelberg (2016). https://doi.org/10.1007/978-3-662-49851-4_1
2. Reinkemeyer, L. (ed.): Process Mining in Action. Principles, Use Cases and Outlook. Springer, Heidelberg (2020). https://doi.org/10.1007/978-3-030-40172-6
3. de Leoni, M., van der Aalst, W.M.P., Dees, M.: A general process mining framework for correlating, predicting and clustering dynamic behavior based on event logs. Inf. Syst. **56**, 235–257 (2016)
4. Marquez-Chamorro, A.E., Resinas, M., Ruiz-Cortes, A.: Predictive monitoring of business processes: a survey. IEEE Trans. Serv. Comput. **11**(6), 962–977 (2018)
5. Aalst, W.M.P.: Object-centric process mining: dealing with divergence and convergence in event data. In: Ölveczky, P.C., Salaün, G. (eds.) SEFM 2019. LNCS, vol. 11724, pp. 3–25. Springer, Cham (2019). https://doi.org/10.1007/978-3-030-30446-1_1
6. Carmona, J., van Dongen, B.F., Solti, A., Weidlich, M.: Conformance Checking - Relating Processes and Models. Springer, Switzerland (2018). https://doi.org/10.1007/978-3-319-99414-7
7. Ramezani, E., Fahland, D., van der Aalst, W.M.P.: Where did i misbehave? Diagnostic information in compliance checking. In: Barros, A., Gal, A., Kindler, E. (eds.) BPM 2012. LNCS, vol. 7481, pp. 262–278. Springer, Heidelberg (2012). https://doi.org/10.1007/978-3-642-32885-5_21
8. van der Aalst, W.M.P., de Beer, H.T., van Dongen, B.F.: Process mining and verification of properties: an approach based on temporal logic. In: Meersman, R., Tari, Z. (eds.) OTM 2005. LNCS, vol. 3760, pp. 130–147. Springer, Heidelberg (2005). https://doi.org/10.1007/11575771_11

9. Bezerra, F., Wainer, J.: Algorithms for anomaly detection of traces in logs of process aware information systems. Inf. Syst. **38**(1), 33–44 (2013)

10. Ghionna, L., Greco, G., Guzzo, A., Pontieri, L.: Outlier detection techniques for process mining applications. In: An, A., Matwin, S., Raś, Z.W., Slezak, D. (eds.) ISMIS 2008. LNCS (LNAI), vol. 4994, pp. 150–159. Springer, Heidelberg (2008). https://doi.org/10.1007/978-3-540-68123-6_17

11. Badakhshan, P., Bernhart, G., Geyer-Klingeberg, J., Nakladal, J., Schenk, S., Vogelgesang, T.: The action engine - turning process insights into action. In: ICPM Demo Track. Aachen, Germany, ceur-ws 2019, pp. 28–31 (2019)

12. Conforti, R., de Leoni, M., La Rosa, M., van der Aalst, W.M.P., ter Hofstede, A.H.: A recommendation system for predicting risks across multiple business process instances. Decis. Support Syst. **69**, 1–19 (2015)

13. Fahrenkrog-Petersen, S.A., et al.: Fire now, fire later: alarm-based systems for prescriptive process monitoring. arXiv:1905.09568 [cs, stat] (2019)

14. Agostinelli, S., Marrella, A., Mecella, M.: Towards intelligent robotic process automation for BPMers. arXiv:2001.00804 [cs] (2020)

Analyzing Comments in Ticket Resolution to Capture Underlying Process Interactions

Monika Gupta[1(✉)], Prerna Agarwal[1], Tarun Tater[1], Sampath Dechu[1], and Alexander Serebrenik[2]

[1] IBM Research, New Delhi, India
{mongup20,preragar,ttater24,sampath.dechu}@in.ibm.com
[2] Eindhoven University of Technology, Eindhoven, The Netherlands
a.serebrenik@tue.nl

Abstract. Activities in the ticket resolution process have comments and emails associated with them. Process mining uses structured logs and does not analyze the unstructured data such as comments for process discovery. However, comments can provide additional information for discovering models of process reality and identifying improvement opportunities efficiently. To address the problem, we propose to extract topical phrases (keyphrases) from the unstructured data using an unsupervised graph-based approach. These keyphrases are then integrated into the event log to derive enriched event logs. A process model is discovered using the enriched event logs wherein keyphrases are represented as activities, thereby capturing the flow relationship with other activities and the frequency of occurrence. This provides insights that can not be obtained solely from the structured data.

To evaluate the approach, we conduct a case study on the ticket data of a large global IT company. Our approach extracts keyphrases with an average accuracy of around 80%. Henceforth, discovered process model succinctly captures underlying process interactions which allows to understand in detail the process realities and identify opportunities for improvement. In this case, for example, manager identified that having a bot to capture specific information can reduce the delays incurred while waiting for the information.

Keywords: Process mining · Ticket resolution · Unstructured data

1 Introduction

A lot of structured and unstructured data is generated during the execution of business processes which gets stored in the information systems [3]. The data captures the runtime process behavior which can be analyzed to discover process reality, and support process improvement. Previous studies show that such an analysis can be based on process mining [2]. Process mining consists of mining

© Springer Nature Switzerland AG 2020
A. Del Río Ortega et al. (Eds.): BPM 2020 Workshops, LNBIP 397, pp. 219–231, 2020.
https://doi.org/10.1007/978-3-030-66498-5_17

event logs generated from business process execution supported by information systems. Every entry in the event log is an event referring to a case, activity, time stamp, and optional attributes such as actor (resource), associated cost, and duration.

Ticket resolution process also has corresponding logs captured in the ticketing system. Also some of the activities in the ticket resolution process have comments associated with them. Existing process mining techniques leverage structured event logs for discovering process model and identifying process inefficiencies [2]. However, comments can provide additional information for effective process improvement decisions. In this study, we aim at discovering the detailed process model for ticket resolution process, using the information present in the comments. The discovered model can then be used to identify the inefficiencies.

To model the detailed process, we extract the topical phrases (keyphrases) from the comments generated during the process execution, using an unsupervised graph-based approach [8]. These keyphrases are then integrated into the event log to derive enriched event logs. A process model is discovered using the enriched event logs wherein keyphrases are represented as activities, thereby capturing the flow relationship with other activities and the frequency of occurrence. This provides insights that could not be obtained solely from the structured data (i.e., activities), and these insights could be used to perform the ticket resolution process more efficiently.

To evaluate the approach, we conduct a case study on the ticket data of a large global IT company. We first extract the keyphrases from the comments associated with the ticket activities with an average accuracy of around 80%. This enables us to succinctly capture the additional information about the activities influencing the ticket resolution process and often causing delays, such as extra information required, priority, and severity. The model allows the managers to understand in detail the process realities and identify opportunities for improvement. In this case, for example, the manager identifies that having a bot to capture the information or adding a mandatory field in the initial ticket template, so as to reduce the delays incurred while waiting for information, can reduce the time (he subsequently had his team implement the bot).

2   Usefulness of Information in Comments

A lot of rich information is present in the comments generated during the process execution, which needs to be integrated into the discovered process model for in-depth process understanding. The in-depth unstructured data-driven (e.g., comments) insights help effectively identify the inefficiencies and make informed process improvement decisions.

Figure 1 shows the snapshot of a real example of a discovered process model for the ticket resolution process of a large global IT company. As part of the ticket resolution process, an analyst (person responsible for servicing the ticket) can ask the user to provide additional information by writing a comment, which gets captured in the information system as an event, *Need Info - Client.*

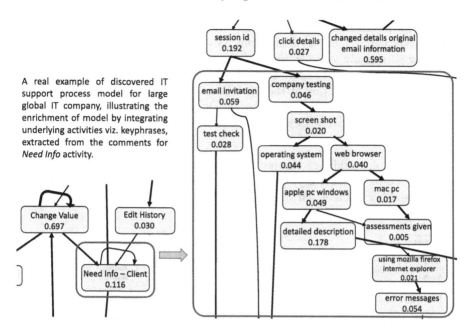

Fig. 1. A real example that compares process model from structured logs against the model capturing underlying activities as per the keyphrases extracted from comments. Here, node corresponds to the activity and edge represents the flow relationship

An analyst can ask for different information, such as error messages and operating system, which gets recorded in the comments. A process model is discovered using only a structured event log where activity, *Need Info - Client* is not further decomposed (refer to Fig. 1, left panel). We extracted the keyphrases from all the comments corresponding to the activity *Need Info - Client*, using an unsupervised graph-based keyphrase extraction approach. The extracted keyphrases represented information typically asked by analysts, using which an enriched event log was derived where the activity, *Need Info - Client*, was mapped to relevant keyphrases on the basis of the comment. The process model discovered using the enriched event log (refer to Fig. 1 - right panel) presented the underlying interactions of the process with activities such as email invitation, screenshot, web browser, operating system and error message, each corresponding to an information asked by the analysts (highlighted in Fig. 1, right panel). This allowed discovering the in-depth reality, which cannot be observed from Fig. 1, left panel. Thus, the following informed improvement decisions could be made to mitigate the delays incurred while waiting for information from the user:

– As *detailed description* (relative frequency is 0.178) is asked more often, the IT company should deploy a system such that a user can be preempted at the time of ticket submission to provide the same upfront [6]. Advantage of having a preemptive model is that a user is preempted selectively based on the ticket requirement as learnt from the historical data [6].

- A user is typically asked to provide an error *screenshot* after asking for *company testing*; therefore, analysts can be preempted to ask both the screenshot and the company testing at the same time.
- As information about the *operating system* and *web browser* is asked in the comments, a bot can be designed to automatically detect this information at the time of ticket submission.

This example highlights the potential of our approach for effective process improvement, by deriving keyphrases from comments corresponding to activities and representing them as part of the discovered process model.

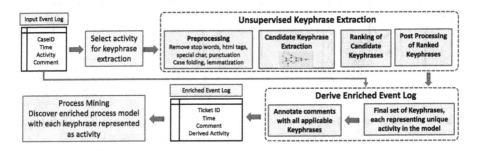

Fig. 2. Proposed approach to discover the underlying process using comments.

3 Proposed Approach

To achieve the objective of integrating knowledge captured in unstructured data, namely comments into the discovered process model, we presented an approach consisting of multiple steps, as shown in Fig. 2. First, a process analyst can select the activity for which the comments should be analyzed for in-depth process understanding. Performing this selection is important because a granular view of every activity can make the discovered process model look like spaghetti. Also, it needs to be decided on the basis of analysis to be performed, such that activities not captured in the structured logs are inferred from the comments. Thereafter, the comments corresponding to selected activities are preprocessed and used for candidate keyphrase extraction in an unsupervised manner. Extracted candidates are ranked and processed to select most relevant keyphrases which in turn are used to annotate the comments, thus, deriving an enriched event log. Finally, using the enriched event log, the process model capturing the flow relationship and frequency is discovered.

Algorithm 1: Unsupervised Keyphrase Extraction

---

1 **Input:** Initial Event Log EL (CaseID, timestamp, Activity, Comments)
2 **Output:** Keyphrase List K for Selected Activity A
3 **Variables:** $Candidate\ keyphrase\ list : M \leftarrow [], lookup_{table} \leftarrow [], score \leftarrow [],$
  $Global\ Graph : G \leftarrow []$
4 $comments \leftarrow SelectActivity(EL, A)$
5 $sentences \leftarrow Preprocess(comments)$
6 $G \leftarrow GlobalGraph(sentences)$
7 **for** $sentence$ in $sentences$ **do**
8   $G_c \leftarrow Graph(sentence)$
9   $G_u \leftarrow G - edges(G_c)$
10   $phrases \leftarrow G_u \cap G_c$
11   $M \leftarrow M \cup phrases$
12   $lookup_{table}[sentence] \leftarrow phrases$
13 **for** m in M **do**
14   $pf = PhraseFrequency(phrase)$
15   $cf = CommentFrequency(phrase)$
16   $score[m] \leftarrow pf \times -log(1 - cf)$
17 $score_n \leftarrow Sort(score, n)$
18 $K \leftarrow PostProcess(score_n)$
19 **return** K

3.1 Unsupervised Keyphrase Extraction

Keyphrase extraction aims at automatic selection of important and topical phrases from the body of documents [16]. Broadly, keyphrases are extracted using two approaches: supervised and unsupervised. In the supervised approach, a model is trained to classify a candidate keyphrase, requiring human-labeled keyphrases as training data. It is impractical to label training data (in this case, process execution comments), given the effort required for manual labeling. Thus, we focused on the unsupervised approach for our purpose. Unsupervised approaches can be grouped as follows [9]: graph-based ranking, topic-based clustering, simultaneous learning, and language modeling. Graph-based ranking methods are state-of-the-art methods [12], based on the idea of building a graph from the input document. Nodes in the graph are ranked based on their importance to select the most relevant keyphrases. Therefore, we used CorePhrase [8], a graph-based algorithm for topic discovery, that is, keyphrase extraction from multidocument sets based on frequently and significantly shared phrases between documents. The algorithm is domain independent and thus suitable for our purpose with some adaptations. The algorithm first identifies a list of candidate keyphrases from the set of documents and then selects top n-ranked keyphrases for the output by using a ranking criterion. The ranked keyphrases are then postprocessed to be adapted according to the domain.

Preprocessing of Comments: The comments from event logs are preprocessed (*Preprocess* in Algorithm 1) including stemming, case folding, removal of HTML tags, stop words, and special characters. Further, we created a set of unique sentences across all the comments to improve the scalability of the approach. The sentences in the comment were demarcated by a period. A unique set was created such that if a sentence was present in multiple comments then it was considered only once. This significantly reduced the number of sentences to be processed in further steps because the sentences were repeated across various comments (emails). This preprocessing did not affect the final set of extracted keyphrases because the keyphrase for a comment was a set of keyphrases extracted for its constituting sentences.

Candidate Keyphrase Extraction: To extract candidate keyphrases, the algorithm compares every pair of sentences to identify the common phrases. If there are n sentences in the corpus, comparing every pair is inherently $O(n^2)$. However, as highlighted in the CorePhrase [8] algorithm, using a data structure called the *Document Index Graph* (DIG), the comparison can be done in approximately linear time [7]. For our purpose, the DIG stored a cumulative graph representing the entire set of unique sentences (e.g., *GlobalGraph* function in Algorithm 1). When the keyphrase for a sentence has to be extracted, its subgraph is matched (by performing graph intersection) with the cumulative graph (viz. Global Graph) except for the sentence (Line 9 and 10 in Algorithm 1). It gives a list of matching phrases between the sentence and the rest of the sentences. This process generates matching phrases between every pair of sentences in near-linear time with varying length phrases. A master list M is maintained that contains unique matched phrases for all sentences that will be used as a list of candidate keyphrases. A *lookuptable* is also maintained that contains all sentences and the corresponding matching phrases (Line 12 in Algorithm 1) irrespective of whether the phrase gets selected after ranking or not (in the follow up steps of algorithm). Therefore, if a comment remains unannotated as phrases for none of its constituting sentences are in the selected set then the *lookuptable* is referred for annotation (discussed in Sect. 3.2).

Ranking of Candidate Keyphrases: Quantitative phrase metrics are used to calculate the *score* representing the quality of the extracted candidate keyphrase. The *score* is computed as $pf \times -\log(1 - cf)$, where cf is the comment frequency and pf is the average phrase frequency. Inspired by term frequency-inverse document frequency (TF-IDF) [8], the *score* rewards the phrases that appear in more documents (high cf) rather than penalizing them. For a phrase p, the comment frequency $cf(p)$ is the number of comments in which p appears, normalized by the total number of comments: $\frac{|\text{comments containing } p|}{|\text{all comments}|}$.

The average phrase frequency pf is the average number of times p appears in one comment, normalized by the length of the comment in words: $\arg \text{avg}[\frac{|\text{occurrences of } p|}{|\text{words in comment}|}]$

Postprocessing of Ranked Keyphrases: Selected top-ranked keyphrases contain some nonrelevant phrases, that is, phrases that fall out of the domain but still are common in most of the comments. Examples of such phrases are *thank you for contact,* and *contact helpdesk.* Such keyphrases are removed by creating a common domain dictionary that contains unwanted words to be removed from the keyphrases (function *PostProcess* in Algorithm 1). This domain dictionary thus postprocesses the keyphrases to obtain the final set of keyphrases. The following is an example of how postprocessing is applied to the keyphrases:

Extracted Phrase: Please tell unemployment benefits
Postprocessed Phrase: Unemployment benefits

Here words *please* and *tell* belong to an unwanted dictionary, as they are not relevant in keyphrases and thus removed. Also, if a keyphrase is a proper substring of any other selected keyphrase, then it is removed to resolve the spurious multilabel assignment.

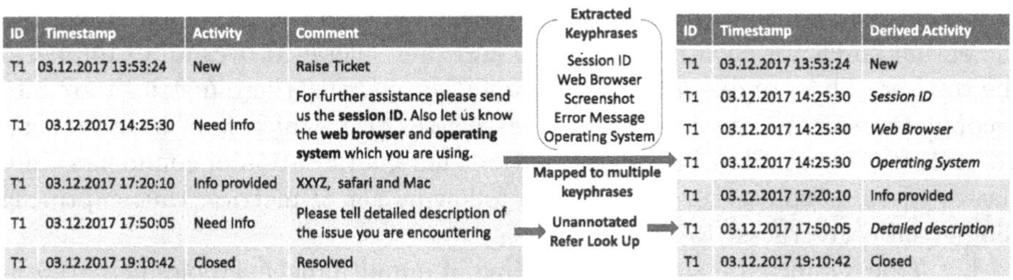

ID	Timestamp	Activity	Comment	Extracted Keyphrases	ID	Timestamp	Derived Activity
T1	03.12.2017 13:53:24	New	Raise Ticket	Session ID Web Browser	T1	03.12.2017 13:53:24	New
T1	03.12.2017 14:25:30	Need Info	For further assistance please send us the **session ID**. Also let us know the **web browser** and **operating system** which you are using.	Screenshot Error Message Operating System	T1	03.12.2017 14:25:30	Session ID
					T1	03.12.2017 14:25:30	Web Browser
				Mapped to multiple keyphrases	T1	03.12.2017 14:25:30	Operating System
T1	03.12.2017 17:20:10	Info provided	XXYZ, safari and Mac		T1	03.12.2017 17:20:10	Info provided
T1	03.12.2017 17:50:05	Need Info	Please tell detailed description of the issue you are encountering	Unannotated Refer Look Up	T1	03.12.2017 17:50:05	Detailed description
T1	03.12.2017 19:10:42	Closed	Resolved		T1	03.12.2017 19:10:42	Closed

Fig. 3. Example to illustrate comment annotation for deriving an enriched event log.

3.2 Annotating Comments with Keyphrases to Derive Enriched Event Log

The initial event log, *EL*, contains activities and corresponding comments. Once the keyphrases are extracted, each comment in the dataset is analyzed to determine whether one of the keyphrases matches with it. To make the matching consistent, we performed the same preprocessing as mentioned earlier. If a comment contained a keyphrase, it was annotated with the corresponding keyphrase. As we only extracted the top n most relevant keyphrases, some comments might be annotated by one or more keyphrases, while other comments might not be annotated at all.

To tackle the latter cases, we referred to the *lookup* table and retrieved the keyphrases for that comment. These keyphrases were added as labels to the comment. Therefore, maintaining the *lookup* table helped in assigning labels to otherwise unannotated comments.

Figure 3 depicts a real example of an event log where the first comment for the activity, *Need Info*, is mapped to three keyphrases. However, the second comment is not annotated with any of the selected top-ranked keyphrases and, therefore, is mapped to *detailed description* after referring to the lookup table. The enriched event log is generated with a new attribute, *derived activity*, replacing *activity* and *comment*, and representing the extracted keyphrases. To capture the ordering relationship of activities in the enriched event log, keyphrase(s) derived for a particular comment are assigned the timestamp of original comment.

3.3 Evaluating Keyphrase Extraction and Annotation

There are two aspects for evaluation, that is, the identification of informative keyphrases and correct annotation of comments with keyphrases. We evaluated the quality of extracted keyphrases manually, that is, checked whether they conveyed the required information. Further, we needed to evaluate the annotation of comments. As discussed in Sect. 3.2, a comment can be mapped to multiple keyphrases. Therefore, we used multilabel evaluation metrics that could be *example-based* or *label-based* [15]. We chose the *example-based evaluation metrics* that could capture the average difference between the predicted labels and the actual labels for each test example, and then averaged over all examples in the test set. Thus, unlike *label-based evaluation metrics*, these metrics took into account the correlations among different classes [17], which is of interest here. To evaluate the quality of the classification (here, annotation of comments) into classes (here, keyphrases), we used the following set of metrics, thus capturing the partial correctness [15]:

Let T be a multilabel dataset consisting of n multilabel examples $(x_i, Y_i), 1 \leq i \leq n, (x_i \in X, Y_i \in Y = \{0,1\}^k)$, with a labelset $L, |L| = k$. Let h be a multilabel classifier (here, annotator in Sect. 3.2) and $Z_i = h(x_i) = \{0,1\}^k$ be the set of label memberships predicted by h for the data point (i.e., comment) x_i.

$$Accuracy, A = \frac{1}{n} \sum_{i=1}^{n} \frac{|Y_i \cap Z_i|}{|Y_i \cup Z_i|} \quad (1) \qquad Recall, R = \frac{1}{n} \sum_{i=1}^{n} \frac{|Y_i \cap Z_i|}{|Y_i|} \quad (3)$$

$$Precision, P = \frac{1}{n} \sum_{i=1}^{n} \frac{|Y_i \cap Z_i|}{|Z_i|} \quad (2) \qquad F_1 = \frac{1}{n} \sum_{i=1}^{n} \frac{2|Y_i \cap Z_i|}{|Y_i| + |Z_i|} \quad (4)$$

$$HammingLoss, HL = \frac{1}{kn} \sum_{i=1}^{n} \sum_{l=1}^{k} [I(l \in Z_i \wedge l \notin Y_i) + I(l \notin Z_i \wedge l \in Y_i)], \quad (5)$$

where I is the indicator function which is equal to 1 if $Z_i = Y_i$, else 0. Since HL is a loss function, it should be minimum for better performance.

4   Case Study: IT Support Ticket Resolution Process

To illustrate the value of integrating knowledge from unstructured data into the discovered process model, we performed a case study on the IT support process data of a large global IT company. The dataset represented interactions between the users and the support team (analysts), and thus, comments were present with relevant activities. While the IT support process was continuously monitored by the process analyst, the unstructured data, for example, comments, were not taken into account.

Data extracted from the organization's ticket system includes the required information about a ticket starting from the time of ticket submission until it is closed. Downloaded data consists of 2620 tickets with 15,819 events in total. We observed from the dataset that two activities (out of 19), *Change Value* and *Need Info*, existed where analysts wrote comments. The number of events with the activities *Change Value* and *Need Info* was 4036 and 280, respectively.

In *Change Value*, changes in the ticket attributes were captured by a descriptive comment as shown below with an anonymized example (for confidentiality): *Changed Category from to "Y". Changed Sub-Category from to "test reset". Changed Severity from to "Sev 4". Changed Summary from to "reset the test". Changed Support Contract from None to Contract1.*

An analyst asks information from the user (here, customer) by writing a comment, which is captured as activity *Need Info* in the database. For example,

Dear ABC, Thank you for contacting the Support Center. In order to assist you more effectively we ask that you please provide the following information: Are you using a Macintosh Computer (Apple) or a PC (Windows)?: What web browser are you using (Internet Explorer, Mozilla Firefox, Safari, Google Chrome)?: Website you were directed to access: Session ID/login info: Detailed description of the issue you are encountering: Screen shot of error message: Thank you in advance!

To enrich the event logs, we performed keyphrase extraction for these two activities. This allowed us to precisely capture what values were changed and what information was typically asked by the analysts, in coherence with the complete process flow. IT support data were not made publicly available for confidentiality reasons; however, examples and results were included for an explanation.

Table 1. Experimental results where **K**: Total extracted keyphrases in final set, **L**: average number of labels for each comment, **P**: Precision, **R**: Recall, **A**: Accuracy, **F1**: F1 measure, and **HL**: Hamming Loss.

Data	K	L	P	R	A	F1	HL
Change value	33	3.63	84.74 %	81.27%	80.26%	82.12%	8.15%
Need info	16	4.28	84.71%	89.97%	80.21%	86.57%	6.21%

4.1 Unsupervised Keyphrase Extraction and Enriched Event Log Derivation

Comments for the selected activities, namely, *Change Value* and *Need Info*, were preprocessed. All the preprocessing steps as discussed in Sect. 3.1 were performed and the final set of preprocessed unique sentences was used for candidate keyphrase extraction. As per Algorithm 1, a set of candidate keyphrases is extracted for both the activity sets. Extracted keyphrases were ranked using the scoring function. We selected top 50 keyphrases from the ranked list for *Change Value* and *Need Info* respectively which were postprocessed as per the data properties. These postprocessed set of final keyphrases were used for annotating the respective comments as discussed in Sect. 3.2,

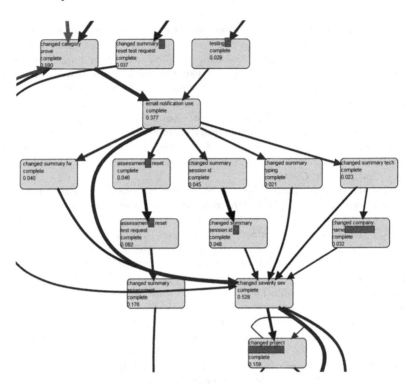

Fig. 4. A real example of discovered IT support process model for a large global IT company, illustrating enrichment of model by integrating keyphrases extracted from the comments for the *Change Value* activity. Some words are masked for confidentiality.

thus generating enriched event logs. Some comments for both the activities were left unannotated for which we referred to the lookup table, and hence assigned keyphrase. Effectively, the total number of unique keyphrases (K) in the resulting enriched event log was 33 and 16 for *Change Value* and *Need Info*, respectively (refer to Table 1). The average number of keyphrases, $L \simeq 4$ for *Change Value* and *Need Info* indicated that multiple important topics were present in a comment, that is, multiple ticket attributes were changed and multiple information was asked in the same comment.

4.2   Visualizing and Analyzing an Enriched Process Model

Enriched event logs are used for process discovery using ProM. We presented and compared *Need Info* for the original and enriched process model of IT support process in Fig. 1. Here, we presented the process model snapshot for IT support process, specifically highlighting the *Change Value* derived activities (refer to Fig. 4).

As shown in Fig. 4, the individual activity *Change Value* was replaced with more specific activities such as *changed category*, *changed company name*, and specific instances of *changed summary*, each corresponding to extracted keyphrases.

As the information captured in comments is integrated into the model, it is possible to derive insights as follows:

- The category was changed for a high percentage of tickets (as the relative frequency was 0.615), highlighting the need for a system to automatically assign a category based on the content of the initial ticket, thus optimizing the time spent for category assignment.
- The summary was changed for various tickets, and some of the most frequent instances were captured as keyphrases in our approach. Hence, we observed many states with a changed summary (suffixed with specific terms), although they all indicated some change in the summary.
- The company name was changed for a small percentage of tickets, which usually happened after the summary was changed in a specific manner. Therefore, the analysts could be preempted in those instances to change the company name in parallel with the summary, thus eliminating the delay.

We showed the process model discovered using structured logs and the enriched discovered process model side-by-side to the manager. He acknowledged that the enriched model helped in making effective process improvement decisions. One of the actionable insights he decided to take forward was to design a robotic process automation solution for the automatic category assignment to a ticket. This could not have been possible without integrating knowledge from the comments into the discovered process model.

4.3   Evaluation of Keyphrase Extraction and Annotation

Establish the Ground Truth: To evaluate keyphrase extraction (as discussed in Sect. 3.3), we established a ground truth for comments corresponding to the selected activities. Thus, we needed to first manually identify a set of keyphrases for comments corresponding to selected activities and annotate comments with the same. First, we identified the ticket attributes typically changed (as part of *Change Value* activity) and information typically asked by the analysts (as part of *Need Info* activity) on the basis of managers' domain knowledge and manual inspection of the comments. Manual inspection was performed by two authors for a disjoint set of comments (random sample of around 25% comments for each) to identify lists of changed attributes and asked information, respectively. Lists by both of them were compared to create a consolidated list. Both the authors used different terms to represent the same information, which were made consistent. Both of them identified the same list with a few exceptions (i.e., rarely occurring content), which were resolved. The final list was verified with the manager. Each item in the list was considered as a keyphrase for the data set.

Now that the list of ground truth keyphrases was identified, to establish the ground truth, comments were annotated with keyphrases using a keyword-based dictionary [14]. A list of keywords corresponding to each keyphrase was prepared iteratively, for example, the keyword "summary" for the keyphrase "changed summary". If the comment contained keywords, it was annotated with the corresponding keyphrase. Thereafter, authors of the paper manually investigated the disjoint set of randomly selected comments to distill the wrongly annotated comments. This process was repeated two to three times until very few/no updates were made in the set of keywords.

As an example, ground truth keyphrases for *Change Value* and *Need Info* were {Changed Category, Changed Sub-Category, Changed Severity, Changed Summary, Changed Support Contract}, and {mac pc, web browser, website directed, session id, detailed description, screen shot}, respectively.

Analysis of Results: Automatically extracted keyphrases can be structurally different from the human-identified ones, although both represent the same topical information.

To avoid spurious penalty on the metrics, we took this into account by manually creating a mapping between the two. Table 1 shows that the proposed approach performed with an accuracy of around 80% and had a low hamming loss. High F_1 measure ensured that the approach achieved a good balance between precision and recall. Hence, the proposed approach efficiently derived an enriched event log for the given data set.

5 Related Work

Unstructured text can be analysed to accrue benefits for process understanding and improvement at different levels however, sort with various challenges [1]. Some of the example use cases are discovering process models from natural language text [4,5] and searching textual as well as model-based process descriptions [11]. Automatic keyphrase extraction is used for a wide range of natural language processing and information retrieval tasks such as text clustering and summarization [13,18], text categorization [10], and interactive query refinement [16]. However, the application of keyphrase extraction to process model enrichment is not explored which is the focus of this work.

6 Conclusion

Process mining techniques are activity focused and do not consider comments generated during process execution. We presented a multistep approach to integrate hidden knowledge captured in unstructured text, namely comments, into the discovered process model. This was achieved by extracting keyphrases in an unsupervised manner and using them to annotate the comments thus deriving enriched event logs. We observed that the keyphrase extraction and annotation approach performed with an average accuracy of around 80% across different activities (*NeedInfo* and *ChangeValue*) for a data set. Further, we discovered the process model using a derived enriched event log and highlighted the value of enhanced process model in deriving actionable insights.

Our future plan is to extend the keyphrase extraction approach, such that the semantics is leveraged, and compare it with the proposed approach to analyze whether the insights derived from the discovered business processes are further enhanced.

References

1. Van der Aa, H., Vargas, J.C, Leopold, H., Mendling, J., Padró, L.: Challenges and opportunities of applying natural language processing in business process management. In: COLING 2018: The 27th International Conference on Computational Linguistics: Proceedings of the Conference: August 20–26, 2018 Santa Fe, New Mexico, USA, pp. 2791–2801. Association for Computational Linguistics (2018)
2. van der Aalst, W.M.P.: Process mining - discovery, conformance and enhancement of business processes (2011)
3. van der Aalst, W.M.P., La Rosa, M., Santoro, F.M.: Business process management. Bus. Inf. Syst. Eng. **58**, 1–6 (2016). https://doi.org/10.1007/s12599-015-0409-x/
4. Chen, Y., Ding, Z., Sun, H.: PEWP: Process extraction based on word position in documents. In: Ninth International Conference on Digital Information Management (ICDIM 2014), pp. 135–140. IEEE (2014)

5. Friedrich, F., Mendling, J., Puhlmann, F.: Process model generation from natural language text. In: Mouratidis, H., Rolland, C. (eds.) CAiSE 2011. LNCS, vol. 6741, pp. 482–496. Springer, Heidelberg (2011). https://doi.org/10.1007/978-3-642-21640-4_36

6. Gupta, M., Asadullah, A., Padmanabhuni, S., Serebrenik, A.: Reducing user input requests to improve it support ticket resolution process. Empir Softw. Eng. **23**, 1–40 (2017). https://doi.org/10.1007/s10664-017-9532-2

7. Hammouda, K.M., Kamel, M.S.: Efficient phrase-based document indexing for web document clustering. IEEE TKDE **16**(10), 1279–1296 (2004). https://doi.org/10.1109/TKDE.2004.58

8. Hammouda, K.M., Matute, D.N., Kamel, M.S.: CorePhrase: keyphrase extraction for document clustering. In: Perner, P., Imiya, A. (eds.) MLDM 2005. LNCS (LNAI), vol. 3587, pp. 265–274. Springer, Heidelberg (2005). https://doi.org/10.1007/11510888_26

9. Hasan, K.S., Ng, V.: Automatic keyphrase extraction: a survey of the state of the art. In: ACL, vol. 1, pp. 1262–1273 (2014)

10. Hulth, A., Megyesi, B.B.: A study on automatically extracted keywords in text categorization. In: Proceedings of the 21st International Conference on Computational Linguistics and the 44th annual meeting of the Association for Computational Linguistics, pp. 537–544. Association for Computational Linguistics (2006)

11. Leopold, H., van der Aa, H., Pittke, F., Raffel, M., Mendling, J., Reijers, H.A.: Searching textual and model-based process descriptions based on a unified data format. Softw. Syst. Model. **18**(2), 1179–1194 (2017). https://doi.org/10.1007/s10270-017-0649-y

12. Liu, Z., Huang, W., Zheng, Y., Sun, M.: Automatic keyphrase extraction via topic decomposition. In: EMNLP, pp. 366–376 (2010)

13. Manning, C.D., Manning, C.D., Schütze, H.: Foundations of Statistical Natural Language Processing. MIT Press, Cambridge (1999)

14. Pletea, D., Vasilescu, B., Serebrenik, A.: Security and emotion: sentiment analysis of security discussions on GitHub. In: MSR, pp. 348–351 (2014)

15. Sorower, M.S.: A literature survey on algorithms for multi-label learning. Oregon State University, Corvallis 18 (2010)

16. Turney, P.D.: Learning algorithms for keyphrase extraction. Inf. Retrieval **2**(4), 303–336 (2000). https://doi.org/10.1023/A:1009976227802

17. Zhang, M.L., Zhang, K.: Multi-label learning by exploiting label dependency. In: KDD, pp. 999–1008 (2010)

18. Zhang, Y., Zincir-Heywood, N., Milios, E.: World wide web site summarization. Web Intell. Agent Syst. Int. J. **2**(1), 39–53 (2004)

Automated Business Process Discovery from Unstructured Natural-Language Documents

Alexander J. Chambers[1]([✉]), Amy M. Stringfellow[1], Ben B. Luo[1],
Sophie J. Underwood[1], Tony G. Allard[1], Ian A. Johnston[1], Sarah Brockman[2],
Leslie Shing[2], Allan Wollaber[2], and Courtland VanDam[2]

[1] Defence Science and Technology, Edinburgh, SA 5111, Australia
{alexander.chambers,amy.stringfellow,ben.luo,sophie.underwood,
tony.allard,ian.johnston12}@defence.gov.au
[2] MIT Lincoln Laboratory, Lexington, MA 02421, USA
sbrockman@cs.umass.edu
{leslie.shing,allan.wollaber,courtland.vandam}@ll.mit.edu

Abstract. Understanding the processes followed by organizations is important to ensure business outcomes are achieved in an optimal, efficient and compliant manner. Process mining techniques rely on the existence of structured event logs captured by process management systems. These systems are not always employed and may not capture all process steps, leaving out those that occur through emails and chat software or edits to documents and knowledge-management systems.

Here we present an algorithm for the automated extraction of processes from unstructured natural-language documents. Action and topic analysis is used to generate an event log, from which process models are mined using standard techniques. We show the algorithm is capable of generating consistent software-development processes from an Apache Camel email dataset.

Keywords: Natural language · Machine learning · Process discovery

1 Introduction

Organizations follow systematic patterns of work to achieve outcomes in an optimal, efficient and compliant manner. Modeling these processes requires structured event logs captured by process-logging software. Many organizations do not employ this software however, as costs and potential risks may be prohibitive. Additionally, these event logs may be incomplete, missing events that occur outside of the logging context. For example, an employee may be required to email

DISTRIBUTION STATEMENT A. Approved for public release. Distribution is unlimited. This material is based upon work supported by the Under Secretary of Defense for Research and Engineering under Air Force Contract No. FA8702-15-D-0001. Any opinions, findings, conclusions or recommendations expressed in this material are those of the author(s) and do not necessarily reflect the views of the Under Secretary of Defense for Research and Engineering.

© Springer Nature Switzerland AG 2020
A. Del Río Ortega et al. (Eds.): BPM 2020 Workshops, LNBIP 397, pp. 232–243, 2020.
https://doi.org/10.1007/978-3-030-66498-5_18

their supervisor for "approval in principle" for travel before costing a trip to obtain full approval. The generation of event logs from unstructured data would allow wider application of business process management (BPM) techniques.

Our previous research and experimentation with the Event Labeling and Sequence Analysis (ELSA) package [39] has suggested that activities and processes can be learned, at least in part, from unstructured natural-language documents (NLDs) such as emails, chat logs, and document edits. ELSA used latent semantic indexing (LSI) [26] and density-based spatial clustering of applications with noise (DBSCAN) [16] to perform topic clustering, and association rule learning [17] to extract association rules between topics and participants. These rules were evaluated against a ground-truth dataset of human-annotated emails from the Apache Camel software project [4,5]. Several shortcomings were identified, including limitations of the natural-language processing (NLP) and association rule learning algorithms, and the absence of an activity-discovery step.

There has been relatively little work performed in the area of semi-automated extraction of business processes from NLDs to build complete end-to-end process models. Di Ciccio et al. [12] propose an approach similar to our own for their "MailOfMine" approach, but did not make use of recent NLP developments and did not validate their results against a ground truth. Jlailaty et al. [24] used recent developments to generate business activities from emails, although no processes were built or analyzed from these models. Dredze et al. [13] examined how to group emails by their contents. Du Four et al. [14] and Schumacher et al. [38] used the verbiage inherent in procedural texts to sequence tasks, but both used semi-structured data as input. Van der Aalst et al. [2] presented the EmailAnalyzer tool for deriving event logs from email data, requiring possible tasks to be pre-defined, and for process cases and tasks to be determined from specific metadata fields.

Outside of the BPM field, some researchers have generated activities from email content, with the aim of automatically constructing 'to-do' lists and similar. Corston-Oliver et al. [11] presented the SmartMail tool for generating tasks from email messages. Bellotti et al. [6] presented Taskmaster, a tool for managing email from a task-centric viewpoint by identifying 'thrasks'– threads of message files, links and drafts related to a particular task.

In this work, we implement action-labeling, case-assignment/clustering and process modeling components of the process-discovery algorithm, and explore alternative algorithms for the NLP steps, presenting a method for automated process discovery from unstructured natural-language data. Our approach demonstrates that, with minimal human-assisted labeling of activities, it is possible to generate meaningful process models from natural-language data.

2 Process-Discovery Algorithm

We draw on our previous work [39] for the structure of the process-discovery algorithm, The algorithm (Fig. 1) can be summarized as follows:

1. Ingest NLDs into a database.
2. Analyze content of NLDs.
   (a) Classify the action implied by each NLD.
   (b) Cluster NLDs by topic.
3. Construct event logs.
   (a) Generate events and associated activities from action-labeled NLDs.
   (b) Assign events to cases, clustering by associated topic and timeframe.
   (c) Cluster cases by process, based on common trace patterns.
4. Model processes from full event logs.

Step 1 is not elaborated on here. Steps 2–4 are described in Sects. 2.1 and 2.2. The algorithm can applied to any timestamped NLDs (emails in this work).

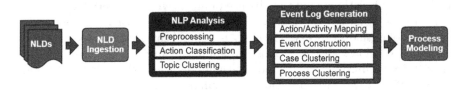

Fig. 1. Pipeline diagram for the process-discovery algorithm.

2.1 Natural-Language Processing

Preprocessing. The NLP algorithms used to perform action-classification and topic-clustering require a preprocessing step. We explored a range of commonly-used preprocessing models, which can be described in terms of their cleaning, vectorization, feature-scaling and feature-selection stages.

Cleaning removes noise and formatting from text, to increase the 'signal-to-noise' ratio of content. This involves stemming, stop-word filtering and punctuation removal. We made use of the Porter stemming algorithm [34] and stop-word filters with differing English dictionaries [21].

Vectorization transforms the natural-langauge content of NLD into a vector of weighted features for numerical analysis. We employed a variety of word-based tokenization methods: a unigram count vectorization method, word-sense induction (WSI) [33], Rapid Automatic Keyword Extraction (RAKE) [37], and Tex-tRank [31]. We also considered the following sequence-based methods: an n-gram vectorizer, a sentence vectorizer with count-based tokens, and a sequence-based ASGD Weight-Dropped LSTM (AWD-LSTM) language model encoding [30].

Feature scaling rescales the matrix representation of NLDs, with the intent of upweighting features which most differentiate samples. We compared no feature scaling, Z-score standardization of the space, and a term frequency-inverse document frequency (TF-IDF) weighting [40].

Feature selection is a dimension-reduction step intended to select features which most differentiate samples, so as to further reduce noise and computational complexity. In addition to testing no feature selection, we implemented a number of common feature-space reduction methods: singular value decom- position (SVD) [15], principle component analysis (PCA) [32], latent Dirichlet allocation (LDA) [35], and top-k maximum-count-based selection.

The output of the preprocessing stage is a matrix representation of the NLD corpus, which is used for the action-classification and topic clustering-steps.

Action Classification. This process assigns an abstract verb-object pair identifying the action implied by an NLD. These actions may not always be directly stated, and extracting them requires a level of abstract understanding of the purpose of the text. While our aim for automated process extraction is to build an entirely unsupervised pipeline, this is most challenging for the action-classification task. In this work, supervised methods using the ground-truth dataset as a training set were employed to ensure acceptable accuracy for further processing steps. See Sect. 4 for further discussion.

We selected a variety of machine learning approaches as candidate classification models: a support vector machine with radial basis function kernel [10,41], a linear kernel with stochastic gradient descent [44], a multinomial naive Bayes classifier [25], multilayer perceptrons, depth-wise separable convolutional network [8], explicit semantic analysis (ESA) [20], and gradient boosting [19]. NLP steps for these methods follow the structure discussed in Sect. 2.1, with a final classification step. Additionally, we implemented a sequential recurrent neural network (RNN) classification process used in association with the AWD-LSTM language model encoding [30], and a depth-wise separable convolutional neural network [8] in association with a simple sentence embedding. An important factor in selecting these models was their ability to handle small datasets well.

Topic Clustering. Here we use a combination of NLP and clustering techniques to group NLDs by shared key concepts and topics to enable the clustering of process cases. Hierarchical density-based spatial clustering of applications with noise (HDBSCAN) [29] and DBSCAN [16] were the primary clustering methods considered, as density-based clustering methods are useful when the number of clusters is unknown *a priori*. Choosing a meaningful distance threshold for DBSCAN can be difficult without an understanding of the underlying data and scale, and it can struggle with dataset clusters of varying densities and/or nested clusters. HDBSCAN's algorithm does not encounter either of these issues. We also investigated k-means [28], affinity-propagation [18] and mean-shift [9] clustering methods.

Many clustering algorithms allow for specification of a distance or similarity measure. In addition to conventional Euclidean and cosine-similarity measures, we also implemented the similarity measure for text processing (SMTP) [23].

2.2   Event Log Generation and Process Modeling

Event and Activity Generation. "Action" and "activity" are distinguished in this work, to generalize to contexts where natural-language actions may be distinct from event log activities. Here however, action labels assigned to the NLD corpus are used to form the set of observed activities. Events are generated by combining the activity associated with each NLD with date and time information. Each action label becomes a possible activity in this case. Although this mapping between actions and activities is trivial, we distinguish the two as in the future we wish to generalize our approach to other data sources. In general, one could consider inferring any number of events from an NLD.

Case Clustering. The assignment of events to process cases here is tailored to emails, however it is generalizable to any NLDs with 'creator' and 'audience' roles. More broadly, case clustering could be performed using any NLD features.

The assignment of events to specific process cases occurs through a three-part algorithm. In the first part, each event is considered a 'seed' of its own case. We initially grouped events by identical associated email subject lines (after removing common signifiers such as "Re:" or "Fwd:") that contain at least one common sender or recipient.

Secondly, we associate each case with a dominant topic determined by summing the counts of each topic for each email and choosing the most frequently-occurring topic. A graph is then constructed with edges between cases that have the same dominant topic, are within a pre-defined time interval of each other, and share at least one sender or recipient. Each connected component of the graph is completely merged into a new case. This process is repeated until no new merges are selected. This second stage is intended to counteract errors due to 'drift' in the topic of conversation for a given subject line or related emails having differing subject line.

For topic-clustering stages which produce vector representations of documents, we proceed to a third part. This involves the same graph construction and merging of connected components as the second stage, except that a cosine-similarity threshold is applied to the addition of each edge instead of the dominant-topic criterion. The equivalent 'case vector' is simply the average of all the topic vectors in the case.

Process Clustering. To identify cases arising from the same process, we used the HDBSCAN algorithm discussed in Sect. 2.1 to group similar traces. For the distance metric, we used the SequenceMatcher class provided by the Python standard library in the difflib module. This is an implementation of an algorithm similar to that of Ratcliff and Metzener [36].

Process Modeling. Given a cluster of similar traces, we construct processes by generating directly-follows graphs (DFGs) [27]. In addition, as a simple place-holder approach, we identified the start node as the activity that most commonly occurs in the first 25% of all traces in the cluster and similarly identify the end node using the last 25%.

3   Algorithm Results

To assess the performance of the process-discovery algorithm, we measured the output of each stage and the output of the whole process against a known ground truth [4], curated from open-source data from the Apache Camel software development project [5]. The ground truth consists of 250 emails, annotated with keywords and actions, and assigned to one of 65 cases. Each case is identified as an instance of one of six processes (e.g. "bugfix"). We use email as an example of an NLD, as it is a commonly used form of NLD, and likely to allude to complementary activities in contexts where logging is employed.

3.1   Individual Component Evaluation

Action Classification. To assess the action classification models of Sect. 2.1, we trained and verified using stratified four-fold cross-validation on 90% of the ground-truth dataset and subsequently evaluated a micro-averaged F_1 score on the remain 10% of the data. To partition labeled data for cross-validation, activities with fewer than four sample events were aggregated into an "other" class.

The best performing model type, found through a grid-search of the parameter space for each model type, used the unigram embedding, with Z-score scaling, top-k feature-selection and gradient boosting. Figure 2a shows the confusion matrix for this model, where we note that "provide support", which accounts for only 5% of the training set, has been misclassified as "git commit", which accounts for around 35%. As these classes are highly imbalanced, this misclassification is hard to avoid with limited training data. The micro-averaged F_1 validation score for this pipeline is 0.96, where the misclassification of the underrepresented "provide support" class has minimal impact.

Topic Clustering. The Apache ground-truth dataset does not include explicit topic information; thus, we can only assess topic clustering indirectly. Topic clustering is intended to enable the clustering of events into cases, which are completely contained by topic clusters. In an ideal case, the topic clusters and ground-truth trace clusters correspond exactly, enabling us to estimate topic-clustering performance by comparing topic and trace partitioning of events. This is only a proxy measure, however, as we intend topic clustering only as a coarse, first-pass grouping of NLDs.

We used the adjusted Rand index (ARI) [22] and the adjusted mutual in-formation (AMI) score [42] to compare topic and trace clusters, using the same stratified four-fold cross-validation model optimization step described in Sect. 2.1. The best model used RAKE vectorization, with TF-IDF scaling, PCA-based feature selection and the HDBSCAN clustering algorithm. We obtained an ARI of 0.15, and an AMI score of 0.20. While these scores are relatively low, topic clustering is intended to provide an initial coarse grouping of emails, which are further refined in the case-clustering step.

Case Clustering. As cases are constrained to be time-ordered, we can roughly estimate the performance of the case clustering algorithm by using the best-performing topic clustering model, and calculating the clustering similarity scores used in Sect. 2.1. We specified a 3-day time interval for this component (longer than the duration of most processes in the ground truth). Taking this approach, we obtained an ARI of 0.43, and an AMI score of 0.49. Figure 2b shows the confusion matrix of event assignment to cases between the ground truth and the process-discovery algorithm.

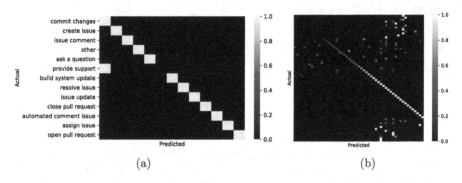

(a)   (b)

Fig. 2. (a) Row-normalized validation confusion matrix for the best performing action classification model. (b) Row-normalized confusion matrix for event assignment to cases between the ground truth and Sect. 2.2 algorithm.

Process Clustering. To measure the performance of the process-clustering step, we treat the set of traces associated with each process graph in the ground-truth dataset as process clusters and compare them to algorithmic process clusters. We obtained an ARI of 0.24, and an AMI score of 0.37.

Process Modeling. The performance of the process-construction component was evaluated using the ground-truth traces clustered using the process clustering methods of Sect. 2.2 as input. This minimized the flow-on effects of inaccuracy earlier in the pipeline.

The Apache Camel dataset contains two dominant processes, "bugfix" and "user support", that encompass 90% of the traces. Figure 3 shows the ground-truth "user support" process and an example trace. We compared the ground-truth processes for both "user support" and "bugfix" with those produced by the trace-clustering and process-modeling algorithms, using van der Aalst's fitness metric [1]. We used this metric to quantify how well a set of traces matches a

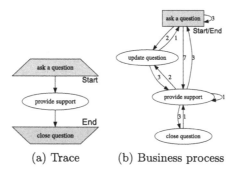

(a) Trace   (b) Business process

Fig. 3. Ground-truth "user support" trace and process. Counts on edges indicate the number of times a particular transition is observed.

process graph. To find the best matching trace of a process graph for a particular case, we used the $A^\star$ approach [3].

The algorithm is able to identically reproduce the "user support" process of Fig. 3b, and both algorithmic and ground-truth traces have fitness scores of 1.0 in this model. The fitness values for the ground-truth model are 0.797 for the ground-truth data and 0.940 for the algorithmically-constructed traces while the generated model produced values of 0.797 for the ground-truth data and 0.980 for the algorithmically-constructed traces. The generated "bugfix" process is well-supported by the ground-truth traces and self-consistent, achieving a fitness score close to 1.0 when measured against its own instances. In practical applications, the discovery pipeline will not have access to ground-truth process information, so the shared fitness consistency between ground truth and model information is a promising result.

3.2 Overall Evaluation

In assessing the overall performance of the process-discovery algorithm, we used the best topic clustering and action classification models and configured subsequent stages as described in Sect. 3.1. The only ground-truth input to this assessment was the action classification training dataset.

Figure 4 shows a comparison between the ground-truth "bugfix" process, and a process discovered by a full-run of our best algorithm. These are substantially similar, demonstrating that our approach is capable of generating processes that appear reasonable and interpretable to an analyst. For the ground truth, traces have a global fitness of 0.87 in the process graphs. For the full run of the process-discovery pipeline, traces have a fitness of 0.85.

4 Discussion

The pipeline we have presented is capable of extracting reasonable processes from unstructured NLDs, requiring only an action classification training set.

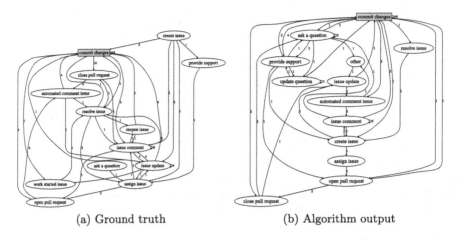

(a) Ground truth   (b) Algorithm output

Fig. 4. Comparison of "bugfix" process models. Counts indicate the number of times a transition is observed in the ground truth, or inferred by the algorithm.

Given the same dataset, two subject matter experts (SMEs) are unlikely to derive strictly identical processes. Similarly, while ground-truth and generated processes are not identical, they share many commonalities, including activities involved and allowed transitions between activities. There is plenty of room for future development of this research and improvements to the algorithm. Possible improvements include investigating more descriptive performance measures, reducing the supervised learning required and incorporating noteworthy recent developments in the NLP and BPM fields. Developments of interest include jointly extracting topical and discussive acts [43] and a new framework for combining data science and business process mining [7] which includes more involved process modeling methods. However, this research has shown a large step forward in capabilities of generating a business process end to end from natural-language documents with only one supervised step in the process.

While the process-discovery algorithm has been demonstrated here using email data, it can be employed with other sources of NLDs, such as chat logs, edits to Office documents, or edits to internal knowledge-management systems. In future, we intend to incorporate these additional data sources, which will allow greater visibility of processes that occur across multiple communication domains. Testing the algorithm on other datasets introduces another challenge due to the level of detailed SME labeling required. We plan to investigate methods to assist with this labeling process or ways of potentially bootstrapping existing datasets. We are also investigating integration of ancillary data sources such as corporate directories and project-management software, which will enrich the organizational information available in the final process construct.

5   Conclusion

In this work we have presented an algorithm for process discovery from NLDs, capturing events missed by process management software. The algorithm can demonstrably recover processes resembling those identified by SMEs.

In future, we intend to continue to refine the process-discovery algorithm, include other sources of NLDs such as chat logs and edits to documents and knowledge-management systems, and enrich process models with ancillary data, in order to provide a comprehensive overview of organizational processes.

References

1. van der Aalst, W., Adriansyah, A., van Dongen, B.: Replaying history on process models for conformance checking and performance analysis. Wiley Interdisc. Rev. Data Mining Knowl. Discov. **2**(2), 182–192 (2012). https://doi.org/10.1002/widm. 1045
2. van der Aalst, W., Nikolov, A.: EMailAnalyzer: an E-Mail mining plug-in for the ProM framework. BPM Center Report-07-16 (2007)
3. Adriansyah, A., van Dongen, B., van der Aalst, W.: Conformance checking using cost-based fitness analysis. In: 2011 IEEE 15th International Enterprise Distributed Object Computing Conference, pp. 55–64, August 2011. https://doi.org/10.1109/ EDOC.2011.12
4. Allard, T., Alvino, P., Shing, L., Wollaber, A., Yuen, J.: A dataset to facilitate automated workflow analysis. PLoS ONE **14**(2), 1–22 (2019). https://doi.org/10. 1371/journal.pone.0211486
5. Apache Camel (2019). https://camel.apache.org/. Accessed 31 Oct 2019
6. Bellotti, V., Ducheneaut, N., Howard, M., Smith, I.: Taskmaster: recasting email as task management. In: CSCW Workshop on Re-designing E-mail for the 21st Century (2002)
7. Berti, A., van Zelst, S.J., van der Aalst, W.M.P.: Process mining for python (PM4PY): bridging the gap between process- and data science. In: 2019 International Conference on Process Mining (ICPM) (2019)
8. Chollet, F.: Xception: deep learning with depthwise separable convolutions. In: Proceedings of the IEEE Conference on Computer Vision and Pattern Recognition, pp. 1251–1258 (2017). https://doi.org/10.1109/CVPR.2017.195
9. Comaniciu, D., Meer, P.: Mean shift: a robust approach toward feature space analysis. IEEE Tran. Pattern Anal. Mach. Intell. **24**(5), 603–619 (2002). https://doi. org/10.1109/34.1000236
10. Cooley, R.: Classification of news stories using support vector machines. In: Proceedings 16th International Joint Conference on Artificial Intelligence Text Mining Workshop. Citeseer (1999)
11. Corston-Oliver, S., Ringger, E., Gamon, M., Campbell, R.: Task-focused summarization of email. In: Text Summarization Branches Out, Barcelona, Spain, pp. 43–50. Association for Computational Linguistics, July 2004
12. Di Ciccio, C., Mecella, M., Scannapieco, M., Zardetto, D., Catarci, T.: MailOfMine – analyzing mail messages for mining artful collaborative processes. In: Aberer, K., Damiani, E., Dillon, T. (eds.) SIMPDA 2011. LNBIP, vol. 116, pp. 55–81. Springer, Heidelberg (2012). https://doi.org/10.1007/978-3-642-34044-4_4

13. Dredze, M., Lau, T., Kushmerick, N.: Automatically classifying emails into activities. In: Proceedings of the 11th International Conference on Intelligent User Interfaces, IUI 2006, New York, NY, USA, pp. 70–77. ACM (2006). https://doi.org/10.1145/1111449.1111471

14. Dufour-Lussier, V., Ber, F.L., Lieber, J., Nauer, E.: Automatic case acquisition from texts for process-oriented case-based reasoning. Inf. Syst. **40**, 153–167 (2014). https://doi.org/10.1016/j.is.2012.11.014

15. Eckart, C., Young, G.: The approximation of one matrix by another of lower rank. Psychometrika **1**(3), 211–218 (1936). https://doi.org/10.1007/BF02288367

16. Ester, M., Kriegel, H.P., Sander, J., Xu, X., et al.: A density-based algorithm for discovering clusters in large spatial databases with noise. Kdd **96**, 226–231 (1996)

17. Fournier-Viger, P., Tseng, V.S.: TNS: mining top-k non-redundant sequential rules. In: Proceedings of the 28th Annual ACM Symposium on Applied Computing, SAC 2013, New York, NY, USA, pp. 164–166, ACM (2013). https://doi.org/10.1145/2480362.2480395

18. Frey, B.J., Dueck, D.: Clustering by passing messages between data points. Science **315**(5814), 972–976 (2007). https://doi.org/10.1126/science.1136800

19. Friedman, J.H.: Greedy function approximation: a gradient boosting machine. Ann. Stat. **29**(5), 1189–1232 (2001). https://doi.org/10.1214/aos/1013203451

20. Gabrilovich, E., Markovitch, S.: Overcoming the brittleness bottleneck using wikipedia: enhancing text categorization with encyclopedic knowledge. In: Proceedings of the 21st National Conference on Artificial Intelligence - Volume 2, AAAI 2006, pp. 1301–1306. AAAI Press (2006)

21. Honnibal, M., Johnson, M.: An improved non-monotonic transition system for dependency parsing. In: Proceedings of the 2015 Conference on Empirical Methods in Natural Language Processing, Lisbon, Portugal, pp. 1373–1378. Association for Computational Linguistics, September 2015. https://doi.org/10.18653/v1/D15-1162

22. Hubert, L., Arabie, P.: Comparing partitions. J. Classif. **2**(1), 193–218 (1985). https://doi.org/10.1007/BF01908075

23. Jiang, J.Y., Cheng, W.H., Chiou, Y.S., Lee, S.J.: A similarity measure for text processing. In: 2011 International Conference on Machine Learning and Cybernetics, vol. 4, pp. 1460–1465, July 2011. https://doi.org/10.1109/ICMLC.2011.6016998

24. Jlailaty, D., Grigori, D., Belhajjame, K.: Email business activities extraction and annotation. In: Kotzinos, D., Laurent, D., Spyratos, N., Tanaka, Y., Taniguchi, R. (eds.) ISIP 2018. CCIS, vol. 1040, pp. 69–86. Springer, Cham (2019). https://doi.org/10.1007/978-3-030-30284-9_5

25. Kibriya, A.M., Frank, E., Pfahringer, B., Holmes, G.: Multinomial Naive Bayes for text categorization revisited. In: Webb, G.I., Yu, X. (eds.) AI 2004. LNCS (LNAI), vol. 3339, pp. 488–499. Springer, Heidelberg (2004). https://doi.org/10.1007/978-3-540-30549-1_43

26. Landauer, T.K., Foltz, P.W., Laham, D.: An introduction to latent semantic analysis. Discourse Process. **25**(2–3), 259–284 (1998). https://doi.org/10.1080/01638539809545028

27. Leemans, S.J.J., Poppe, E., Wynn, M.T.: Directly follows-based process mining: exploration & a case study. In: 2019 International Conference on Process Mining (ICPM), pp. 25–32, June 2019. https://doi.org/10.1109/ICPM.2019.00015

28. Lloyd, S.P.: Least squares quantization in PCM. IEEE Trans. Inf. Theory **28**(2), 129–137 (1982). https://doi.org/10.1109/TIT.1982.1056489

29. McInnes, L., Healy, J., Astels, S.: HDBSCAN: hierarchical density based clustering. J. Open Source Softw. **2**(11), 205 (2017). https://doi.org/10.21105/joss.00205

30. Merity, S., Keskar, N.S., Socher, R.: Regularizing and optimizing LSTM language models. In: International Conference on Learning Representations (2018)
31. Mihalcea, R., Tarau, P.: TextRank: bringing order into text. In: Proceedings of the 2004 Conference on Empirical Methods in Natural Language Processing, Barcelona, Spain, pp. 404–411. Association for Computational Linguistics, July 2004
32. Pearson, K.: LIII: on lines and planes of closest fit to systems of points in space. Philos. Mag. J. Sci. **2**(11), 559–572 (1901). https://doi.org/10.1080/14786440109462720
33. Pelevina, M., Arefiev, N., Biemann, C., Panchenko, A.: Making sense of word embeddings. In: Proceedings of the 1st Workshop on Representation Learning for NLP, Berlin, Germany, pp. 174–183. Association for Computational Linguistics, August 2016. https://doi.org/10.18653/v1/W16-1620
34. Porter, M.F.: An algorithm for suffix stripping. Program **14**(3), 130–137 (1980). https://doi.org/10.1108/00330330610681286
35. Pritchard, J.K., Stephens, M., Donnelly, P.: Inference of population structure using multilocus genotype data. Genetics **155**(2), 945–959 (2000)
36. Ratcliff, J.W., Metzener, D.E.: Pattern-matching-the gestalt approach. Dr Dobbs J. **13**(7), 46 (1988)
37. Rose, S., Engel, D., Cramer, N., Cowley, W.: Automatic Keyword Extraction from Individual Documents, vol. 1, pp. 1–20. Wiley, Hoboken (2010). https://doi.org/10.1002/9780470689646.ch1
38. Schumacher, P., Minor, M., Walter, K., Bergmann, R.: Extraction of procedural knowledge from the web: a comparison of two workflow extraction approaches. In: Proceedings of the 21st International Conference on World Wide Web, New York, NY, USA, pp. 739–747. WWW 2012 Companion, ACM (2012). https://doi.org/10.1145/2187980.2188194
39. Shing, L., et al.: Extracting workflows from natural language documents: a first step. In: Daniel, F., Sheng, Q.Z., Motahari, H. (eds.) BPM 2018. LNBIP, vol. 342, pp. 294–300. Springer, Cham (2019). https://doi.org/10.1007/978-3-030-11641-5_23
40. Sparck Jones, K.: A statistical interpretation of term specificity and its application in retrieval. J. Documentation **28**(1), 11–21 (1972). https://doi.org/10.1108/eb026526
41. Vapnik, V.N.: The Nature of Statistical Learning Theory. Springer, Heidelberg (1995). https://doi.org/10.1007/978-1-4757-3264-1
42. Vinh, N.X., Epps, J., Bailey, J.: Information theoretic measures for clusterings comparison: variants, properties, normalization and correction for chance. J. Mach. Learn. Res. **11**, 2837–2854 (2010)
43. Zeng, J., Li, J., He, Y., Gao, C., Lyu, M.R., King, I.: What you say and how you say it: joint modeling of topics and discourse in microblog conversations. Trans. Assoc. Comput. Linguist. **7**, 267–281 (2019). https://doi.org/10.1162/tacl_a_00267
44. Zhang, T.: Solving large scale linear prediction problems using stochastic gradient descent algorithms. In: Proceedings of the Twenty-first International Conference on Machine Learning, ICML 2004, New York, NY, USA, pp. 116–123. ACM (2004). https://doi.org/10.1145/1015330.1015332

Workshop on Business Process Management in the Era of Digital Innovation and Transformation (BPMinDIT)

Workshop on Business Process Management in the era of Digital Innovation and Transformation (BPMinDIT)

The fundamental nature of many organizations is being rapidly transformed with the ongoing diffusion of digital technologies. In this era, organizations in many domains are challenged to question their existing business models and to improve or revolutionize them using new technologies such as Artificial Intelligence (AI), Internet of Things (IoT), 4D printing, or Blockchain technology. Many IT-based initiatives, such as Uber, Car2Go, DriveNow, Udacity, or Airbnb have emerged, and disrupt traditional markets as such. To stay ahead of their competitors, even ICT giants (e.g., Google or Amazon) face the need to constantly evaluate and improve the value they propose to their end customers.

These developments are also challenging the role of BPM. For instance, advances in data analytics and AI, uptake of emerging technologies, increased adoption of cloud and mobile technologies, and new business paradigms such as service-dominant logic, open the path for more innovative business processes and new opportunities that arise for the application of BPM. We see, for example, how automated business process management can be used to tightly link business analytics and business execution in short process-based iterations to follow quickly changing markets, how real-time data from physical entities ('things' in the IoT sense) is directly injected into decision making in business processes, or how agile and IT-reliant business models are directly mapped to executable business processes.

However, the traditional role of BPM in structuring and optimizing (mostly but not limited to operational) processes often falls short in making use of the above-mentioned opportunities. This can risk the position of the BPM discipline to act as the driving force in digital innovation and transformation initiatives. New BPM capabilities that reflect an explorative-dominant (instead of exploitation-dominant) view may help in addressing the emerging opportunities and challenges of digitalization.

In this workshop, we question and investigate the new role of BPM in the digital era. The goal is to advance our understanding of the BPM capabilities that organizations require to explore emerging opportunities of digital innovation and transformation, and cope with the related challenges.

We are excited to receive high-quality submissions in BPMinDIT's second edition. Each submission was reviewed by at least three members of the Program Committee. From these submissions, the top-two was accepted for presentation at the workshop. These papers feature highly relevant and novel research ideas.

Kirss and Milani provide a good example of how a recent technology can be used as a leverage for process innovation. The authors report on their investigation of how the blockchain technology can be used to redesign the Know-Your-Customer process in banks. Secondly, Alam, Dallasega, Marengo, Nutt, and Rahman focus on constructions processes, which are acknowledged to have the potential for benefiting significantly from digitalization as well. The authors propose an approach and demonstrator tool that can help organizations operating in the construction domain to improve control over their processes.

The second edition of the workshop features also a keynote talk by Paul Grefen. His talk challenges the position of BPM with respect to the Service-Dominant Logic (SDL) as a new business paradigm. He coins the term 'Service-Dominant Process Engineering' and discusses how BPM can be (re)positioned in a broad context of business engineering in the era of SDL.

We hope that the reader will find the selected papers relevant and interesting. Wishing you a nice reading experience!

September 2020   Oktay Turetken
                                                                     Amy Van Looy

Organization

Workshop Chairs

Oktay Turetken   Eindhoven University of Technology,
The Netherlands

Amy Van Looy   Ghent University, Belgium

Program Committee

Banu Aysolmaz   Eindhoven University of Technology,
The Netherlands

Wasana Bandara   Queensland University of Technology,
Australia

Vesna Bosilj-Vukšić   University of Zagreb, Croatia

Marco Comuzzi   Ulsan National Inst. of Science and Tech.,
South Korea

Mahendrawathi Er   Institut Teknologi Sepuluh Nopember,
Indonesia

Peter Fettke   Saarland University, Germany

Renata Gabryelczyk   University of Warsaw, Poland

Paul Grefen   Eindhoven University of Technology,
The Netherlands

Thomas Grisold   University of Liechtenstein, Liechtenstein

Mojca Indihar Stemberger   University of Ljubljana, Slovenia

Andrea Kő   Corvinus University of Budapest, Hungary

Rob Kusters   Open University, The Netherlands

Peter Loos   Saarland University, Germany

Monika Malinova   Vienna University of Economics and Business,
Austria

Charles Møller   Aalborg University, Denmark

Baris Ozkan   Eindhoven University of Technology,
The Netherlands

Maximilian Röglinger   University of Bayreuth, Germany

Estefania Serral   KU Leuven, Belgium

Service-Dominant Process Engineering: A New Reality for BPM?

Paul Grefen ⓘ

Eindhoven University of Technology and Atos Netherlands
p.w.p.j.grefen@tue.nl

Extended Abstract. Traditionally, business process engineering (BPE) follows a development approach from specification to realization of business processes that resembles the well-known waterfall approach in information systems engineering. Even though the BPM life cycle approach in all its variations emphasizes the cyclical nature of BPE, it is based on a sequence of life cycle phases. In the modern world of agile business engineering, this sequencing is often not flexible enough anymore. Traditional BPE also focuses heavily on the 'what' and 'how' questions of business processes: what should be done in which way? In a world of service-dominant business models [1, 2], the emphasis is much more on the perspective of value delivery than on the perspective of process flow. In other word, the 'why' question of business processes becomes dominant. To address the advent of both agility and service-dominance, we advocate an approach of bi-speed business process engineering shown in the figure. The left, relatively slow development cycle (evolution) creates the basis for business processes in the form of business services. The right, relatively fast cycle (revolution) creates and operationalizes business models (answering the 'why' question) in the form of executable business processes based on the business services.

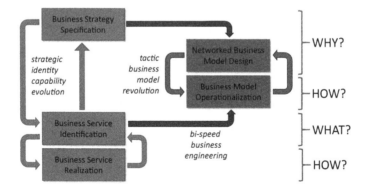

Fig. 1. Bi-speed service-dominant business process engineering

References

1. Grefen, P.: Service-Dominant Business Engineering with BASE/X: Business Modeling Handbook. CreateSpace Independent Publishing Platform, USA (2015)
2. Grefen, P, Turetken, O.: Achieving Business Process Agility through Service Engineering in Extended Business Networks. BPTrends (2018)

Using Blockchain Technology to Redesign Know-Your-Customer Processes Within the Banking Industry

Kristin Kamilla Kirss and Fredrik Milani[(⊠)]

University of Tartu, Narva Mantee 18, 51009 Tartu, Estonia
kristin@juhe.ee, milani@ut.ee

Abstract. Blockchain has emerged as a technology with the potential to innovate business processes. The potential benefits of this technology have attracted the attention of banks. In particular, banks are interested in redesigning compliance processes to reduce costs and risks. One of the most crucial compliance processes is that of the *Know-Your-Customer* (KYC) process. The KYC process serves to ensure that banks assess the risk of doing business with a client and adopt appropriate mitigation strategies. The KYC process is mandatory, expensive, reduces customer satisfaction, and does not add value to the banks. Therefore, banks seek to use technologies to redesign the KYC process. In recent years, blockchain technology has been considered for the redesign of the KYC process. However, history has taught us that merely substituting existing technology has limited value. Rather, it is necessary to use the capabilities of new technologies, such as blockchain, to enable redesign of the KYC process. In this paper, we examine how blockchain technology can enable the redesign of the *Know-Your-Customer* (KYC) process. To this end, we conduct a case study and discuss how the capabilities of blockchain technology can enable banks to redesign their KYC processes.

Keywords: Blockchain · Business process redesign · Know-Your-Customer

1 Introduction

The banking industry is one of the most regulated in the world, forcing them to comply to with a growing number of regulatory demands. Therefore, banks invest in improving their compliance processes. However, there are challenges. Firstly, compliance processes are difficult to standardize as different governmental agencies have different requirements [1]. Secondly, the costs for compliance are relatively high. For instance, compliance costs for banks in US alone, doubled from 2007 to 2013 [2]. Thirdly, extensive and rigorous compliance processes, such as the Know-Your-Customer (KYC) process, negatively affect customer onboarding and satisfaction [3]. Forth, banks are held responsible for identifying and stopping fraudulent financial behavior, such as money laundering [4]. Failure to fulfil this responsibility leads to hefty fines.

In addressing the above issues, banks have relied on technologies to improve efficiency. However, existing technologies support banks in increasing automation of compliance processes but not necessarily enable transformational redesign. In recent years,

© Springer Nature Switzerland AG 2020
A. Del Río Ortega et al. (Eds.): BPM 2020 Workshops, LNBIP 397, pp. 251–262, 2020.
https://doi.org/10.1007/978-3-030-66498-5_19

blockchain has emerged as a technology with the potential to re-think and redesign banks' compliance processes. Blockchain enable independent parties to maintain a safe, tamper-proof, and permanent ledger of transactions without relying on a central authority [5]. Transactions are not kept centrally but instead, each party records and maintains a copy of the ledger once a majority have reached consensus about the validity of the transaction. This capability i.e., a replicated append-only transactional storing of data, enables tamper-proof and secure data interactions among independent parties [5].

Banks have closely followed the applicability of blockchain to better understand how its use can redesign for instance, cross-border payments and settlement of financial instruments [6]. However, the focus has been on the technology rather than using blockchain for innovative redesign of compliance processes [7]. Therefore, it is relevant to examine how blockchain, given its capabilities, can enable redesign of compliance processes. In particular, we explore how the KYC process can be redesigned as it is mandatory for all banks, directly affects the customers, and is the basis for other compliance processes. To this end, this paper explores the research question of "*how can blockchain technology enable redesign of banks' Know-Your-Customer process*".

In this paper, we illustrate how banks' KYC process can be redesigned to address the issues of differing requirements, high costs, customer experience, and quality in detecting suspicious entities. Thus, the contribution of this paper is relevant for analysts involved with designing or implementing blockchain use cases for compliance within the banking sector. Understanding of how blockchain can redesign compliance processes and the benefits thereof, can help analysts move beyond substituting existing technology when redesigning compliance processes. We use a case study to study the KYC process within the setting of a multi-national bank operating in northern Europe.

The remainder of the paper is structured as follows. Section 2 introduces background on blockchain, smart contracts, and KYC. Then, in Sect. 3, we present the case study followed by a discussion in Sect. 4. Finally, Sect. 5 concludes the paper.

2 Background

In this section, we present background on blockchain technology and smart contracts. We, furthermore, briefly describe the mandatory KYC process.

2.1 Blockchain Technology and Smart Contracts

Blockchain is based on the principle that data is replicated across a number of ledgers, each of which keeps a full or partial copy. Its architecture enables recording and sharing of data across multiple ledgers. New transactions are gathered in blocks and once verified, added to the ledger. Thus, each block is linked to the previous one. To change a transaction would require modifying other blocks, an action requiring a high computational effort. The transactions become, therefore, immutable and tamper-proof [8].

The distributed architecture of blockchain enables multiple independent parties to interact and share data with each other without necessarily relying on a central authority. Such blockchain platforms are commonly divided into two main categories, *public* or *permissioned* [9]. A public blockchain is fully decentralized i.e., open for all to join

[10] while a permissioned one has restrictions on joining and data access. As such, permissioned blockchain is preferred for consortiums with shared interests such as cross-organizational settlements of records [11] and compliance processes of banks.

Blockchain enable deploying code with predefined rules that automatically execute when conditions are met [12] called smart contracts [13]. Smart contracts automatically execute across all ledgers [14]. As smart contracts inherit the capabilities of blockchain, they become tamper-proof, efficient, precise, and transparent once deployed. Smart contracts are tamper-proof and cannot be modified, manipulated, or deleted [15]. The efficiency and precision are gained as the smart contracts self-execute when pre-defined conditions are met and as such, execute what has been agreed upon. Transparency is achieved as the pre-defined conditions are verifiable by involved parties [5].

2.2 KYC

The purpose of compliance is to ensure that banks identify existing risks and implement mitigation strategies. At its core, compliance processes aim at balancing the risk appetite of banks with their abilities to manage the business without endangering the bank, clients, or the funds of involved counterparties. The essence of banking is the customer.

Customers can expose banks to risks of money laundering or terrorist funding. As such, banks must assess the extent of such risks by knowing their customers. This process, which is a legal requirement for banks, is called *Know-Your-Customer* (KYC). The KYC aims at establishing the customers' identity, ensure that the source of the funds is legitimate, and monitor risks for money laundering.

The KYC process consists of three main steps. The first is customer identification and serves to verify that the customer is who he/she claims to be. The next is customer due diligence. The aim of the second step is to determine if the potential customer can be trusted. This is achieved by collecting and examining data, such as occupation, type of transactions, expected financial activity patterns, and if the person is politically exposed. Upon completion of the KYC, customers can be onboarding as a new customer. In addition, banks also have the responsibility to renew the KYC and regularly screen the customers' financial transactions. The KYC process is generally considered to be cumbersome and to have negative implications for both banks and their customers [16].

3   Case Study

In this section, we first present the case study methodology and research questions. We also present the case study setting and design. Then, we present the case study execution that includes the current KYC process and how KYC could be conducted if leveraged by blockchain and smart contract technology.

3.1   Case Study, Setting, and Design

Case study research method employs a qualitative methods to investigate a particular reality within its real-life context [17], in particular when the boundaries distinguishing the object of study and its context are not clear [18]. Case studies can be used for confirming a hypothesis [17, 19], method evaluation [20], and for exploratory purposes [17, 19]. Yin [18] argues for the necessity of defining a research question when designing a case study. Hence, we use the case study method to explore the research question of *"how can blockchain technology enable redesign of banks' Know-Your-Customer process?"*. The research question is relevant given that if new technology is used to substitute existing ones, limited value can be derived. Rather, new technologies should be used to redesign processes [21]. Therefore, it is relevant to explore how new technologies, such as blockchain, can enable redesign of processes.

Given the above defined research question, we sought a banking case that (1) is examining blockchain usage for KYC processes and (2) provides access to information. To this end, we selected the compliance department of a mid-sized European bank. Currently, the bank conducts KYC with limited support from IT systems. The management of the KYC process is costly, does not add value to the bank, and demands high quality as failure to identify and report suspicious entities results in fines. The bank, therefore, is exploring how blockchain can enable redesign of the KYC process.

The design of the case studies consists of three steps. The first step is data collection. Data was collected from primarily four sources. The first source was bank internal documentation on KYC available at the bank. These documents including instructions, templates, guidelines, and relevant documents published by regulatory agencies. The second source of information came from attending banking association group meetings (22 in total). During these meetings, held in the domestic language, specialists discussed various compliance issues such as KYC. The third source was meetings with 4 KYC specialists, 4 client managers, and 2 blockchain specialists. The fourth source was workshops conducted with domain experts at the bank. The condition for accessing the above sources of data was anonymity. Thus, cannot reveal the identity of the bank. We took notes as we did not have permission to record the interviews.

The second step of the design was the mapping of the current state. This was achieved by conducting workshops with domain experts. The objective was to understand and model the current state of how KYC is conducted. In so doing, special attention was given to the general requirements. As such, the current state does not focus on specifics for the bank but for KYC. The workshops resulted in a set of models representing the main components and actors of the KYC process.

The third step, design of a blockchain-based KYC process, was conducted in the same manner as step 2 but with two iterations. The input from the first and second step were used to develop a blockchain-based KYC process. The first proposal was verified with blockchain specialists who provided feedback on the feasibility of the solution. Then, a second workshop was conducted with the domain experts to finalize and verify the blockchain-based KYC process. The workshops conducted in steps two and three lasted, on average, 90 min. The participants were domain experts but not experts in blockchain technology. Therefore, the discussions were on conceptual solutions and not on technical implementation.

3.2 Case Study Execution

This section describes the execution of the case study. We first provide an overview of how compliance is conducted as-is followed by a conceptual solution.

Current Know-Your-Customer Processes

There are recommendations that stipulates banks to identify, assess, and understand the risk levels they are exposed to, and implement appropriate measures to mitigate risks. However, the details and methods are left to discretion of the banks. RBA stipulates banks to assess the risk of a potential client before onboarding. This is commonly achieved by cross-checking the documentation provided by the potential clients with external sources such as state registries, trusted parties, and different watchlists. The main information required for onboarding potential clients are depicted in Fig. 1. The data required are either public, private, or authority data. The role of the bank becomes a secretarial one where banks act as collectors and validators of data. In the end, all the documents are preserved and, if requested, presenting to regulators.

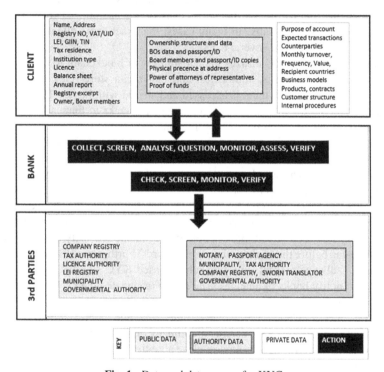

Fig. 1. Data and data source for KYC

Most of the data that clients present to the banks, are already recorded in state-related or controlled entities. About a third of the necessary information comes from the client. Such data include information about, for instance, place of employment or estimated revenue streams, counterparty (customer) info, and contract data (for corporate clients). Banks also evaluate the risks of clients' ownership structure considering the industry, origin of assets, business plans, and counterparties. All data are analyzed and assessed against internal risk policies. For internationally active clients, additional verification and control methods are required as data from clients' overseas operations is not easy to gather and validate. For such cases, banks rely on third parties.

Based on the data collected and verified, KYC involves investigating, monitoring, reporting, closing, and re-opening of accounts. Typically, the KYC process begins with the onboarding of a new client. When onboarding a new client, data is collected and examined. If needed, the case is escalated for detailed assessment. Banks are also required to regularly update their knowledge on clients. Thus, KYC is conducted regularly, the frequency of which, is dependent on the risk assessment made. Banks can, at any time, based on their analysis, close a client's account. The KYC, both for onboarding and for update, rely mostly on collecting and verifying data from various sources.

Blockchain-Based Know-Your-Customer Process

In the current KYC solutions, each participating entity is responsible for the data they capture and mange. For instance, each bank has processes to capture, verify, store, and update client data. Such data management is time consuming and error prone as it is mostly conducted manually. In many cases, the exact same client data is stored (duplicated) and managed by several banks.

A blockchain-based KYC utilizes the capabilities of blockchain to act as a shared data storage. As such, clients, authorities, and banks can store relevant data on the blockchain. The stored data, then, is accessed and shared with banks. The data, therefore, is no longer stored at multiple sources, requiring banks to manually capture and validate them. However, such data is confidential and should not be accessed by all. Furthermore, not all parties should access, enter, or update all types of data. Therefore, permissioned blockchain solutions are required. Functionalities for access rights and account management are commonly accommodated by permissioned blockchain platforms, such as Hyperledger Fabric.

Clients can add new and update data. Likewise, authorities can do the same. For instance, events such as bankruptcy, acquisitions, mergers, or buyouts require data updating. In existing solutions, given that the data management is mostly manual at banks, there is a time-lag before the banks update their records. Furthermore, the risk of missing such updates is always present. In a blockchain-based KYC solution, smart contracts can be used to notify banks of changes. The newly or updated data is validated against existing data and stored on the ledger. Then, smart contracts, according to predefined rules, send notifications about the updates to the banks the client is engaged with. An example is when authorities put a person or company on special watchlists (lists of individuals and corporate entities who require close attention due to suspicious or abnormal financial behavior). With notifications, sent by smart contracts, banks do not run the risk of missing such vital information, reduce errors of onboarding clients on the watchlist, and effectively prevent processing transactions to clients or their subsidiaries recently added to the watchlist.

Banks have to regularly compile and submit reports to authorities. These reports include information on, for instance, declined customers as a result of KYC, terminated accounts, and reason for termination. Such reports are time consuming to produce and has no value for the banks. A blockchain-based solution enables banks to add such data to the blockchain from where it is accessible to all relevant authorities. The authorities will gain access to the information in real-time. Furthermore, authorities do not have to compile and compare the reports received by different banks.

Clients who wish to open an account with a bank, can grant the bank access to their data stored on the blockchain. Banks can, thereby, get the required data for KYC from one source. Although such access reduces the time, costs, and errors associated with manual data management, banks can still conduct their analysis to determine risk level and appropriate measures for monitoring and mitigation.

At the core of the redesigned KYC solution lies data. Banks, in conducting their analysis, spend most of the effort in collecting and verifying the data to ensure that the potential client does not pose a risk for the bank. Likewise, clients have to compile all the data required, submit it to various banks, and inform about changes. Third parties also provide data and authorities rely on banks to report data. Blockchain can redesign the process by facilitating data sharing among clients, banks, authorities, and third parties. Clients enter their data to the blockchain. Likewise, authorities and third parties enter and verify data on the clients. Thus, banks can rely, to a larger extent, on one source of data. The data, when updated, is propagated to all relevant parties by means of smart contracts. As illustrated in Fig. 2, blockchain becomes a secure hub for data sharing. Authorities can retrieve the data they need from the blockchain. Clients can, when wishing to become clients at a new bank, grant access to the bank to retrieve their data. The KYC process becomes redesigned in that activities, for all parties involved, that capture, verify, and update data, are eliminated or reduced.

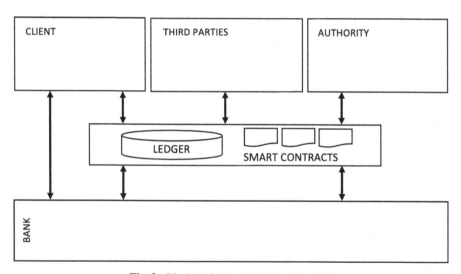

Fig. 2. Blockchain-based solution for KYC

4   Discussion

In this section, we address our research question of how blockchain technology can enable redesign of banks' Know-Your-Customer processes by moving KYC tasks to the authorities, having a single source of truth, and creation of an eco-system. We also discuss potential challenges with implementing blockchain-based KYC solutions. Finally, we conclude this section with a discussion on threats to validity.

4.1 Redesign of KYC Processes

Blockchain enables shared ledger for data storage, automatic rule-based computation by means of smart contracts, and data communication between different stakeholders [22]. These capabilities enable banks, authorities, and clients to replace manual data management with decentralized data capture with automated sharing and processing of data. Regulatory mandated tasks, such as the KYC process, are costly but does not deliver value for banks. Automation of such tasks saves banks time and costs. Likewise, banks reduce the risk of errors that might lead to fines. Furthermore, when banks and authorities can access and update, given permission rights, important client data, the process becomes faster. Clients also gain benefits as they update their information at one place rather than communicating the changes to different banks. Clients also benefit from less work when opening new accounts. Instead of providing comprehensive documentation and verification, they grant banks permission to access their client data.

Currently, banks perform the KYC process while the authorities are the consumers of the results. As such, those who need the output and have the best expertise (authorities), rely on those who neither have a strong incentive nor the expertise (banks) for the execution of KYC processes. It has been demonstrated that process efficiency can be significantly improved if those who use the output, perform the process [23, 24]. This principle builds on enabling consumers of the output, to perform the process (self-service) by granting them access to required data. Hammer [25] exemplifies this principle using the procurement process of a company. If all departments have to turn to the central department for procurement of for instance pencils, the processing of such an order will cost more than the items ordered, making the solution inefficient.

Similarly, the authorities are the users of the output of the KYC process performed at every bank and as such, should perform the process themselves. Such a set-up is more efficient due to several reasons. First of all, the analysis is conducted by the party with best expertise (authorities who need and process the data). Currently, authorities issue regulatory documents that are in turn, interpreted by banks and used as requirements for building and/or adapting software solutions. Therefore, the reports submitted to the authorities are based on processed data without guarantees that all banks have analyzed the data in the same consistent manner. Secondly, as the data analysis and report compilation differ from banks to bank, it is fully possible for faulty data to be included which is difficult, if not impossible, to detect.

The capability of blockchain to share sensitive data with restricted access right, enable banks to reduce costs and risks by becoming providers of data rather than executors of KYC processes. Authorities, on the other hand, reduce the time of compiling reports received from banks and can focus on developing better methods for identifying suspicious individuals and corporate entities. Furthermore, authorities rely on the blockchain rather than banks who might vary in their rigor when executing the KYC processes.

With blockchain, a single source of truth is created where the data is continuously updated by clients, banks, and authorities. Data is recorded once and at its source

[23, 24], which enable data to be updated in a tamper-proof manner and made available for relevant stakeholders in near real-time. Banks can, thereby, reduce their costs for data collection and risk of processing transactions to clients on sanction lists. Authorities gain efficiencies as they reduce time for managing issues born of poor data integrity.

A blockchain-based solution involves work conducted by clients, authorities and banks. These independent and in some cases competing entities, are connected by a common blockchain-solution. The blockchain can, thus, connect collaborating and competing entities by enabling secure sharing, transportation, and processing of data. Thus, blockchain can enable the emergence of a collaborative eco-system [26]. The eco-system is possible as blockchain provides secure, immutable, and tamper-proof transaction records in an untrusted decentralized ecosystem. As such, the blockchain system becomes foundational for treating dispersed entities (clients, banks, and authorities across several countries) in an ecosystem as operating on the same platform.

A blockchain-based execution of KYC processes reduces the overall redundancy and duplication of effort. Currently, every bank conducts KYC, often for the same client. With a blockchain system, KYC can be performed once and used when the client opens an account at another bank. Smart contracts can be used to automated some of manual and repetitive tasks, such as verification of data from different sources.

A blockchain-based eco-system can also facilitate innovation driven by new companies that focus on using technology for KYC. In the current structure, access to data and regulatory requirements pose a barrier for new entrants within this domain [27]. However, with an eco-system, this barrier can be significantly reduced. For instance, if a new company wishes to offer financial services, they have to ensure compliance to KYC. Such a process is costly to set up and pose as a barrier market entry. In an eco-system, the barrier is reduced as new entrants need to provide data to authorities rather than build up full KYC capabilities.

4.2   Challenges for Adopting Blockchain-Based KYC Solution

Implementation of a blockchain-based KYC solution has technical, legal, governance, and adoption challenges. Technical challenges primarily concern scalability. Blockchain has limitations. For instance, public blockchain solutions provide full decentralization but at the expense of speed and scalability. On the other hand, consortium solutions remove much of the scalability issues but cannot offer fully decentralized solutions. While the technology is improving and investments are being made, such as Corda, it is not yet fully mature for industry-wide implementation.

A blockchain-based KYC solution would require a thorough examination of current legal frameworks and, most likely, new laws [28]. For instance, smart contracts are considered as smart as they are automated and contracts in that they execute their rules once conditions are met. However, according to current law, they are not legally binding [29]. Furthermore, a blockchain solution for compliance will most likely extend across national boundaries, requiring legal aspects of several countries to be examined. As such, one of the main challenges will be to secure legally binding a blockchain-based KYC solution.

Another challenge that must be addressed is that of governance [30]. A blockchain-based KYC solution, specifically as it most likely cannot be fully public and self-governed like Bitcoin, requires an entity to manage the application. Finally, the adoption is a challenge. By this is meant the degree and speed by which independent parties agree to be onboarded on such a solution. The value of a blockchain-based KYC solution depends on number of participating stakeholders. The more, the greater the benefits from automation, data quality, and prevention of illegal financial activity.

4.3 Threats to Validity

Case studies have threats to validity such as reliability and external validity [17] that should be considered. Reliability refers to the level of dependency between the results and the researcher or in other words, would the same redesigns be produced with other researchers. This threat was to an extent tackled by having verifications by the domain experts and peer debriefing [31]. External validity concerns the extent to which findings can be generalized beyond the setting of the study. Case studies have the inherent limitation of being restricted in the extent results can be generalized. Although our findings suffer from this limitation, the results might be valuable for analysts working with introducing compliance solutions based on blockchain technology. In this regard, it should be noted that implementation of a blockchain-based KYC solution will require a set of technical, legal, governance, and new technology adoption challenges to be overcome. However, these aspects are beyond the scope of this paper to discuss.

5 Conclusion

Blockchain technology has been positioned as a game changer for banks. Particularly, banks are interested in redesigning compliance processes, such as the KYC process, as they are expensive but does not provide added value for them. In this paper, we explored how blockchain technology can enable redesign of the KYC processes. Blockchain can enable banks to become providers of data while authorities who use the outputs, assumes a greater role in analyzing data for the purpose of identifying suspicious financial entities and activities. Furthermore, blockchain can enable creation of a shared platform and a single source of truth for KYC related data. Such a solution will reduce the work with managing data inconsistency and quality for all involved stakeholders. Finally, smart contracts can automate many of the repetitive tasks that require manual validation of data. It should be noted that our study, like other case studies, have limitations. In particular, regarding the extent by which the results can be generalized. However, KYC is a highly regulated and mandatory for all banks and, thereby, somewhat standardized. Furthermore, while our study cannot be prescriptive, it can provide concepts to consider when designing blockchain-based KYC processes.

In this paper, we focused on KYC whereas there are other compliance processes, such as anti-money laundering, that needs to be studied as part of our future work.

Acknowledgments. This research is funded by the European Research Council (PIX project).

References

1. Belinsky, M., Rennick, E., Veitch, A.: The fintech 2.0 paper: rebooting financial services (2015)
2. Federal Financial Analytics: The regulatory price-tag: cost implications of post-crisis regulatory reform, Washington (2014)
3. Harrop, M., Mairs, B.: Know your customer survey. https://www.thomsonreuters.com/en/press-releases/2016/may/thomson-reuters-2016-know-your-customer-surveys.html. Accessed 20 May 2020
4. Kaminski, P., Robu, K.: A best-practice model for bank compliance (2016)
5. Xu, X., et al.: A taxonomy of blockchain-based systems for architecture design. In: Proceedings - 2017 IEEE International Conference on Software Architecture, ICSA 2017, pp. 243–252 (2017)
6. Mori, T.: Financial technology: blockchain and securities settlement. J. Secur. Oper. Custody. **8**, 208–227 (2016)
7. Milani, F., García-Bañuelos, L., Dumas, M.: Blockchain and business process improvement. BPTrends, vol. 4 (2016)
8. Deshpande, A., Stewart, K., Lepetit, L., Gunashekar, S.: Distributed ledger technologies/blockchain: challenges, opportunities and the prospects for standards (2017) https://doi.org/10.7249/RR2223
9. Zheng, Z., Xie, S., Dai, H., Chen, X., Wang, H.: An overview of blockchain technology: architecture, consensus, and future trends. In: Proceedings - 2017 IEEE 6th International Congress on Big Data, BigData Congress 2017, pp. 557–564 (2017). https://doi.org/10.1109/BigDataCongress.2017.85
10. Tschorsch, F., Scheuermann, B.: Bitcoin and beyond: a technical survey on decentralized digital currencies. IEEE Commun. Surv. Tutorials **18**, 2084–2123 (2015). https://doi.org/10.1109/COMST.2016.2535718
11. Santo, A., Minowa, I., Hosaka, G., Hayakawa, S., Kondo, M.: Applicability of distributed ledger technology to capital market infrastructure, Tokyo (2016)
12. Kosba, A., Miller, A., Shi, E., Wen, Z., Papamanthou, C.: Hawk: The blockchain model of cryptography and privacy-preserving smart contracts. In: IEEE Symposium on Security and Privacy, pp. 839–858 (2016). https://doi.org/10.1109/SP.2016.55
13. Szabo, N.: Smart contracts: building blocks for digital free markets. Extropy J. Transhuman Thought 1–10 (1996). https://doi.org/10.1200/JCO.2011.40.6546
14. Bartoletti, M., Pompianu, L.: An empirical analysis of smart contracts: platforms, applications, and design patterns. In: Brenner, M., et al. (eds.) FC 2017. LNCS, vol. 10323, pp. 494–509. Springer, Cham (2017). https://doi.org/10.1007/978-3-319-70278-0_31
15. Zhang, K., Jacobsen, H.A.: Towards dependable, scalable, and pervasive distributed ledgers with blockchains. In: Proceedings - International Conference on Distributed Computing Systems, pp. 1337–1346 (2018)
16. Gill, M., Taylor, G.: Preventing money laundering or obstructing business? Financial companies' perspectives on "Know your customer" procedures. Br. J. Criminol. **44**, 582–594 (2004). https://doi.org/10.1093/bjc/azh019
17. Runeson, P., Höst, M.: Guidelines for conducting and reporting case study research in software engineering. Empir. Softw. Eng. **14**, 131–164 (2009). https://doi.org/10.1007/s10664-008-9102-8
18. Yin, R.K.: Case Study Research: Design and Methods. SAGE Publications, New York (2009)
19. Flyvbjerg, B.: Five misunderstandings about case-study research. Qual. Inq. **12**, 219–245 (2006)

20. Kitchenham, B., Pickard, L., Pfleeger, S.L.: Case studies for method and tool evaluation. IEEE Softw. **12**, 52–62 (1995). https://doi.org/10.1109/52.391832
21. Brynjolfsson, E., Mcafee, A.: The Second Machine Age. WW Norton & Company, New York City (2014)
22. Xu, X., Pautasso, C., Zhu, L., Lu, Q., Weber, I.: A pattern collection for blockchain-based applications. In: Conference: 23rd European Conference on Pattern Languages of Programs (EuroPLoP 2018), New York, NY, USA, pp. 1–20. ACM (2019)
23. Hammer, M., Champy, J.: Reengineering the Company - A Manifesto for Business Revolution. HarperCollins Publishing Inc., New York City (2001). https://doi.org/10.5465/AMR.1994.9412271824
24. Milani, F., García-Bañuelos, L.: Blockchain and principles of business process re-engineering for process innovation. ArXiv (2018)
25. Hammer, M.: Reengineering work: don't automate, obliterate. Harv. Bus. Rev. **68**, 104–112 (1990)
26. Glaser, F.: Pervasive decentralisation of digital infrastructures: a framework for blockchain enabled system and use case analysis. In: Proceedings of the 50th Hawaii International Conference on System Sciences (HICSS-50), pp. 1543–1552 (2017). https://doi.org/10.1145/1235
27. Porter, M.E.: The five competitive forces that shape strategy. Harv. Bus. Rev. **86**, 25–40 (2008)
28. Wright, A., De Filippi, P.: Decentralized blockchain technology and the rise of lex cryptographia. SSRN Electron. J. (2015). https://doi.org/10.2139/ssrn.2580664
29. Weber, S.: Blockchain and the Law. The Rule of Code. Zeitschrift für Bankr. und Bankwirtschaft **31**, 414–420 (2019). https://doi.org/10.15375/zbb-2019-0609
30. Beck, R., Müller-Bloch, C., King, J.L.: Governance in the blockchain economy: a framework and research agenda. J. Assoc. Inf. Syst. **19**, 1020–1034 (2018)
31. Runeson, P., Host, M., Rainer, A., Regnell, B.: Case Study Research in Software Engineering: Guidelines and Examples. Wiley, Hoboken (2012)

Increasing Control in Construction Processes: The Role of Digitalization

Arif Ur Rahman[1], Syed Mehtab Alam[2], Patrick Dallasega[3], Elisa Marengo[2(✉)], and Werner Nutt[2]

[1] Department of Computer Science, Bahria University, Islamabad, Pakistan
arif.buic@bahria.edu.pk
[2] Faculty of Computer Science, Free University of Bozen-Bolzano, Bolzano, Italy
{mehtabalam.syed,werner.nutt,elisa.marengo}@unibz.it
[3] Faculty of Science and Technology, Free University of Bozen-Bolzano, Bolzano, Italy
patrick.dallasega@unibz.it

Abstract. Digitalization could support the execution of construction processes resulting in a decrease of *waste* and a better synchronization with the supply chain. However, in construction the level of digitalization is low, also compared to other sectors. This is due to some characteristics of the sector, such as the limited margins of *Small and Medium-size Enterprises* to be invested in digitalization, the little standardization of construction processes, and the imponderabilities characterising their execution and making the execution on-site less controllable. Common practice of the companies is to define rough long-term plans as Gantt chart, which however cannot be used to manage the daily work on-site and that are rarely aligned with the real progress on-site.

In this paper we discuss the challenges and the benefits in increasing digitalization to better support construction execution processes. Digitalization enables a new way of supporting *process modeling*, short- and medium-term *planing*, and *monitoring* of construction processes. By means of this approach, companies can, among other benefits, reduce waste on site and improve synchronization with the supply chain. In the paper we also discuss the characteristics of suitable digital tools.

Keywords: Construction execution processes · Process modeling · Medium- and short-term planning · Process monitoring and analysis

1 Introduction

According to [1], digitalization is not only about digital infrastructures, technologies or tools. Digital innovation is about facilitating "new pathways of actions"

Research carried out as part of the research project "COCkPiT - Collaborative Construction Process Management - FESR1008", funded by the European Regional Development Fund (ERDF) of the Autonomous Province of Bolzano-South Tyrol.

© Springer Nature Switzerland AG 2020
A. Del Río Ortega et al. (Eds.): BPM 2020 Workshops, LNBIP 397, pp. 263–275, 2020.
https://doi.org/10.1007/978-3-030-66498-5_20

or, in other terms, digital innovation enables *new processes*. In this paper, we discuss how digitalization would enable a new approach for construction execution processes. The idea is that by means of digitalization it is possible to consider and to properly handle details when managing construction execution processes, which results in better performance and waste reduction on-site. First steps towards increasing digitalization, indeed, showed that it would allow companies to have better control over the execution process and, in particular, to understand the current status of the process and plan with some advance what to perform next. This would allow companies to reduce waste like idle time of the workers, and to *synchronize the execution with the supply chain*, supporting a just-in-time delivery of the material and the prefabrication of the components.

As shown in [2], construction companies have to orchestrate several processes at the same time (e.g., administrative, organizational, execution, procurement). In this paper we consider the execution process, in its phases of *defining* the activities to be performed, *scheduling* and *monitoring* their execution. The current adoption of digitalization in construction is known to be low compared to other sectors, such as manufacturing, tourism, telecommunications [3].

To understand the reasons for a limited digitalization and before advancing proposals to increase it, one has to look at the characteristics of the sector. It is important to consider that most companies are Small and Medium-size Enterprises (SMEs), which usually are more fragile than bigger ones. In particular, digitalization in process management requires an initial effort from the companies in abstracting, standardizing and planning their processes. To do this, companies have to identify and allocate resources. Although companies would benefit from this during the execution of the process, not all SMEs can or are willing to afford this initial investment, also because amortization of such costs will take long. Existing technologies, such as BIM-based technologies, do not represent a solution either, since they are usually expensive and quite complex to use. In this respect we can say that these companies can count on *low organizational resources* and that the culture is *medium-* or *non-supportive* of BPM [4].

Additionally, abstraction and standardization of construction processes are not trivial tasks. It is not hard to see that construction processes are less uniform than processes in other sectors (such as manufacturing), compared to which construction projects can be seen as one-of-a-kind, both in terms of technologies adopted for the construction and elements to be built. This makes it difficult to find a good level of abstraction that suits all possible projects, which instead have to be considered individually. Also, other than with manufacturing processes, the work is performed by several companies (SMEs), each one responsible for a different part. For the process to be successful, their activities need to be coordinated and synchronized. However, the consortium of companies changes from project to project, therefore the synchronization cannot be standardized. Construction processes are *non-repetitive, creative* and highly *interdependent* [4].

Another not negligible aspect is that construction processes are inevitably subject to imponderabilities which impact on the execution of activities causing delays, affecting the outcome of the activities or even preventing some activities to be executed at all (the environment is highly *uncertain* [4]). Bad weather conditions are an example of such imponderabilities, which may increase the time needed for some activities (e.g., painting), and may prevent other activities from being executed (e.g., building the roof or excavating). As a consequence, when managing construction processes one has to take into account that the sequence of work performed on-site may differ from what was planned.

Given these aspects, the kind of support needed to assist companies in the management of the execution should *i)* require as little effort as possible from the companies (for instance in understanding new technologies), and should rather be close to the common practice and tools they are already familiar with; *ii)* allow for the definition of the elements characterizing a process, at the desired level of abstraction; *iii)* support a flexible definition and adjustment of the execution plans, so as to react to imponderabilities; *iv)* reflect the current state of a project, considering that the project execution may deviate from what was planned (imponderabilities).

In this paper we discuss challenges and advantages in increasing digitalization in construction execution processes which results from a *design science research* methodology. The paper is organized as follows. In Sect. 2 we present what is the current practice in construction process management, which leads to the proposal that we present in Sect. 3. In Sect. 4 we present CoSMos, a demonstrator developed to support scheduling and monitoring of construction processes. Sect. 5 ends the paper.

2   Current Practice and Motivations

Common practice in managing construction processes is to define a long-term plan as a Gantt chart, covering the entire duration of the project. Gantt charts, however, are general purpose tools and thus fail in capturing some elements of construction processes (such as, locations or teams). Moreover, they do not provide support for changes and adaptation of the plans [5,6]. This however is very important due to the frequent unexpected events happening on-site. As a result, these plans are defined in a rough way, mainly with the purpose of showing to customers how intermediate milestones could be achieved or how the construction is planned to be carried out. They, however, are not suitable to manage the operative work on-site, nor to represent the execution in a detailed way. One can therefore see that, although having a digital format, Gantt Charts are limited in the digital support they provide for construction process management. Support that is limited to computations such as the expected usage of resources or the Critical Path.

Currently, Gantt chart plans are defined at the beginning of projects and rarely updated to reflect the real progress on-site. Therefore, is the foreman on-site who, leveraging on his experience and on the construction strategy agreed with his company, defines the activities to be performed weekly by his workers. This is done with very limited or no support. Also, when unexpected events happen, he is the only person capable of rescheduling the activities based on the strategy he has in mind. Since there is no explicit *process model* representing the requirements and the elements of the process, the foreman represents a *single point of failure*, making the success of the process just depending on him. Additionally, this approach prevents the identification of execution alternatives and the selection of the most promising one. Control of an ongoing project is also very limited. The status of a project (on-time, late, within budget and so on) can only be estimated roughly by the foreman and by the project manger. This latter, in particular, frequently bases his estimates on indirect measures, such as the material delivered on-site.

Digitalization of the entire process and the adoption of suitable digital tools can improve these aspects, opening up the way for a more detailed way of planning, and offering the possibility of synchronizing plans with the actual progress on-site. Over the years a number of tools, technologies and methodologies to increase digitalization in construction have been proposed. Among these, the most prominent one is probably BIM (Building Information Modeling) and the plethora of software tools supporting it [7]. The BIM philosophy is to maximize the exchange of data and support the management of a building in all of its phases, from the design to its maintenance. BIM tools are on the one hand very powerful, but on the other hand they are complex to understand and use in a profitable way. For this reason, they would require a dedicated and knowledgeable person working with it, and a change in the way companies are used to manage their processes. Many companies, especially SMEs, cannot afford or are not willing to perform this change [8,9]. This also justifies why the very promising adoption of BIM technologies is nowadays rather limited in reality.

Our proposal is to bring digitalization in construction execution processes using a bottom up approach, that is starting from how companies manage their construction processes. To this aim, as a result of the collaboration with construction companies, we decompose a construction process into four sub-processes: *modeling, scheduling, monitoring* and *analysis*. In summary, the digital innovation we are proposing consists of *i)* introducing a representation of the process requirements with a declarative and explicit representation of the process model which would allow one to overcome the single point of failure problem and to identify the execution alternatives; *ii)* introducing a detailed but flexible short and medium-term planning approach; *iii)* combining the planning with the monitoring of the process; and *iv)* using the monitored data to give a better picture to the management about the actual progress of the project and to support future projects. To be successful, digital innovation need the support of adequate digital tools. We decided to start with what companies are already familiar with and to use these tools to show concepts to the companies and acquire feedback and requirements to then build more sophisticated tools. This is in line with the *agile* approach in software development. In the following, for each of the sub-processes we explain challenges, benefits and support needed for digitalization.

3 Digitalization to Support Execution Processes

In this section we describe the challenges and the advantages of increasing the level of digitalization in construction process *models* (Sect. 3.1), in short- and medium-term *planning* (Sect. 3.2), and in *monitoring* and *analysis* of processes (Sect. 3.3).

3.1 Execution Process Modeling in Construction

Typically, a process is constituted by *what* needs to be done (activities), *who* is responsible for it and preferred *orders* in performing the activities. In construction an additional dimension needs to be considered: one also needs to define *where* activities are executed. These elements can be seen as the requirements of a process and in the business process literature constitute what is known as a *process model*. A process model basically defines *what* needs to be done and differs from an *execution plan*, which additionally specifies when the activities should be performed, assigning explicit days for their execution, resources, deciding the level of parallelism where possible and so on. Anyway, an execution plan should satisfy the requirements specified in a model.

In construction, the common practice is to define *execution plans* directly, neglecting the creation of process models. Having a process model, however, would allow one to identify the possible alternatives of how the work can be performed, which would be beneficial for several reasons. For instance, before starting the execution, it would allow one to select a preferred way of performing the work, such as the one that minimizes the waste. During the execution, when the plan needs to be modified to react to unexpected events, it would support the identification of the best way to limit the consequences of such events. Defining directly an execution plan, instead, means committing too early to one possible way of performing the activities, preventing the possibility of identifying and analyzing other ways. In this case, indeed, the process requirements are hidden inside an execution plan and are hard to distinguish from preferences or choices of the planner (e.g., the project manager).

Example 1. Suppose for instance that in an execution plan Put Window occurs before Lay Floor. Suppose that the windows are late, can the two activities be switched? Is there some requirement which makes the switch impossible? This type of requirements are explicit in a process model, while by looking at the plan it is not clear whether this order is a requirement or a preference of the planner.

In order to take advantage from a process model, suitable support is needed. General purpose tools usually fail in capturing some relevant aspects of the domain. Among the known drawbacks of adopting Gantt charts in construction, for instance, there is the fact that there is no proper abstraction for representing the locations. These are, therefore, improperly represented as macro-activities (*summary tasks* in MS Project), composed of the activities to be performed in the location. So, for instance, one can specify that Lay Floor and Put Window are sub-activities of the macro-activity `Floor One`. However, one would have to repeat such activities for each location in which they appear. One can immediately see that this will *i)* make the overall plan less readable, *ii)* increase the chances of introducing mistakes, *iii)* make updates and changes more difficult. To limit these drawbacks, the solution adopted in reality is to keep activities at an abstract level, neglecting the details and thus keeping a plan compact. Given these characteristics, it is not surprising that execution plans based on Gantt charts are used at the beginning of the project for managerial purposes but are not suitable to manage the operative work on site. To do this, the foreman when scheduling the daily or weekly activities to be performed on-site, will have to interpret the Gantt chart end determine additional details, such as more information on the locations, requirements on the execution of activities, expected productivity and so on.

A first step towards digitalization of construction process models is defined in [6], where a declarative approach for construction process modeling is formally introduced. The requirements for such approach were collected in collaborative meetings with companies, considering real construction projects. Out of those requirements, we defined the modeling language CoPMod, its formal semantics and a software tool supporting the definition of CoPMod models [10]. Please refer to [6] for an in depth contextualization of CoPMod in the BPM literature. The benefit of this digital innovation in construction process modeling, allows for *i)* the selection of the desired level of detail, *ii)* using the right abstractions and *iii)* support automatic checks (such as the implemented satisfiability check of a process model). The possibility of choosing the desired level of detail in which to represent a process model is possible thanks to the digitalization of the approach. Without it, only an abstract level would be reasonably manageable, however, making it not suitable for reality.

3.2   Short- and Medium-Term Execution Planning

When defining the requirements, one challenge is the identification of the right level of abstraction for activities and locations. Choice that, as discussed, is possible thanks to digitalization that would allow to capture and handle also detailed representations. Similarly, when it comes to the definition of an execution plan, one needs to identify the right time granularity (e.g., day or week). The choice of a suitable abstraction in the process model and the choice of the time granularity in the execution plan are tightly connected. Specifically, the amount of work to be performed for the activities in the locations should be easy to define and to measure in the chosen unit of time. Let us consider an example.

Example 2. A common phase in construction processes is the skeleton phase. Common activities are `Excavate`, `Pour Concrete`, `Erect Scaffolding` and so on. The execution of these activities usually takes several days in one location. Therefore, scheduling and monitoring in terms of days would require a lot of effort to measure the quantity of work performed and would not be very informative. In this case a weekly granularity is more appropriate and a representation of the locations in terms of floors would be detailed enough. Indeed, subdividing the floors into smaller units would not increase the control over the process, but would just require more effort for scheduling and measuring. Another phase is the interior construction. Here, activities have usually shorter duration (Put Window, Lay Floor), and several activities are performed in parallel in the same floor. As a consequence, a finer representation of the locations (such as in terms of rooms per floor) is more adequate and is compatible with a time unit of weeks (expressing how many windows can be installed in a week) or even days.

Note that, since the process model abstractions need to be compatible with the time granularity of the plans, this granularity needs to be determined already when defining the process model.

Besides choosing the time granularity, one also needs to define the period of time that the execution plan should cover. Usually, project managers in construction define a long-term schedule covering the entire duration of the project. Alternatively, one can think of defining short-term ones (e.g., one week) with the drawbacks of having to define schedules more frequently but with the advantage of being able to consider and react to the real progress on-site. However, the risk is that short-term plans are too specific, and thus not useful to guide the execution towards the achievement of some goal (such as a milestone) or so as to guarantee the alignment with the supply chain, which in some cases requires some weeks to produce the material.

Our approach consists in defining a detailed execution plan, considering a week (or day) as time unit, and planning for more than one week in the future. The number of weeks to be planned in advance, called *look-ahead window*, is another aspect that has to be decided for each project. Also in this case it is a matter of balance: scheduling too far in the future would very likely mean that the plan will soon be inapplicable in reality. Think, for instance, of the frequent unpredictable events which require to deviate from the expected plan. On the other hand, planning not enough in advance will prevent the synchronization with the supply chain. In our approach, we do not fix the look-ahead window size but we allow it to be decided from project to project (although usually it is between four and six weeks). Also in this case, the possibility of choosing the right level of detail and the right seize of the time window is enabled only if one can rely on proper tools. Without specialized support, the best one can do is to produce coarse schedules for a sufficiently long period, or detailed ones for a duration too short to be practical. To this purpose, we collected requirements from construction companies in real projects and propose a first approach as described in Sect. 4.

3.3   Monitoring and Analysis of Executions

Digitalization not only gives the possibility to improve the scheduling of the activities, but also supports the update of such plans to keep them aligned with the construction site. Due to unforeseeable events, indeed, it is very important to frequently monitor which activities have been performed, in which locations and how much of them has been completed. This information should then be used to adjust the schedule for the subsequent periods. Also in this case, digitalization supports this task, making it applicable to detailed medium-term plans. In particular, it would support the "planned" vs "performed" analysis. Knowing that a foreseen activity has not been performed, indeed, is important to understand what are the consequences. In the most lucky situation, if the activity has no impact on subsequent ones, then the consequences for the overall project are limited. If the activity acts as a precondition for the execution of other activities, or if the activity needs specific material from the supply chain, then this may result in severe delays and wastes on-site (for instance, idle resources which cost money but that cannot perform their activities because the conditions to perform them are not satisfied). The sooner these situations can be identified, the greater the chances to implement countermeasures, such as notifying the supply chain, rescheduling the resources and the like. Not only could a proper tool support the identification of such problems, but it could also be used to estimate delays and cost overruns. Without a proper tool, this analysis would become difficult to perform and time consuming, making it less likely to be performed frequently.

Besides supporting the "planned" vs "performed" analysis, digitalization would also allow the execution of automated analysis over the data collected from the monitoring. In particular, the aim is to support companies when defining execution plans. Consider, indeed, that planning in more detail would require more effort from the companies. Therefore, by considering and analyzing how the work was scheduled (or performed) in past or finished (similar) projects one can provide hints and suggestions to the foreman on what to schedule next. This will make it easier for the companies to adopt the approach, allowing to collect more and more data on construction projects that can in turn be used to improve the suggestions. These aspects are currently under investigation and part of the research project Confucius [11].

4   CoSMos: Scheduling and Monitoring of Construction Processes

To support the approached described in the previous sections we developed CoSMos, a tool for Scheduling and Monitoring construction processes. The aim of the tool is mainly to present the digital innovation to the companies in an effective way, in order to show them the advantages, collect feedback and additional requirements that can be used to improve the approach and the tools. To this aim we applied an agile approach, leveraging on technologies and tools companies are already familiar with, that is MS Excel. The development of enhanced and ad-hoc tools replacing Excel is part of current work. In Sect. 4.1, we present first applications of the approach that then lead to the implementation of CoSMos (Sect. 4.2).

4.1   Preliminary Applications of the Approach

The approach described above for a detailed and medium-term scheduling was developed and tested in the context of a collaboration with a leader company in fabricating and installing facades [12]. The approach consisted in a first *modeling* phase, for the identification of the process requirements, and a second *scheduling* phase for the scheduling and monitoring of the execution process. The process requirements for a construction project were discussed during a one day workshop organized by the company and involving the project manager, the foreman and the architect. During this meeting, the participants agreed on a suitable representation of the activities to be performed, how to organize the locations so has to have enough detail and, at the same time, being able to schedule and monitor the work in an effective way. They also provided the quantities of work to be performed for each activity in each location and an estimate of the productivity for each activity.

These requirements were then elaborated by the coauthor from the Faculty of Science and Technology, who manually produced a number of tables to support both the scheduling and the monitoring of the work on-site. Figure 1 shows a section of such a table, for a specific week and a part of the building. In the table, activities are represented as rows and locations as columns. The intersection between them is used to represent the quantity of work scheduled in that week. Similar tables were manually created for the monitoring, with the difference that the quantities reported for activities in a location represent the quantities of work actually performed on-site.

To implement the medium-term planning, the foreman defined a schedule for the current week and for a few upcoming weeks (usually three). At the end of the week, the activities actually performed on-site were measured and the quantities reported in the monitoring table corresponding to that week. This data was then used to update the schedule for the (three) upcoming weeks and perform the schedule for a new week (the fourth), keeping the look-ahead window size.

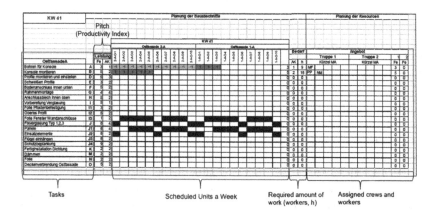

Fig. 1. Excel tables for scheduling and monitoring of construction processes.

The tables were created in MS Excel. This offered several advantages, such as the possibility to link them by means of formulas (for instance to automatically compute the remaining quantity of work for an activity in a location), the possibility to compute man-hours spent or to make forecasts to completion. Non negligible is the fact that companies are already familiar with MS Excel. This makes them less hesitant to apply the new approach. After trying it, they were convinced of the increased level of control over the project. However, one of the shortcoming is that tables and all the analysis that can be performed on the data have to be defined anew and individually for each project. This demand of resources and time is a severe restriction for applying the approach in new projects. To overcome it we implemented CoSMos.

4.2 Functionalities Supported by CoSMos

In the context of the research projects MoMaPC [13] and COCkPiT [14] we generalized the approach described in Sect. 4.1, with the aim of digitalizing it. To this end, we developed a Java-based demonstrator called CoSMos (Scheduling and Monitoring of Construction processeS).

Modeling. A process model consists of the following information: *i)* a list of activities to be performed on-site; *ii)* a representation of the locations; *iii)* and a list of workers available on-site. Additionally, we give the possibility of specifying several companies, aiming at using the approach to orchestrate the work of more than one company. To this aim, the input also foresees *iv)* the list of companies and *v)* the relation activity-company, to represent which company is responsible of which activity. Each of these is specified in Excel as a list in a dedicated sheet. As a next step in the development, this information will be automatically extracted from a CoPMod model [10] (mentioned in Sect. 3.1 and described in [6]). Concerning the representation of the locations, in the current version we define them in terms of a two-level hierarchy, for instance representing floors and locations inside the floors (e.g., rooms, corridors and so on), or orientations (N, S, E and W for each floor). Part of a process model is also the definition of the quantity of work for each activity, which can be done by filling in Excel tables that CoSMos automatically generates after inserting the initial information.

Scheduling. As sketched in Fig. 2, after defining the process model and the quantity of work per location, CoSMos can be used to generate the scheduling and monitoring tables. The tables generated are similar to those tested in the preliminary phase (Fig. 1). The first time, the tool generates a number of scheduling tables equal to the look-ahead window size (parameter that can be given as input). Later, only one additional scheduling table will be generated. The tables are linked together, and this allows to propagate the quantities from the process model to the scheduling tables, but also update them based on what is scheduled or monitored in previous weeks. This ensures that quantities are always updated, showing the quantity of work remaining.

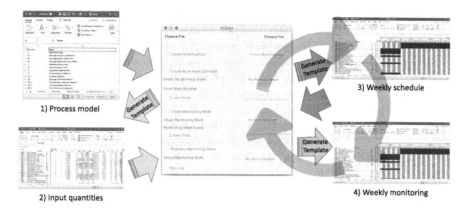

Fig. 2. Generation of the excel tables for scheduling and monitoring using CoSMos.

Monitoring tables can be generated similarly and can be filled-in with the quantity of work performed per activity in location. By means of Excel formatting rules, the tool is able to compare the quantities recorded with the quantities scheduled. Differences are then highlighted, both in the scheduling and the monitoring tables, with a color coding: *i)* yellow for those activities and locations where more work than scheduled has been performed; *ii)* orange when less work than scheduled has been performed; and *iii)* red when more work than remaining has been scheduled or performed. All these colors indicate a warning. Note, indeed, that also being ahead of schedule could be a problem (for instance, for the synchronization with the supply chain or with other companies).

5 Conclusion and Future Work

In this paper we discussed how digitalization enables a new process for managing construction executions. This new way consists in increasing the level of details in which construction processes are modeled, scheduled and monitored. All this results in better control of processes for companies, thus reducing wastes and better supporting the synchronization with the supply chain. Without digitalization and the support of adequate tools, this would not be possible.

We also introduced CoSMos, a first step toward supporting digitalization of short- and medium-term scheduling and monitoring of the execution on-site. Currently, we rely on Excel for this phase as a tool that companies already know. This allows us to convey the advantages of increasing digitalization in construction. It also allows us to collect a number of requirements and feedback that we are using in the implementation of a web application that will support this phase better and that will eventually combine and enhance CoPMod and CoSMos.

Currently we are testing CoSMos on real projects with the collaboration of a company responsible for the installation of Heating, Ventilation and Air

Conditioning systems (HVAC). We are also working in the implementation of the web application and on the analysis of monitoring data as described in Sect. 3.3. The analysis concerns both on-going projects, such as the estimation of budget to completion or the computation of the current status of a project in a reliable way. Additionally, as part of the project Confucius, we will use monitoring (and potentially also scheduling) data to extract frequent patterns in construction process execution. Such patterns will then be used to suggest future scheduling steps. First steps in this direction allow to suitable parse and store monitoring data collected in Excel files or, alternatively, directly collected via the web application by means of an interface developed for the purpose. Companies could decide which way for inserting monitoring data is less demanding for them.

First tests in the adoption of the approach and of the tools, revealed that the monitoring activity is very interesting for the companies, which have a simple but reliable tool to assess the status of a project.

References

1. Mendling, J., Pentland, B., Recker, J.: Building a complementary agenda for business process management and digital innovation. Eur. J. Inf. Syst. **29**, 208–219 (2020)
2. Bhargav, D.: Business process management - a construction case study. Constr. Innov. **17**(1), 50–67 (2017)
3. Calvino, F., Criscuolo, C., Marcolin, L., Squicciarini, M.: A Taxonomy of digital intensive sectors. OECD Science, Technology and Industry Working Papers 2018, vol. 14 (2018). https://doi.org/10.1787/f404736a-en
4. vom Brocke, J., Zelt, S., Schmiedel, T.: On the role of context in business process management. Int. J. Inf. Manage. **36**(3), 486–495 (2016)
5. Shankar, A., Varghese, K.: Evaluation of location based management system in the construction of power transmission and distribution projects. In: International Symposium on Automation and Robotics in Construction, vol. 30 (2013)
6. Marengo, E., Nutt, W., Perktold, M.: CoPModL: construction process modeling language and satisfiability checking. Inf. Syst. (2019)
7. Hardin, B., McCool, D.: BIM and Construction Management: Proven Tools, Methods, and Workflows. Wiley, Hoboken (2015)
8. Forsythe, P., Sankaran, S., Biesenthal, C.: How far can BIM reduce information asymmetry in the Australian construction context? Project Manage. J. **46**(3), 75–87 (2015)
9. Dallasega, P., Marengo, E., Revolti, A.: Strengths and shortcomings of methodologies for production planning and control of construction projects: a systematic literature review and future perspectives. Prod. Plann. Control (2020)
10. Marengo, E., Nutt, W., Perktold, M.: CoPMod: support for construction process modeling. In: Proceedings of BPM 2018 DEMO track. vol. 2196 of CEUR (2018)
11. Confucius: Study the past if you would define the future: discovering patterns in scheduling and monitoring data (2020). Project financed by UNIBZ
12. Marengo, E., Dallasega, P., Montali, M., Nutt, W., Reifer, M.: Process management in construction: expansion of the Bolzano hospital. In: vom Brocke, J., Mendling, J. (eds.) Business Process Management Cases. MP, pp. 257–274. Springer, Cham (2018). https://doi.org/10.1007/978-3-319-58307-5_14

13. MoMaPC: Modeling and managing processes in construction (2015). Project financed by the Free University of Bozen-Bolzano (UNIBZ)
14. COCkPiT: COllaborative Construction process Management (2017). Project financed by the European Regional Development Fund (ERDF)

16th International Workshop on Business Process Intelligence (BPI)

16th International Workshop on Business Process Intelligence (BPI)

Business Process Intelligence (BPI) is a growing area both in industry and academia. BPI refers to the application of data- and process-mining techniques to the field of Business Process Management. In practice, BPI is embodied in tools for managing process execution by offering several features such as analysis, prediction, monitoring, control, and optimization.

The main goal of this workshop is to promote the use and development of new techniques to support the analysis of business processes based on run-time data about the past executions of such processes. The workshop aims at discussing the current state of research and sharing practical experiences, exchanging ideas and setting up future research directions that better respond to real needs. We aim to bring together practitioners and researchers from different communities such as business process management, information systems, business administration, software engineering, artificial intelligence, process mining, and data mining who share an interest in the analysis of business processes and process-aware information systems. In a nutshell, it serves as a forum for shaping the BPI area.

The 16th edition of this workshop attracted 9 international submissions. Each paper was reviewed by at least three members of the Program Committee. From these submissions, the top 5 were accepted as full papers for presentation at the workshop, which was held in an online format due to the COVID-19 pandemic. The papers presented at the workshop provide a mix of novel research ideas, this year mainly focused on process discovery and predictive process monitoring.

Fani Sani, Boltenhagen and van der Aalst focus on the process discovery task by proposing a new method to select prototypes based on clustering and conformance metrics. By incrementally selecting prototypes and providing those in a final step to a process discovery technique, the data variability of event logs is reduced, which has a significant positive effect on the precision and simplicity of discovered process models. Next, *Janssenswillen, Depaire and Faes* address the problem that typical discovered process models do not convey any notion of probability and uncertainty. Therefore, they propose an approach based on Bayesian inference and Markov Chain Monte Carlo to build a statistical model on top of a process model, using event data. The approach is capable to generate probability distributions of choices in a process' control-flow. In a third contribution, *van Zelst and Cao* turn to the problem of trace clustering. They specifically focus on including data-attribute-driven clustering by proposing a hierarchical trace clustering framework. They provide an implementation in PM4Py. *Weytjens and De Weerdt* report on a comparative analysis of CNN and LSTM models for process outcome prediction. They show that the application of CNNs might be advantageous given that they are not outperformed in terms of predictive accuracy, while the model training times are an order of magnitude faster than those of LSTMs. Finally, *Mannel, Epstein and van der Aalst* put forward an improvement of their eST-Miner. The problem addressed relates to the fact that the number of candidate places is exponential in the number of activities, which makes the standard eST-miner slow compared to other discovery techniques. Two approaches are introduced, both relying on the organization of candidate places in a tree-like search structure.

As with previous editions of the workshop, we hope that the reader will find this selection of papers useful to keep track of the latest advances in the BPI area. We are looking forward to keeping bringing new advances in future editions of the BPI workshop.

October 2020

Organization

Workshop Chairs

Andrea Burattin   Technical University of Denmark, Denmark
Jochen De Weerdt   KU Leuven, Belgium
Marwan Hassani   Eindhoven Univ. of Technology,
     The Netherlands

Program Committee

Ahmed Awad   Cairo University, Egypt
Johannes De Smedt   KU Leuven, Belgium
Benoit Depaire   Universiteit Hasselt, Belgium
Claudio Di Ciccio   Vienna Univ. of Economics and Business,
     Austria
Chiara Di Francescomarino   Fondazione Bruno Kessler – IRST, Italy
Diogo R. Ferreira   Universidade de Lisboa, Portugal
Luciano García-Bañuelos   Tecnologico de Monterrey, Mexico
Gianluigi Greco   University of Calabria, Italy
Gert Janssenswillen   Universiteit Hasselt, Belgium
Michael Leyer   University of Rostock, Germany
Fabrizio Maggi   Free University of Bozen/Bolzano, Italy
Jorge Munoz-Gama   Pontificia Universidad Católica de Chile, Chile
Pnina Soffer   University of Haifa, Israel
Seppe vanden Broucke   Ghent University, Belgium
Eric Verbeek   Eindhoven Univ. of Technology,
     The Netherlands
Matthias Weidlich   Humboldt-Universität zu Berlin, Germany
Wil van der Aalst   RWTH Aachen University, Germany

Prototype Selection Using Clustering and Conformance Metrics for Process Discovery

Mohammadreza Fani Sani[1(✉)], Mathilde Boltenhagen[2], and Wil van der Aalst[1]

[1] RWTH Aachen University, Aachen, Germany
{fanisani,wvdaalst}@pads.rwth-aachen.de
[2] LSV, Université Paris-Saclay, ENS Paris-Saclay, CNRS, Inria, Cachan, France
boltenhagen@lsv.fr

Abstract. Automated process discovery algorithms aim to automatically create process models based on event data that is captured during the execution of business processes. These algorithms usually tend to use all of the event data to discover a process model. Using all (i.e., less common) behavior may lead to discover imprecise and/or complex process models that may conceal important information of processes. In this paper, we introduce a new incremental prototype selection algorithm based on the clustering of process instances to address this problem. The method iteratively computes a unique process model from a different set of selected prototypes that are representative of whole event data and stops when conformance metrics decrease. This method has been implemented using both ProM and RapidProM. We applied the proposed method on several real event datasets with state-of-the-art process discovery algorithms. Results show that using the proposed method leads to improve the general quality of discovered process models.

Keywords: Process mining · Process discovery · Prototype selection · Trace clustering · Event log preprocessing · Quality enhancement

1 Introduction

Process Mining bridges the gap between traditional data science techniques and business process management analysis [1]. Process discovery, one of the main branches of this field, aims to discover process models (e.g., Petri nets or BPMN) that describe the underlying processes captured within the event data. Event data that is also referred to as *event logs*, readily available in most current information systems [1]. Process models capture choice, concurrent, and loop behavior of activities.

To measure the quality of discovered process models, four criteria have been presented in the literature, i.e., *fitness*, *precision*, *generalization*, and *simplicity* [2]. *Fitness* indicates how much of the observed behavior in data is described

© Springer Nature Switzerland AG 2020
A. Del Río Ortega et al. (Eds.): BPM 2020 Workshops, LNBIP 397, pp. 281–294, 2020.
https://doi.org/10.1007/978-3-030-66498-5_21

by the process model. In opposite, *Precision* computes how much modeled behavior exists in the event log. Generalization represents the ability of a model to correctly capturing parts of the system that have not been recorded [3]. Simplicity measures the understandability of a process model by limiting the number of nodes and complex structures of the resulting model.

Several automated process discovery algorithms have been proposed in the literature that work perfectly on synthetic event logs. However, when dealing with real event logs, many of them have difficulties to discover proper models and generate spaghetti-like process models, i.e., the discovered models contain too many nodes and arcs. Such structures are too complex for human analysis. Therefore, the quality of discovered process models depends on the given event log which can be noisy or very complex [4]. Moreover, sometimes the discovered models are unacceptably imprecise and describe too much behavior compared to the given event log. Thus, many state-of-the-art process discovery algorithms have difficulties to balance between these four quality criteria.

The mentioned problems are usually caused by high data variability of event logs and the existence of infrequent behavior in them. Therefore, by applying *data preprocessing* techniques, e.g., noise reduction [5,6], we are able to decrease the data variability of event logs and consequently improve the resulting process models. Using this approach, the preprocessed event log is given to process discovery algorithms instead of the original event log.

In this paper, we aim to improve the results of process discovery algorithms by proposing a new preprocessing method that incrementally selects prototypes in event logs. Our main motivation is to get the most representable trace instances. For this purpose, the method uses trace clustering. Each cluster has a representative instance that we consider as a *prototype*. The selection of prototypes is incremental and depends on the moderate use of conformance checking artifacts. By using prototypes we reduce the data variability of event logs and consequently improve the precision and simplicity of discovered models.

Using `RapidProM` [7], we study the usefulness of the proposed method by applying it on several real event logs while using different process discovery algorithms. The experimental results show that applying our method improves the balance between the quality metrics of discovered process models.

The remainder of this paper is structured as follows. We first provide a motivating example in Sect. 2. Then, in Sect. 3, we discuss related work. Section 4 defines preliminary notations. We present the prototype selection method in Sect. 5. The evaluation and its results are given in Sect. 6. Finally, Sect. 7 concludes the paper and presents some future work.

2   Motivating Example

Research like [8,33] has shown that by using only a small subset of traces for process discovery we sometimes can improve the quality of process models. The main challenge faced this research is which traces should be selected as input for process discovery algorithms. Some methods, e.g., [8,9], propose to use sampling methods for this purpose without considering the quality of discovered model during the selection phase. We aim to find the most representative process instances of a log, i.e., referred to prototypes, using a clustering method. To motivate our approach, in Fig 1, we show discovered models based on selected traces of an event log (i.e., Fig. 1e) by the inductive miner [10] in conjunction with three preprocessing methods.

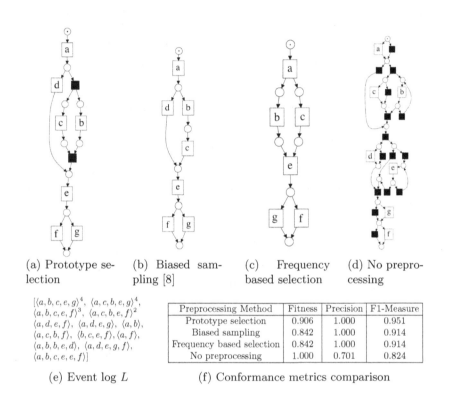

(a) Prototype selection   (b) Biased sampling [8]   (c) Frequency based selection   (d) No preprocessing

$[\langle a, b, c, e, g\rangle^4,\ \langle a, c, b, e, g\rangle^4,$
$\langle a, b, c, e, f\rangle^3,\ \langle a, c, b, e, f\rangle^2$
$\langle a, d, e, f\rangle,\ \langle a, d, e, g\rangle,\ \langle a, b\rangle,$
$\langle a, c, b, f\rangle,\ \langle b, c, e, f\rangle, \langle a, f\rangle,$
$\langle a, b, b, e, d\rangle,\ \langle a, d, e, g, f\rangle,$
$\langle a, b, c, e, e, f\rangle]$

Preprocessing Method	Fitness	Precision	F1-Measure
Prototype selection	0.906	1.000	0.951
Biased sampling	0.842	1.000	0.914
Frequency based selection	0.842	1.000	0.914
No preprocessing	1.000	0.701	0.824

(e) Event log L (f) Conformance metrics comparison

Fig. 1. Comparison of different trace selection methods using the inductive miner

Note that the statistical sampling method [9] returns all the traces to have a high confidence of not loosing information (and because of the small size of the log). The biased sampling method [8] takes as input the percentage of desired traces. In this example, we used 30% of the entire log to get the same number of traces as our method. Moreover, the frequency based selection method returns the top most frequent traces in the log. Considering conformance metrics of discovered models using different preprocessing methods that is presented in Fig. 1f, we found that choosing the right instances in the log is a key factor to discover high quality models.

3 Related Work

Reducing event logs to only significant behaviors is a common practice to improve model quality. Some variants of process discovery algorithms, e.g., the inductive miner [10], directly incorporate filters to remove infrequent behavior. In [11], infrequent traces are qualified as outliers and suggested to be filtered out. Independent to process discovery algorithms, filtering methods like [5,6,12] remove outlier behaviors in event logs. These works show possibilities of improvement when using reduced event logs as the input for process discovery algorithms.

Another way to reduce log variability that causes simpler models is to extract only a small set of traces. Authors in [8] and [9] present different trace selection methods that improve the performance of process discovery algorithms using random and biased sampling. These works are close to the present paper's method as the size of the reduced log is considerably smaller than the original one. We aim to select the most representative traces for process discovery.

As our prototype selection uses a clustering method, we recall that trace clustering has been used in process mining to get several sub-models according to clustered sub-logs [13–16]. The quality of the sub-models is better than a single process model discovered on the whole event log. However, getting several process models may be a barrier for decision-makers who need a single overview of each process.

Finally, in [2] a genetic process discovery algorithm is proposed that benefits from conformance artifacts. However, this method is time-consuming for large real event logs and it is impractical using normal hardware.

4 Preliminaries

In this paper, we focus on sequences of activities, also called traces, that are combined into event logs.

Definition 1 (Event Log). *Let $\mathcal{A}$ be a set of activities. An event log is a multiset of sequences over $\mathcal{A}$, i.e., $L \in \mathbb{B}(\mathcal{A}^*)$ that is a finite set of words. A word, i.e., a sequence of activities, in an event log is also called a trace.*

Figure 1 shows an example of event log L. The occurrence of trace-variant $\langle a, b, c, e, g \rangle$ in this event log is four.

A sub-log L_1 of a log L is a set of traces such that $L_1 \subseteq L$. A trace clustering method aims to find disjoint sub-logs according to the similarity between traces.

Definition 2 (Trace Clustering). *Given a log L, a trace clustering $\xi(L, n)$ is a partitioning of L in a set of sub-logs $\{L_1, L_2 \ldots, L_n\}$ such that $\forall_{i \neq j} (L_i \cap L_j = \emptyset)$ and $\biguplus_{i=1:n} L_i = L$.*

We usually need a distance metric to cluster objects. One distance metric that is widely used to cluster words is Edit distance.

Definition 3 (Edit Distance). *Let* $\sigma, \sigma' \in \mathcal{A}^*$, *edit distance function* $\triangle(\sigma, \sigma') \rightarrow \mathbb{N}$ *returns the minimum number of edits that are required to transform* σ *to* σ'.

We assume that an edit operation can only be a deletion or an insertion of an activity in a trace. To give an example, $\triangle(\langle a, c, f, e, d \rangle, \langle a, f, c, a, d \rangle) = 4$ corresponding to two deletions and two insertions.

Some clustering algorithms return a medoid for each cluster that is a representative object of that cluster. In this paper, we also return prototypes as medoids which have the closest distance with other objects in their cluster.

Definition 4 (Prototypes). *Let* $\delta \colon \mathbb{B}(\mathcal{A}^*) \rightarrow \mathcal{A}^*$ *be a function that for each sub-log* L_i *returns* $p_i \in L_i$ *which has the minimum distance with other traces in that sub-log, i.e.,* $\sum_{\sigma \in L_i} (\triangle(p_i, \sigma))$. *For a trace clustering* $\xi(L, n) = \{L_1, L_2, \ldots, L_n\}$, *prototypes are a set* $P = \{p_i = \delta(L_i) : L_i \in \xi(L, n)\}$.

In other words, a prototype is a unique trace-variant that represents a sub-log.

A *process model*, commonly Petri net or BPMN, describes a set of traces. As the present paper does not propose a specific notation for process models, we define a process model by its describing behavior.

Definition 5 (Runs of Process Model). *Let* M *be a process model with a set of activities* $\mathcal{A}$. *We define a set of all possible traces that can be executed by* M *as* $Runs(M) \subseteq \mathcal{A}^*$. *In case of having loop in the model, this set is infinite.*

For example, Fig. 1a describes six traces; therefore, we have $Runs(M) = \{\langle a, c, b, e, g \rangle, \langle a, b, c, e, g \rangle, \langle a, d, e, g \rangle, \langle a, c, b, e, f \rangle, \langle a, b, c, e, f \rangle, \langle a, d, e, f \rangle\}$.

A process model and an event log may have some deviations. For instance, $\langle a, f \rangle$ is not in the described behavior of the model that is presented in Fig. 1a). It is shown in [17] that using the following formula, we can measure the fitness of a model and a traces based on the edit distance function.

$$trace\_fitness(\sigma_L, M) = 1 - \frac{\min\limits_{\sigma \in Runs(M)} \triangle(\sigma_L, \sigma)}{|\sigma_L| + \min\limits_{\sigma \in Runs(M)} |\sigma|} \tag{1}$$

The fitness of an event log and a model is a weighted average of the *trace-fitness* of trace logs. Thus, log traces with a higher frequency have higher weights.

In contrast, precision shows how much behavior in a model exists in occurs in the log. In this paper, we refer by $Precision(L, M)$ to ETC [18] as it has a high performance computation; however any other precision metrics can be used. To balance between the two main metrics, we use the F-Measure [19]:

$$F\text{-}Measure = 2 \times \frac{Precision \times Fitness}{Precision + Fitness} \tag{2}$$

5   Incremental Prototype Selection for Process Discovery

In this section, we explain the details of our approach to use the selected *Prototypes*, i.e., a subset of traces, for representing the entire log as a process model. As explained, we use a clustering approach to select the representative prototypes for process discovery. The schematic view of the proposed method is presented in Fig. 2. The method contains the following four main steps:

1. *Clustering for prototype selection*: to select prototypes using a clustering method.
2. *Model discovery*: discovering a model based on the selected prototypes.
3. *Quality assessment*: to evaluate the discovered model based on the original event log, conformance artifacts are computed.
4. *Iteration over deviating traces*: while quality metrics improve, we iterate the method (from Step 1) on the deviating traces of the last discovered model.

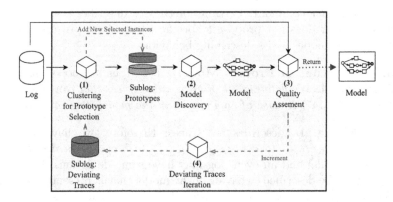

Fig. 2. Structure of the prototype selection approach

The prototype selection is an iterative process; then, the sub-log of selected prototypes gently grows in each iteration. During an iterative process, we expect that fitness increases while the precision value decreases. Here, we explain each step in more detail.

1- Clustering for Prototype Selection. By applying process discovery algorithms directly on a complete event logs, we usually obtain complex and imprecise process models. As presented in [8,20] by modifying and sampling event logs we are able to improve results of process discovery algorithms in terms of F-measure. We apply clustering to extract a very small set of representatives traces, i.e., prototypes. In this regards, we use K-medoids [21] to cluster traces in K sub-logs by considering their similarity (using the Edit distance function). This algorithm works as follows:

1. Select randomly K (i.e., the number of clusters) traces in L as medoids.
2. Create, or update, clusters by associating each trace to its closest medoid based on the Edit distance metric.
3. For each cluster, redefine the medoid as the prototype of the cluster according to Definition 4 using δ function. If medoids haven't changed, return the K prototypes. Otherwise, do Step 2 again.

The prototypes are then added to the set of *selected prototypes* which is empty at the first iteration of our method (see Fig. 2). For example, for the event log that is presented in Fig. 1e, applying the clustering with $K=3$ in the first iteration gives $\langle a, b, c, g \rangle$, $\langle a, c, b, e, f \rangle$ and $\langle a, d, e, f \rangle$ as prototypes.

2- Model Discovery. After selecting a set of prototypes, we discover a descriptive view of it, i.e., a process model. In this regard, we are flexible to use any process discovery algorithm. However, it is recommended to use methods that guarantee to return sound process models as it is necessary for fitness computation. By discovering a process model from the selected prototypes, we will have a general view of what is going on in the process and position different log traces w.r.t, this model.

3- Quality Assessment. To ensure that the discovered process model via prototypes conforms to the whole event log, we incorporate quality assessment evaluations in our method. We use the F-measure to get a good balance between fitness and precision. The metric is computed by considering the original event log and the process model created based on the selected prototypes.

4- Iteration over Deviating Traces. The F-measure is computed for the first time after the initialization step that selects a first set of prototypes. Thereafter, the proposed method starts an iterative procedure. In each iteration, the method first finds the deviating traces that are formally defined as follows.

Definition 6 (Deviating Traces). *Let M be a process model and L be an event log. The Deviating Traces is* $L_d=\{\sigma\in L : \sigma\notin Runs(M)\}$.

After finding the deviating traces, we look for other representative traces among them like what we did in Step 1, i.e., using clustering. Then, the new prototypes will be added to the previous ones (see Fig. 2). Here, for clustering of deviating traces, we are able to use similar or different K compared to the first step. Thereafter, we apply again the process discovery algorithm to find a new process model and so on. Afterward, we compare the previous and current F-measure values. The iterative procedure stops when the quality of the new discovered process model is lower than the previous one. By increasing in the number of the selected prototypes, we expected that the fitness of discovered model is increased, but its precision is decreased. So we use F-measure that balances the metrics. We make the hypothesis that process discovery algorithms tend to approach high fitness and adding traces in the input raises the fitness of the whole log and decreases the precision. This hypothesis is commonly true (as also assumed in [22]). Therefore, the algorithms stops when there is no improvement in F-measure of the discovered process model of prototypes.

Table 1. Some information of the real-life event logs used in the experiments.

Event log	Activities#	Traces#	Variants#	DF relations#
BPIC_2012 [23]	23	13087	4336	138
BPIC_2018_Insp. [24]	15	5485	3190	67
BPIC_2019 [25]	44	251734	11973	538
Hospital [26]	18	100000	1020	143
Road [27]	11	150370	231	70
Sepsis [28]	16	1050	846	115

6 Experiments

In this section, we investigate whether the proposed method results in process models with higher quality. To apply the proposed method, we implemented the *Prototype Selection* plug-in in ProM framework. The plug-in takes an event log as input and outputs the discovered model and the selected prototypes. As presented above, our method uses two parameters, i.e., the number of clusters/prototypes that will be selected in each iteration and the process discovery algorithm. To simplify the plug-in, we consider the same cluster size for both Step 1 and 4. We ported the *Prototype Selection* plug-in into RapidProM that allows us to apply the proposed method on various event logs with different parameters. RapidProM is an extension of RapidMiner that combines scientific workflows with a range of (ProM-based) process mining algorithms.

The experiments have been conducted on six real event logs of different fields, from healthcare to insurance. Event logs have different characteristics which are given in Table 1.

In the following, we first position the proposed method compared to some state-of-the-art preprocessing methods. Later, we analyze the clustering method for selecting prototypes.

6.1 Process Discovery Improvement

Here we aim to find out how the proposed method is able to improve the results of different process discovery algorithms. As the *Prototype Selection* has two parameters, i.e., the number of clusters and the discovery algorithm, we show results over a set of different settings. We repeated the experiments for 2 to 9

Table 2. Average of precision, fitness, and F-Measure for different methods.

Miner	Log	Nothing			Prototype Selection			Sampling			Statistical		
		Fitness	Precision	F-Measure	Fitness	Precision	F-Measure	Fitness	Precision	F-Measure	Fitness	Precision	F-Measure
ILP	BPIC-2012	1,00	0,12	0,21	0,75	0,74	0,65	0,88	0,24	0,37	1,00	0,12	0,22
	BPIC-2018-Ins.	1,00	0,13	0,22	0,88	0,64	0,68	0,96	0,37	0,51	1,00	0,16	0,28
	BPIC-2019	1,00	0,36	0,53	0,89	0,84	0,82	0,98	0,6	0,73	1,00	0,52	0,68
	Hospital	1,00	0,39	0,57	0,87	0,86	0,84	0,98	0,59	0,72	1,00	0,44	0,61
	Road	1,00	0,53	0,69	0,86	0,89	0,85	0,91	0,8	0,84	1,00	0,61	0,76
	Sepsis	1,00	0,2	0,34	0,81	0,68	0,65	0,94	0,39	0,53	1,00	0,22	0,35
IMi	BPIC-2012	0,83	0,59	0,67	0,78	0,78	0,76	0,89	0,56	0,68	1,00	0,54	0,7
	BPIC-2018-Ins.	0,97	0,52	0,67	0,59	0,79	0,66	0,8	0,62	0,65	0,94	0,53	0,66
	BPIC-2019	0,97	0,48	0,64	0,78	0,75	0,74	0,95	0,68	0,76	1,00	0,83	0,9
	Hospital	0,83	0,86	0,83	0,85	0,93	0,88	0,96	0,86	0,9	1,00	0,84	0,91
	Road	0,88	0,65	0,74	0,88	0,9	0,88	0,9	0,95	0,92	1,00	0,82	0,9
	Sepsis	0,9	0,56	0,65	0,84	0,71	0,75	0,91	0,56	0,67	1,00	0,49	0,66
SM	BPIC-2012	0,89	0,72	0,79	0,76	0,89	0,81	0,84	0,69	0,76	0,99	0,59	0,74
	BPIC-2018-Ins.	0,9	0,71	0,79	0,87	0,83	0,83	0,88	0,76	0,81	0,92	0,67	0,77
	BPIC-2019	0,98	0,76	0,85	0,85	0,97	0,9	0,96	0,66	0,76	1,00	0,44	0,61
	Hospital	0,99	0,9	0,94	0,9	1,00	0,94	0,92	0,98	0,94	1,00	0,97	0,98
	Road	0,89	0,9	0,89	0,89	1,00	0,94	0,87	0,99	0,93	1,00	0,95	0,98
	Sepsis	0,87	0,62	0,71	0,85	0,75	0,77	0,86	0,71	0,76	0,99	0,43	0,6

clusters and we used the inductive miner [10], the ILP miner [29], and the split miner [30]. Moreover, we compared our work with two trace sampling methods [8,9], i.e., referred to *Sampling* and *Statistical* respectively. For both of these methods, we ran the experiments with 20 different settings. When we use the preprocessing methods, we used the inductive miner with its filtering mechanism set to 0 and the default setting for the split miner. We also compared our method to normal process discovery algorithms, i.e, discovery without preprocessing which we denote it by *Nothing* in the experiments. For this case, we ran a set of experiments with 50 different settings for the inductive miner (IMi) and 100 for the split miner. For the ILP miner, we just run the experiment without its internal filtering mechanism.

Tables 2 and 3 show the average results of the experiments over the different settings. It is shown in Table 2 that for most of the cases, the F-Measure of discovered process models using the prototype selection method is higher than other preprocessing techniques and generally, the proposed method leads to provide more precise process models.

For simplicity, in Table 3, we consider two metrics that measure the complexity of discovered process models. *Model Size* of process models is a combination of the number of transitions, places, and arcs that connected them. Another metric is the Cardoso metric [31] that measures the complexity of a process model by its complex structures, i.e., *Xor*, *Or*, and *And* components. For both of these measures, a lower value means less complexity and consequently a simpler process model. Results show that we can have much simpler process models using the proposed method. By considering both tables, we see that the presented method helps to get more precise and simpler models in most of the event logs.

Table 3. Comparison of simplicity measures for different preprocessing methods.

Miner	Log	Nothing		Prototype Selection		Sampling		Statistical	
		Cardoso	Model Size	Cardoso	Model Size	Cardoso	Model Size	Cardoso	Model Size
ILP	BPIC-2012	163	33×28×426	83,7	25.7×23.3×172.7	182,5	31×26.6×449.5	234	33.5×27.7×593.7
	BPIC-2018-Ins.	142	26×19×324	123	25.3×17×246	110,2	24×18.1×238.3	112,7	24.3×18.3×234.7
	BPIC-2019	561	61×46×1550	45,7	16×13.3×90.7	379	45.6×39.6×1157.2	365	47.3×39.5×1064.3
	Hospital	120	29×22×440	16,7	10.3×13×38	106,2	22.5×19.8×370.1	103,8	21.5×20.3×386.7
	Road	67	16×15×150	18,5	10.5×12×39.3	52,5	16.4×14.8×115.1	53,2	17×15×109.3
	Sepsis	209	31×20×470	73	19×17×153.3	126,9	25.7×18.7×287.6	187	28.3×20×407.7
IMi	BPIC-2012	27	21×44×90	45,7	34.7×45.7×99.3	47,1	34.7×52.3×111.9	45,3	32.8×55.3×115.7
	BPIC-2018-Ins.	25	19×33×68	16,3	14×23.3×49.3	13,3	11.9×22.8×47.7	13,5	12×23.2×48.3
	BPIC-2019	32	21×65×132	20	15.3×22.3×45.3	34,6	24.5×60×123	38,2	26.2×62.2×128.3
	Hospital	43	31×52×108	24,7	19×26×54.7	33	23×41.5×84.2	35	23.5×43.8×88
	Road	19	19×25×56	10,5	10×15.5×31	18,3	15.7×24.6×51	21,8	18.7×28.2×58.7
	Sepsis	22	17×33×68	17	15×23.3×48.7	27,5	22.4×33.4×72.7	21,7	17.5×32.5×67.7
SM	BPIC-2012	106,8	65.9×101×240.2	41	31×42.3×86	85,8	55×84×179.8	102,7	62.2×98.5×218.8
	BPIC-2018-Ins.	56,5	33.2×56×115.4	39,8	24.8×39.8×80.3	42,6	25.8×43.5×87.3	44,7	26×45.7×91.3
	BPIC-2019	393,3	139×388.1×894.3	19,7	15.7×22×44	255,4	98.5×259.5×526.2	210	89.8×215×430.3
	Hospital	135,7	62.5×137.2×274.3	15	13×16×32	77,7	42.8×79.4×158.8	75,3	42.7×76.8×153.7
	Road	46,2	30.6×41.9×108.7	13,7	12.7×13.7×30	29,8	23×28.6×69.3	29,8	21.5×29.8×68.7
	Sepsis	97,3	49.5×94.5×232	30,8	23.2×32.8×65.7	68,3	34.9×70.6×145.8	95,7	41.2×99×198

6.2 Using Clustering for Prototype Selection

Here, we aim to find how by selecting prototypes using the clustering method we improve the quality of process models. We increased the number of prototypes from 1 to 20 and analyze the quality of the resulted models for Sepsis [28], and Road [27] event logs. We used the inductive miner without the internal filtering for model discovery. In Fig. 3, we compared the results of prototype selection based on clustering and the most frequent variants on the discovered models.

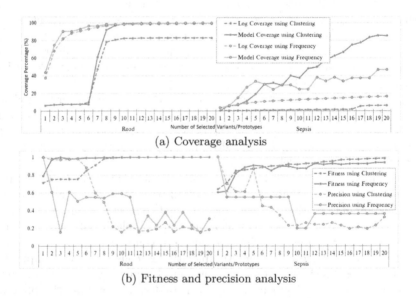

(a) Coverage analysis

(b) Fitness and precision analysis

Fig. 3. Effects of increasing the number of selected prototypes on the quality issues of discovered process models using frequency and clustering methods.

In Fig. 3a, the log coverage shows how many percentage of the traces in the event log, is corresponds to the selected prototypes. Moreover, the model coverages indicates that how many percentages of traces in the event log is replayable (or perfectly fitted) by the discovered process model. For example, in the Sepsis event log, by selecting eight prototypes, i.e., corresponds to 5% of traces, the discovered process model is able to perfectly replay 35% of the traces in the event log. Figure 3a shows that process discovery algorithms depict much behavior in the process model compared to the given event log. For event log with high frequent traces, e.g., *Road*, when we select very few process instances, by selection based on frequency, we usually have higher model coverage. However, when we select more than 10 prototypes, or for event logs with lots of unique variants, e.g., *Sepsis*, the model coverage of clustering method is higher.

In Fig. 3b, we see how by increasing the number of prototypes, generally fitness increases and precision decreases. This reduction is higher when we select based on frequency. Results show that we can discover a high fitted process model without giving just a few prototypes to process discovery algorithms. This experiment shows that using the clustering algorithm we can choose the more representative prototypes specifically if the log has lots of unique behavior.

7   Conclusion

In this paper, we proposed an incremental method to select prototypes of the event logs in order to generate simple and precise process models having a good F-measure. It clusters the traces in the event log based on their control-flow distances. Afterward, it returns the most representative instance for each cluster, i.e., the prototype. We discover a process model of the selected prototypes which is analyzed by common conformance metrics. Then, the method recessively selects new prototypes from deviating traces. A novel set of traces is added to the process discovery algorithm which improves fitness while decreasing precision.

To evaluate the proposed method, we have developed it in ProM and RapidProM, and have applied the proposed prototype selection method on six real event logs. We compared it with other state-of-the-art sampling methods using different process discovery algorithms. The results indicate that the proposed method is able to select process instances properly and help process discovery algorithms to return process models with a better balance between quality measures. Discovered models are less complex and, consequently, easier to understand. Another advantage of our method is that it is more stable in chosen settings of parameters and tends to return process models with higher quality.

As future work, it is possible to find prototypes using more advanced clustering methods and measures that are proposed in the literature. Indeed, the weakness of the F-measure is the fact that it is an average of fitness and precision, which blurs the understanding of the chosen model. Instead, the use of F_β-measure introduced in [32] can help one to balance between the two criteria. Moreover, we aim to use prototypes for other purposes, e.g., conformance checking and performance analysis. One limitation of our method is it may find a local optimum rather than the global optimum. We plan to use an adjustable number of clusters for both initiating phase and incremental steps.

Acknowledgment. We thank Prof. Josep Carmona, Dr. Thomas Chatain and Dr. Sebastiaan J. van Zelst for comments that greatly improved the work. We also thank the Alexander von Humboldt (AvH) stiftung for supporting this research.

References

1. van der Aalst, W.M.P.: Process Mining - Data Science in Action, 2nd edn. Springer, Heidelberg (2016). https://doi.org/10.1007/978-3-662-49851-4
2. Buijs, J.C.A.M., van Dongen, B.F., van der Aalst, W.M.P.: On the role of fitness, precision, generalization and simplicity in process discovery. In: Meersman, R., et al. (eds.) OTM 2012. LNCS, vol. 7565, pp. 305–322. Springer, Heidelberg (2012). https://doi.org/10.1007/978-3-642-33606-5_19
3. Carmona, J., van Dongen, B., Solti, A., Weidlich, M.: Conformance Checking, pp. 241–260. Springer, Cham (2018). https://doi.org/10.1007/978-3-319-99414-7_12
4. Bose, R.J.C., Mans, R.S., van der Aalst, W.M.P.: Wanna improve process mining results? In: IEEE Symposium on Computational Intelligence and Data Mining, CIDM 2013, Singapore, 16–19 April, 2013, pp. 127–134 (2013)
5. Conforti, R., Rosa, M.L., ter Hofstede, A.H.M.: Filtering out infrequent behavior from business process event logs. IEEE Trans. Knowl. Data Eng. **29**(2), 300–314 (2017)
6. Fani Sani, M., van Zelst, S.J., van der Aalst, W.M.P.: Improving process discovery results by filtering outliers using conditional behavioural probabilities. In: Teniente, E., Weidlich, M. (eds.) BPM 2017. LNBIP, vol. 308, pp. 216–229. Springer, Cham (2018). https://doi.org/10.1007/978-3-319-74030-0_16
7. van der Aalst, W.M.P., Bolt, A., van Zelst, S.: Rapidprom: mine your processes and not just your data. CoRR abs/1703.03740 (2017)
8. Fani Sani, M., van Zelst, S.J., van der Aalst, W.M.P.: The impact of event log subset selection on the performance of process discovery algorithms. In: Welzer, T., et al. (eds.) ADBIS 2019. CCIS, vol. 1064, pp. 391–404. Springer, Cham (2019). https://doi.org/10.1007/978-3-030-30278-8_39
9. Bauer, M., Senderovich, A., Gal, A., Grunske, L., Weidlich, M.: How much event data is enough? A statistical framework for process discovery. In: Krogstie, J., Reijers, H.A. (eds.) CAiSE 2018. LNCS, vol. 10816, pp. 239–256. Springer, Cham (2018). https://doi.org/10.1007/978-3-319-91563-0_15
10. Leemans, S.J.J., Fahland, D., van der Aalst, W.M.P.: Discovering block-structured process models from event logs containing infrequent behaviour. In: Lohmann, N., Song, M., Wohed, P. (eds.) BPM 2013. LNBIP, vol. 171, pp. 66–78. Springer, Cham (2014). https://doi.org/10.1007/978-3-319-06257-0_6
11. Ghionna, L., Greco, G., Guzzo, A., Pontieri, L.: Outlier detection techniques for process mining applications. In: An, A., Matwin, S., Raś, Z.W., Slezak, D. (eds.) ISMIS 2008. LNCS (LNAI), vol. 4994, pp. 150–159. Springer, Heidelberg (2008). https://doi.org/10.1007/978-3-540-68123-6_17
12. Fani Sani, M., van Zelst, S.J., van der Aalst, W.M.P.: Applying sequence mining for outlier detection in process mining. In: Panetto, H., Debruyne, C., Proper, H.A., Ardagna, C.A., Roman, D., Meersman, R. (eds.) OTM 2018. LNCS, vol. 11230, pp. 98–116. Springer, Cham (2018). https://doi.org/10.1007/978-3-030-02671-4_6

13. Tax, N., Sidorova, N., Haakma, R., van der Aalst, W.M.P.: Mining local process models. J. Innov. Digit. Ecosyst. **3**(2), 183–196 (2016)
14. Boltenhagen, M., Chatain, T., Carmona, J.: Generalized alignment-based trace clustering of process behavior. In: Donatelli, S., Haar, S. (eds.) PETRI NETS 2019. LNCS, vol. 11522, pp. 237–257. Springer, Cham (2019). https://doi.org/10.1007/978-3-030-21571-2_14
15. Weerdt, J.D., vanden Broucke, S.K.L.M., Vanthienen, J., Baesens, B.: Leveraging process discovery with trace clustering and text mining for intelligent analysis of incident management processes. In: Proceedings of the IEEE Congress on Evolutionary Computation, CEC, pp. 1–8 (2012)
16. De Weerdt, J., Vanden Broucke, S., Vanthienen, J., Baesens, B.: Active trace clustering for improved process discovery. IEEE Trans. Knowl. Data Eng. **25**(12), 2708–2720 (2013)
17. Fani Sani, M., van Zelst, S.J., van der Aalst, W.M.P.: Conformance checking approximation using subset selection and edit distance. In: Dustdar, S., Yu, E., Salinesi, C., Rieu, D., Pant, V. (eds.) CAiSE 2020. LNCS, vol. 12127, pp. 234–251. Springer, Cham (2020). https://doi.org/10.1007/978-3-030-49435-3_15
18. Muñoz-Gama, J., Carmona, J.: A fresh look at precision in process conformance. In: Hull, R., Mendling, J., Tai, S. (eds.) BPM 2010. LNCS, vol. 6336, pp. 211–226. Springer, Heidelberg (2010). https://doi.org/10.1007/978-3-642-15618-2_16
19. Van Rijsbergen, C.J.: Information retrieval (1979)
20. Sani, M.F., van Zelst, S.J., van der Aalst, W.M.P.: Repairing outlier behaviour in event logs using contextual behaviour. Enterp. Model. Inf. Syst. Archit. Int. J. Concept. Model. **14**, 5:1–5:24 (2018)
21. De Amorim, R.C., Zampieri, M.: Effective spell checking methods using clustering algorithms. In: Recent Advances in Natural Language Processing, RANLP 2013, 9–11 September 2013, Hissar, Bulgaria, pp. 172–178 (2013)
22. Augusto, A., Dumas, M., La Rosa, M.: Metaheuristic optimization for automated business process discovery. In: Hildebrandt, T., van Dongen, B.F., Röglinger, M., Mendling, J. (eds.) BPM 2019. LNCS, vol. 11675, pp. 268–285. Springer, Cham (2019). https://doi.org/10.1007/978-3-030-26619-6_18
23. van Dongen, B.F.: BPI Challenge 2012. Eindhoven University of Technology, Dataset (2012)
24. van Dongen, B.F., Borchert, F. (Florian): BPI Challenge 2018 Eindhoven University of Technology. Dataset (2018)
25. van Dongen, B.F.: BPI Challenge 2019. Eindhoven University of Technology, Dataset (2019)
26. Mannhardt, F.: Hospital Billing-Event Log. Eindhoven University of Technology. Dataset. Eindhoven University of Technology. Dataset, pp. 326–347 (2017)
27. De Leoni, M., Mannhardt, F.: Road traffic fine management process. Eindhoven University of Technology, Dataset (2015)
28. Mannhardt, F.: Sepsis cases-event log. Eindhoven University of Technology (2016)
29. van Zelst, S.J., van Dongen, B.F., van der Aalst, W.M.P., Verbeek, H.M.W.: Discovering workflow nets using integer linear programming. Computing **100**(5), 529–556 (2017). https://doi.org/10.1007/s00607-017-0582-5
30. Augusto, A., Conforti, R., Dumas, M., Rosa, M.L., Polyvyanyy, A.: Split miner: automated discovery of accurate and simple business process models from event logs. Knowl. Inf. Syst. **59**(2), 251–284 (2019). https://doi.org/10.1007/s10115-018-1214-x
31. Lassen, K.B., van der Aalst, W.M.P.: Complexity metrics for workflow nets. Inf. Softw. Technol. **51**(3), 610–626 (2009)

32. Chinchor, N.: Muc-4 evaluation metrics. In: ACL (1992)
33. Fani Sani, M., van Zelst, S.J., van der Aalst, W.M.P.: Improving the performance of process discovery algorithms by instance selection. Comput. Sci. Inf. Syst. **17**(3), 927–958 (2020). https://doi.org/10.2298/CSIS200127028S

Enhancing Discovered Process Models Using Bayesian Inference and MCMC

Gert Janssenswillen[1]($\boxtimes$), Benoît Depaire[1], and Christel Faes[2]

[1] Faculty of Business Economics, UHasselt - Hasselt University, Hasselt, Belgium
`gert.janssenswillen@uhasselt.be`
[2] Data Science Institute, Interuniversity Institute of Biostatistics and Statistical Bioinformatics, Leuven, Belgium

Abstract. Process mining is an innovative research field aimed at extracting useful information about business processes from event data. An important task herein is process discovery. The results of process discovery are mainly non-stochastic process models, which do not convey a notion of probability or uncertainty. In this paper, Bayesian inference and Markov Chain Monte Carlo is used to build a statistical model on top of a process model using event data, which is able to generate probability distributions for choices in a process' control-flow. A generic algorithm to build such a model is presented, and it is shown how the resulting statistical model can be used to test different kinds of hypotheses. The algorithm supports the enhancement of discovered process models by exposing probabilistic dependencies, and allows to compare the quality among different models, each of which provides important advancements in the field of process discovery.

Keywords: Process mining · Bayesian statistics · Process model quality

1 Introduction

Due to ever increasing amount of process data recorded by organisations, the list of available techniques to explore, describe and eventually improve business processes has been continually expanding. One of the main tasks within the process mining field has been to discover process models from event data. Since initial attempts to learn process models from event data at the end of the 20th century [2,5], many more advanced process discovery algorithms have been developed in recent years [3,10]. Recent approaches are increasingly better able at tackling challenges such as noisy and incomplete data as well as dealing with unstructured process behaviour.

Notwithstanding the increasing quality of process discovery algorithms, the resulting models offer mostly a non-stochastic representation, without any notion of uncertainty or probabilities. While discovered process models are able to depict when a choice between two or more activities occurs, or when an activity occurs

© Springer Nature Switzerland AG 2020
A. Del Río Ortega et al. (Eds.): BPM 2020 Workshops, LNBIP 397, pp. 295–307, 2020.
https://doi.org/10.1007/978-3-030-66498-5_22

in a loop, they do not provide more statistics about each of these constructs. As a result, it will be unclear how probable a specific choice or loop is, and how this probability relates to other components of the model.

In this paper, we introduce a method to build a statistical model on top of a process model—discovered or handmade—and event data, using Bayesian statistics and Markov Chain Monte Carlo (MCMC) simulation, in order to learn stochastic characteristics and test dependencies between different control-flow constructs. In particular, the contributions of this paper are 1) an algorithm which creates a Bayesian model based on a process model and related event data, and 2) example use cases which illustrate how the output of the Bayesian model can be used for hypothesis testing.

The next section will motivate the contribution and provided context with respect to related techniques. Section 3 describes and illustrates the algorithm to build a statistical model, while in Sect. 4 the approach is demonstrated by way of several use cases. Section 5 concludes the paper.

2   Motivation

Whereas the earlier process discovery algorithms tended to result in spaghetti-like models [1], the focus of more recent and advanced algorithms has been to tackle issues such as long-term dependencies, noise, and duplicate tasks, among others. An overview of existing discovery algorithms can be found in [4]. The output of most process discovery algorithms are *non-stochastic*, in the sense that they show the *possible* sequences of activities, but not how *probable* these are. Nevertheless, Stochastic Petri nets form a well-known extension to regular Petri nets [12]. They are able to mimic stochastic aspects by using probabilistic firing delays. They are therefore very useful for performance analysis [13]. An approach to discover Stochastic Petri nets from event logs, with the aim for performance analysis is described in [15]. However, the use of firing delays only indirectly explains the probabilities of different execution paths in the process, and is therefore less suited to analyse control-flow. Furthermore, the firing rate can only depend on the current marking, thereby making it largely absent of memory. As such, the firing of an activity will always be independent of the history of the current process instance.

Next to performance analysis, understanding stochastic aspects of control-flow is also relevant for predictive process monitoring. Black-box [6] as well as white-box [17] techniques have been developed for this task. While very useful to predict outcomes, cost or cycle-time of running cases, white-box techniques can also provide some insights into the model and rationale behind the predictions. However, they rely strongly on the discovered process model, and are not designed to compare different configurations in terms of statistical distributions and dependencies between them. The latter would allow to see which stochastic model is the best fit with the data, and to examine how the control-flow model itself can be enhanced to better reflect the observed event data.

The objective of this paper is therefore to complement the current state-of-the-art described above, by adding a statistical model on top of discovered process models, which is (1) able to incorporate probabilistic dependencies between different constructs, and (2) which can be easily adjusted, so that different configurations of the model can be compared in a statistically rigorous manner. As a result, insights about obtained probability distributions can be used to enhance the (understanding of the) control-flow of discovered process models.

For example, consider Fig. 1. This Petri net contains two exclusive choice splits: between B and C, and between E and F. What they do not tell us is which of the paths in a choice construct is more probable, and how it might affect the further process execution of a process instance.

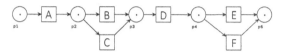

Fig. 1. Petri Net N_1

Models as these only show the extremes of behaviour: something can happen or it cannot. For example, the choice between transitions E and F does not depend on the choice between transitions B and C in this model—i.e. any combination can occur. The other possible extreme is a long-term dependency—e.g., if C is executed, than activity F should always be executed later. Such a constraint can be implemented using additional places and arcs. In reality, often more nuanced patterns can be observed. For instance, execution of C instead of B might lead to a *higher* probability of executing F instead of E, though not one of 100%. Such notions about probability and how they relate to each other are absent in the output of most process discovery algorithms, although they can improve the understanding about the process at hand.

3   Building a Statistical Model

The approach used in this paper to learn probability distributions is that of Bayesian inference, allowing the practitioner to define *prior beliefs* about the different distributions which, together with the data, is converted into posterior distributions. Due to the complexity of typical process models, it is likely not feasible to calculate the posterior distributions analytically [11]. Therefore—as often in Bayesian inference—Markov Chain Monte Carlo sampling will be used. Markov Chain simulation is one of the possible approaches for posterior simulation [8]. In particular, the Gibbs sampler will be used [7], which is the state-of-the-art MCMC technique used for Bayesian inference.

In order to illustrate the approach, consider again the process model in Fig. 1, and the accompanying simplified Event log L_1 in Table 1a. Suppose we denote the probability that B is executed as ϵ_1 and the probability for E as ϵ_2.

Table 1. Motivational example.

(a) Eventlog L_1

Trace	Frequency
ABDE	36
ACDE	37
ABDF	21
ACDF	6

(b) Posteriors for Bayesian Model 1.

Variable	Mean	St. Dev.	2.5 perc	Median	97.5 perc
θ_1	0.4127	0.04327	0.3267	0.4119	0.4954
θ_2	0.314	0.04258	0.2371	0.3131	0.4003
θ_3	0.1554	0.03035	0.101	0.1541	0.219
θ_4	0.1179	0.02357	0.07758	0.1171	0.1673
ϵ_1	0.5682	0.05012	0.4684	0.5684	0.6663
ϵ_2	0.7267	0.04589	0.6391	0.7256	0.8121

Then, $1-\epsilon_1$ and $1-\epsilon_2$ denote the probability that C or F are executed, respectively. The probability of observing each of the four traces which are allowed by the model, denoted by $\theta_1, \ldots, \theta_4$, can then be defined using these probabilities, as shown in Bayesian Model 1 (lines 8–11).

Bayesian Model 1. Specification for Petri net N_1 and Event log L_1.

1:	Data:	
2:	$Y = (36,37,21,6)$	▷Trace frequencies
3:	$N = 100$	▷Number of traces
4:	Model:	
5:	$Y \sim Mult(N, (\theta_1, ..., \theta_4))$	▷Event log as multinomial distribution
6:	$\epsilon_1 \sim Beta(1, 1)$	▷Prior distribution of ϵ_1
7:	$\epsilon_2 \sim Beta(1, 1)$	▷Prior distribution of ϵ_2
8:	$\theta_1 = \epsilon_1 * \epsilon_2$	▷Trace ABDE
9:	$\theta_2 = (1 - \epsilon_1) * \epsilon_2$	▷Trace ACDE
10:	$\theta_3 = \epsilon_1 * (1 - \epsilon_2)$	▷Trace ABDF
11:	$\theta_4 = (1 - \epsilon_1) * (1 - \epsilon_2)$	▷Trace ACDF

Subsequently, we assume that the trace frequencies observed in the event log (line 2–3) are the result of 100 draws from a multinomial distribution with these four trace probabilities (line 5). Finally, two beta distributions with shape parameters $a = 1$ and $b = 1$ are used as prior beliefs for ϵ_1 and ϵ_2 (line 6–7). These are uniform distributions, giving equal weight to all values between 0 and 1. As such, we assume no prior knowledge about the probability related to both choices for this illustration.

Using MCMC we can then generate the posterior probability distributions for the ϵ's and θ's, which are shown in Table 1b. Based on these results, it can for instance be seen that executing activity E is more probable than activity F (mean $\epsilon_2 = 0.7267$), while the choice between B and C is more balanced (mean $\epsilon_1 = 0.5682$). It's important to note that the method does not return point estimates of the probability that—for instance—B occurs, but rather a distribution that also shows how certain we are about this probability. For example, the probability that B happens lies with 95% confidence in the interval $[0.4684, 0.6663]$. As a result, we can use these distributions to test hypotheses, which is one of the main novelties of this approach. In this example, we cannot reject the hypothesis that $\epsilon_1 = \epsilon_2$, as their confidence intervals overlap.

Constructing the model as shown in Bayesian Model 1 can be challenging when the model contains a large, perhaps infinite, amount of traces, or does not perfectly fit the event log. As such, a formal algorithm to do so will be proposed in the remainder of this section.

3.1   Algorithm

The algorithm to create a model to be used for Bayesian statistics based on an event log and a process model consists of eight steps, which are listed below.

1. Split the event log in a fitting and non-fitting part.

2. For the fitting part, construct a prefix automaton.
3. Add escaping arcs based on transitions possible in the model that did not occur in the log, based on [14]. We call these escaping end states.
4. Add additional escaping arcs based on activities in the non-fitting part of the log which were not allowed by the model (i.e. the first deviating activity). We call these unfitting end states.
5. For each state with more than one outgoing arc, define a probability distribution.
6. Add constraints to probabilities based on the use case and hypothesis to be tested.
7. For each normal end state (i.e. not escaping or unfitting end state), calculate the probability to end up in that state.
8. Define a probability to end up in an escaping state.
9. Define a probability to end up in an unfitting end state.

By passing through these steps, a probability will have been defined for each fitting trace, together with an overall probability of finding a non-fitting trace and an overall probability that the model produces an unobserved trace. As such, all ingredients for a model similar as shown in Bayesian Model 1 will be available. Prefix automatons based on the event log are chosen as a way to decompose traces as a series of choices. Restricting this to the event log allows to cope with models of infinite size, while the escaping arcs cover for unobserved as well as unfitting behaviour. In subsequent paragraphs we first illustrate the algorithm with a more advanced example. A formal description of the algorithm and the implementation is available on GitHub.[1] The event log L_2 in Table 2 and the Petri net N_2 in Fig. 2 will be used to illustrate the algorithm.

Split the Event Log. The splitting of the event log into fitting and non-fitting traces is indicated in the penultimate column of Table 2. Of the 12 traces in the event log, 10 are fitting, while 2 are not. In trace 11 activity A or B should have been executed while in trace 12 the activity D is missing.

Mine a Prefix Automaton. Using all the prefixes of the fitting traces as states, the transition system as shown by the black nodes and arrows in Fig. 3 is obtained.

Add Escaping States. There are two states in the model in which additional, unobserved behaviour is possible. Firstly, after the execution of $\langle A, C, E, D \rangle$, both F and H are enabled, but only H was observed in the log. Secondly, after the execution of $\langle B, C, E, D \rangle$, again both F and H are enabled, but only H was observed in the log. Both instances have been annotated with red escaping arcs in Fig. 3.

---

[1] https://github.com/bupaverse/propro.

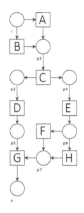

Fig. 2. Model N_2.

Table 2. Event log L_2.

	Trace	Freq	Step 1 Fits?	Step 6 Probability
1	A,C,D,E,H,G	10	✓	$\beta_1\beta_4(1-\beta_8)$
2	B,C,D,E,H,G	8	✓	$\beta_2\beta_4(1-\beta_8)$
3	B,C,E,F,D,G	7	✓	$\beta_2(1-\beta_4)\beta_6$
4	B,C,E,D,H,G	4	✓	$\beta_2(1-\beta_4)\beta_5(1-\beta_8)$
5	B,C,D,E,F,G	4	✓	$\beta_2\beta_4\beta_8$
6	A,C,D,E,F,G	4	✓	$\beta_1\beta_4\beta_8$
7	A,C,E,F,D,G	3	✓	$\beta_1(1-\beta_4)\beta_6$
8	A,C,E,D,H,G	3	✓	$\beta_1(1-\beta_4)\beta_5(1-\beta_8)$
9	B,C,E,H,D,G	2	✓	$\beta_2(1-\beta_4)\beta_7\beta_9$
10	A,C,E,H,D,G	2	✓	$\beta_1(1-\beta_4)\beta_7$
11	C,D,E,H,G	1	✗	β_3
12	B,C,E,H,G	2	✗	$\beta_2(1-\beta_4)\beta_7(1-\beta_9)$

Add Unfitting Escaping States. Two traces in the log, i.e. 11 and 12 do not fit the model. For both traces, we add an additional escaping state, at the moment the first deviation occurs. For trace 11, this means adding an additional arc from the initial state. For trace 12, this means adding an additional arc from the state $\langle B, C, E, H \rangle$. Both instances have been annotated with blue escaping arcs in Fig. 3.

Define Probability Distributions/Add Constraints. [2]There are 10 states in the prefix automaton which have more than one outgoing arc. For each of these states, we can define a probability distribution. Not all of the 10 split states require a unique distribution, as some choices recur more than once in the prefix automaton. For example, the choice between activity D and E occurs both in state $\langle A, C \rangle$ as well as in state $\langle B, C \rangle$. Since these states share the same marking in the Petri net, a strict interpretation of the net requires that both are appointed the same probability distribution. These constraints can be decided upon based on the use case. An overview of the choice probabilities is given in Table 3. Note that for states with only two options, we use β_x and $1 - \beta_x$ as probabilities, while for states with more than two options we will define a β_x for each option, and make sure that the sum of these equals to one. The reason for this is that both require a different approach for setting prior probability distributions—the binary case being an exception.

At this moment, we can also add derived variables to facilitate the analysis. For example, if we want to compare the probability that D occurs after $\langle B, C \rangle$ or $\langle A, C \rangle$ with the probability that it occurs after $\langle B, C, E \rangle$ or $\langle A, C, E \rangle$, we can define the difference between those variables as an additional variable, e.g.

---

[2] For brevity, we directly apply the constraints by giving different prefixes the same probability parameters: there are 10 different splits, but we only define 5 probability distributions. In practice, equality constraints can be added in the final stage, yielding more flexibility in adding and removing specific constraints.

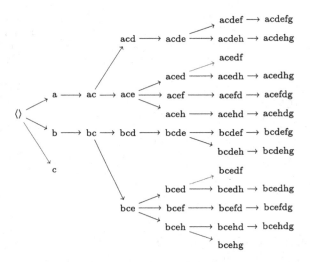

Fig. 3. Prefix automaton of log L_2 based on Petri net N_2.

Table 3. Probability distributions for each split.

Prefixes	Options	Probability
$\langle\rangle$	A	$Pr(A) = \beta_1$
	B	$Pr(B) = \beta_2$
	C	$Pr(C) = \beta_3$
		such that $\beta_1 + \beta_2 + \beta_3 = 1$
bc, ac	D	$Pr(D) = \beta_4$
	E	$Pr(E) = 1 - \beta_4$
bce, ace	D	$Pr(D) = \beta_5$
	F	$Pr(F) = \beta_6$
	H	$Pr(H) = \beta_7$
		such that $\beta_5 + \beta_6 + \beta_7 = 1$
bcde, bced, acde, aced	F	$Pr(F) = \beta_8$
	H	$Pr(H) = 1 - \beta_8$
bceh	D	$Pr(D) = \beta_9$
	G	$Pr(G) = 1 - \beta_9$

$\Delta_1 = \beta_4 - \beta_5$. As a result, we will obtain a posterior distribution and confidence interval for Δ_1 as well, making it easier to test hypotheses.

Compute Probability Normal End States. The probability of arriving in a specific end state, i.e. of observing the corresponding trace, can be established by following the path from the initial state to this end state and combining all the probabilities along the way. For each fitting trace, the probability has been listed in the last column of Table 2.

Compute Probability Escaping End State. In a similar way, the probability of arriving in an escaping end state can be computed. The overall probability θ_{esc} is equal to the sum of all probabilities of arriving in an escaping state, i.e. $\theta_{esc} = \beta_1(1 - \beta_4)\beta_5\beta_8 + \beta_2(1 - \beta_4)\beta_5\beta_8$.

Compute Probability Unfitting End States. Likewise, the probability of arriving in an unfitting end state can be computed. The overall probability θ_{unf} is equal to $\theta_{unf} = \beta_3 + \beta_2(1 - \beta_4)\beta_7(1 - \beta_9)$.

Results. Using the outputs of the algorithm, we can now construct the Bayesian model in the same manor as Bayesian Model 1.[3] This model would define the event log Y as the result of a multinomial distribution with probabilities $\theta_1, ...\theta_{12}, \theta_{esc}$. Here, θ_1 to θ_{12}, which corresponds with the equations in Table 2, refer to the probability of seeing a particular trace in the log. The probability θ_{esc} refers to the situation that an escaping arc is reached. Optionally, θ_{unf} can be added as an additional variable to get an estimate of the probability of an unfitting trace. In this case, $\theta_{unf} = \theta_{11} + \theta_{12}$, and as such is already incorporate in the model.

The specification of priors is no output of the formal argument. In the illustration before, a beta distribution was taken as prior for those splits where only 2 options are possible. For the split where more than 2 options exist, a Dirichlet distribution can be taken, which is the multivariate generalization of the beta distribution. For each of the priors, the shape parameters can be set in such a way that the priors are uninformative—or a domain expert can be asked to give her prior belief on the probabilities of different splits, so that the resulting statistical model takes into account both domain knowledge and input from the event log.

4 Demonstration of Use Cases

In this section, we will discuss two different use case of Bayesian inference and MCMC in the context of process discovery. In order to do so, we will use the

---

[3] Because of space limitations, the Bayesian model and resulting posterior distributions have not been included in this paper. Instead, we will look at two example use cases in the following section.

examples introduced so far. The first use case relates to the detection of choice probabilities and non-deterministic dependencies between choices—which was already briefly introduced in Sect. 2. Secondly, we will discuss how the approach can be used to compare the quality of different process models.

4.1 Detecting Non-deterministic Dependencies

In the prefix automaton in Fig. 3 it can be observed that the activity F is *never* recorded if E is executed before D. This might indicate a deterministic long-term dependency. However, it's not clear whether this is truly the case, as we have only seen a limited number of cases, and the next case we see might prove us wrong. In other situations, dependencies might be more nuanced, and deterministic dependencies might be too strong.

Using the Bayesian model, and different configurations of the probability distributions, both these aspects can be formally tested. In the previous section, we used the same probabilities for choices in different states when those states shared the same marking, thereby strictly taking into account the Petri net. However, we can create an alternative model where we define different probabilities for different states, even though they share a marking. As such, we can build implicit dependencies between choices in our statistical model and test whether they are significant.

Table 4. Prefix-dependent configuration for choice between F and H.

Variable	Mean	St. Dev	2.5 perc	Median	97.5 perc
β_{10}	0.351	0.125	0.136	0.338	0.607
β_{11}	0.17	0.144	0.006	0.132	0.53
β_{12}	0.319	0.114	0.118	0.306	0.556
β_{13}	0.202	0.167	0.006	0.16	0.629
$\delta_1 = \beta_{11} - \beta_{10}$	−0.18	0.192	−0.52	−0.195	0.248
$\delta_2 = \beta_{12} - \beta_{10}$	−0.032	0.169	−0.364	−0.031	0.279
$\delta_3 = \beta_{13} - \beta_{10}$	−0.149	0.206	−0.485	−0.171	0.333
$\delta_4 = \beta_{12} - \beta_{11}$	0.149	0.184	−0.267	0.163	0.468
$\delta_5 = \beta_{13} - \beta_{11}$	0.032	0.225	−0.43	0.02	0.511
$\delta_6 = \beta_{13} - \beta_{12}$	−0.117	0.201	−0.466	−0.14	0.332

(b) Posteriors.

Prefixes	$Pr(F)$	$Pr(H)$
bcde	β_{10}	$1 - \beta_{10}$
bced	β_{11}	$1 - \beta_{11}$
acde	β_{12}	$1 - \beta_{12}$
aced	β_{13}	$1 - \beta_{13}$

(a) Probabilities

For example, instead of a single probability distribution for the choice between F and H, we can define four separate distributions as in Table 4a—one for each state—to see whether there are significant differences between them in terms of probabilities. If we replace each β_8 in Table 2 appropriately and set additional priors for them, the resulting posteriors are as shown in Table 4b.[4] For analytical purpose, we added each pairwise difference between the β's.

---

[4] Note that due to space limitations, we only show the posterior distributions of the variables of interest.

Looking at the mean values for the different β's we see that there are differences in how probable it is that F gets executed and not H. When looking at the δ's, which are the pairwise differences between these distributions, it should be noted that zero is included in the 95% confidence interval for each of these. As such, given the 50 observed traces, we cannot reject the hypothesis that there are no dependencies between the choice of F versus H and earlier choices.

4.2 Comparing the Quality of Multiple Models

Whereas the previous use case mainly looked at *local* characteristics of the process, the output of the procedure can also be used to get a *global* measurement of the quality of a model. By running the Bayesian models, we can obtain a Deviance Information Criterion (DIC), which indicates which of multiple models is preferable. The DIC measures goodness-of-fit, also taking into account the model complexity—denoted by pD, which is a measure of model dimensionality which can be interpreted as the number of *effective* parameters in the model [8].[5]

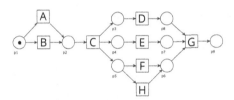

Fig. 4. Petri net discovered from L_2 using inductive miner [10].

Whereas the Petri net in Fig. 2 was constructed by hand, the Petri net in Fig. 4 was discovered using the Inductive miner [10]. It differs from the previous model in that it allows E to happen in parallel with F or H. When we run through the different steps of the algorithm, we will thus find that the final prefix automaton contains more escaping arcs, as this model allows for more behaviour that was not observed compared to the model in Fig. 2.

Both the Deviance Information Criterion (DIC) and the model dimensionality (pD) for the three different models relating to the event log in Table 2 which we have discussed so far are shown in Table 5. With regards to complexity, it can be seen that the original specification (as shown in Bayesian Model ??) has the lowest complexity, while both the specification with probabilities depending on the prefix as well as the specification using the Inductive miner model are more complex. The overall goodness-of-fit measured by the DIC—which takes into

---

[5] By *model*, we are referring to the *Bayesian* models. As such, we can compare the goodness-of-fit among different Petri net models, as well as among a single Petri net model with different probability specifications, such as the different specifications in Table 3 vs Table 4a.

account this complexity—is also the best for the original specification (a lower DIC is better). According to [16], models with a difference in DIC less than or equal to 2 compared to the *best* model should still be considered, whereas models for which the DIC is considerably higher should not. As such, the model with the prefix-dependent probabilities should not necessarily be regarded as less good than the original model, while the goodness-of-fit of the configuration using the Inductive miner model is considerably worse.

Table 5. Deviance Information for different model specifications.

Model	DIC	pD
Original model specification (Table 3)	48.0	5.7
Model specification with prefix-dependent probabilities (Table 4a)	50.3	8.6
Inductive miner model specification (Fig. 4)	67.2	12.2

5   Conclusions and Future Work

In this paper, an algorithm was defined to represent a process as a Bayesian model, based on event data and a process model. The resulting Bayesian model can then be used, together with MCMC, to gather further insights about the process, in terms of probability distributions and dependencies. While the algorithm provides a general approach to do so, two exemplary use cases were used to show the added value of the results in specific contexts. The algorithm is implemented in R and published as the R-package propro[6], which is compatible with the bupaR-framework [9].

The main limitation of the suggested approach is that it requires a model of certain quality as input, as the algorithm is not usable in case for instance all traces in the event log are unfitting. Secondly, defining prior belief about the probabilities might be challenging, even for domain experts. To this end, a more thorough evaluation of the approach using real-life data together with domain experts is needed. It should also be noted that the approach is mostly targeted towards confirmatory analysis, as the use cases assume a certain hypothesis that needs to be tested. Based on the hypothesis, one can focus on the right elements of the model and compare different configurations. Additional research will be needed to automatically extract interesting patterns from the model, without a starting hypothesis - as many different configurations might be possible in a real-life case. Secondly, novel visualisations are needed in order to truly exploit the insights that the Bayesian models produces.

---

[6] https://github.com/bupaverse/propro.

References

1. van der Aalst, W.M.P., Weijters, A.J.M.M., Maruster, L.: Workflow mining: discovering process models from event logs. IEEE Trans. Knowl. Data Eng. **16**(9), 1128–1142 (2004)
2. Agrawal, R., Gunopulos, D., Leymann, F.: Mining process models from workflow logs. In: Schek, H.-J., Alonso, G., Saltor, F., Ramos, I. (eds.) EDBT 1998. LNCS, vol. 1377, pp. 467–483. Springer, Heidelberg (1998). https://doi.org/10. 1007/BFb0101003
3. Augusto, A., Conforti, R., Dumas, M., La Rosa, M.: Split miner: discovering accurate and simple business process models from event logs (2017)
4. Augusto, A., et al.: Automated discovery of process models from event logs: review and benchmark. In: IEEE Transactions on Knowledge and Data Engineering (2018)
5. Datta, A.: Automating the discovery of as-is business process models: probabilistic and algorithmic approaches. Inf. Syst. Res. **9**(3), 275–301 (1998)
6. Di Francescomarino, C., Ghidini, C., Maggi, F.M., Petrucci, G., Yeshchenko, A.: An eye into the future: leveraging a-priori knowledge in predictive business process monitoring. In: Carmona, J., Engels, G., Kumar, A. (eds.) BPM 2017. LNCS, vol. 10445, pp. 252–268. Springer, Cham (2017). https://doi.org/10.1007/978-3-319-65000-5_15
7. Geman, S., Geman, D.: Stochastic relaxation, Gibbs distributions, and the Bayesian restoration of images. IEEE Trans. Pattern Anal. Machine Intell. **6**, 721–741 (1984)
8. Gill, J.: Bayesian Methods: A Social and Behavioral Sciences Approach. Chapman and Hall/CRC (2002)
9. Janssenswillen, G., Depaire, B., Swennen, M., Jans, M., Vanhoof, K.: bupaR: enabling reproducible business process analysis. Knowledge-Based Syst. **163**, 927–930 (2019)
10. Leemans, S.J.J., Fahland, D., van der Aalst, W.M.P.: Discovering block-structured process models from incomplete event logs. In: Ciardo, G., Kindler, E. (eds.) PETRI NETS 2014. LNCS, vol. 8489, pp. 91–110. Springer, Cham (2014). https://doi.org/10.1007/978-3-319-07734-5_6
11. Lesaffre, E., Lawson, A.B.: Bayesian Biostatistics. Wiley, New York (2012)
12. Marsan, M.A.: Stochastic petri nets: an elementary introduction. In: Rozenberg, G. (ed.) APN 1988. LNCS, vol. 424, pp. 1–29. Springer, Heidelberg (1990). https://doi.org/10.1007/3-540-52494-0_23
13. Molloy, M.K.: Performance analysis using stochastic Petri nets. IEEE Trans. Comput. **9**, 913–917 (1982)
14. Muñoz-Gama, J., Carmona, J.: A fresh look at precision in process conformance. In: Hull, R., Mendling, J., Tai, S. (eds.) BPM 2010. LNCS, vol. 6336, pp. 211–226. Springer, Heidelberg (2010). https://doi.org/10.1007/978-3-642-15618-2_16
15. Rogge-Solti, A., van der Aalst, W.M.P., Weske, M.: Discovering stochastic petri nets with arbitrary delay distributions from event logs. In: International Conference on Business Process Management. pp. 15–27. Springer (2013)
16. Spiegelhalter, D.J., Best, N.G., Carlin, B.P., Van Der Linde, A.: Bayesian measures of model complexity and fit. J. Royal Stat. Soc. Ser. B (Statistical Methodology) **64**(4), 583–639 (2002)
17. Verenich, I., Nguyen, H., La Rosa, M., Dumas, M.: White-box prediction of process performance indicators via flow analysis. In: Proceedings of the 2017 International Conference on Software and System Process, pp. 85–94 (2017)

A Generic Framework
for Attribute-Driven Hierarchical Trace
Clustering

Sebastiaan J. van Zelst[1,2(✉)] and Yukun Cao[2]

[1] Fraunhofer Institute for Applied Information Technology (FIT),
Sankt Augustin, Germany
sebastiaan.van.zelst@fit.fraunhofer.de
[2] Chair of Process and Data Science, RWTH Aachen University, Aachen, Germany

Abstract. The execution of business processes often entails a specific process execution context, e.g. a customer, service or product. Often, the corresponding event data logs indicators of such an execution context, e.g., a customer type (bronze, silver, gold or platinum). Typically, variations in the execution of a process exist for the different execution context of a process. To gain a better understanding of the global process execution, it is interesting to study the behavioral (dis)similarity between different execution contexts of a process. However, in real business settings, the exact number of execution contexts might be too large to analyze manually. At the same time, current trace clustering techniques do not take process type information into account, i.e., they are solely behaviorally driven. Hence, in this paper, we present a hierarchical data-attribute-driven trace clustering framework that allows us to compare the behavior of different groups of traces. Our evaluation shows that the incorporation of data-attributes in trace clustering yields interesting novel process insights.

Keywords: Process mining · Trace clustering · Process comparison.

1 Introduction

Modern information systems employed at companies track the execution of business processes in great detail. The activities executed in the context of a business process, i.e., *events*, are stored in *event logs*. Automated analysis of event logs allows one to get a better understanding of the process based on what happened in reality (as recorded in the information system). In the field of *process mining* [1], several techniques have been developed that provide such automated analyses, e.g., methods exist that automatically construct a process model that describes the process as captured by the event data [2].

Ideally, process mining techniques allow us to get instant insights into the execution of a process, based on an event log. However, the direct application of an automated process discovery algorithm on a real event log often leads to

© Springer Nature Switzerland AG 2020
A. Del Río Ortega et al. (Eds.): BPM 2020 Workshops, LNBIP 397, pp. 308–320, 2020.
https://doi.org/10.1007/978-3-030-66498-5_23

a complex process model which is hard or even impossible to comprehend. This is typically caused by incorrect logging of events and low-frequent executions of the process that severely deviate from the process' main flow. Some authors have proposed methods to preprocessing/filter event data [10, 16–18], however, even properly filtered data often yields imprecise and complex process models. In some cases, this is because a company executes the process slightly different for the different *process execution contexts* of its end product and/or service. Even if the differences between the process execution are subtle, e.g., just swapping two activities, state-of-the-art process discovery algorithms tend to discover process models of inferior quality.

Process execution contexts are omnipresent, yet, little to no work focuses on providing end-to-end solutions to exploit event data describing execution contexts. To some degree, e.g., when a company distinguishes between 4 different customer types, manually slicing the data into relevant subsets and subsequently analyzing/comparing the corresponding process models is feasible. However, often, the number of process categories is too large to perform such an analysis manually. For example, consider the WABO/CoSeLoG event data [8], which describes different sub-logs capturing how five Dutch municipalities handle the application of building permits. In its current form, a comparative study is still feasible, however, in case one collects data of the same process among all Dutch municipalities (>300), such a manual analysis is no longer feasible.

Therefore, in this paper, we propose a framework that allows for attribute-driven hierarchical clustering of traces. Given a user-defined case attribute, we construct a hierarchical clustering of the given event data, grouping the most similar executions of the process. As such, the framework allows the user to inspect collections of similar process executions belonging to different process contexts. The clusters allow us to group different process contexts that have a similar process execution. Furthermore, the clusters allow us to improve the overall process performance of specific process execution contexts based on similar process contexts. The accompanying implementation, built on top of the PM4Py framework [4], allows the user to compare different groups in the hierarchy. Using the implementation, we conducted a set of large-scale experiments based on publicly available real event data sets. Our experiments consistently show that the models found based on attribute-driven event logs are of superior quality compared to the models discovered by process discovery algorithms when directly applied on the original event data.

The remainder of this paper is structured as follows. In Sect. 2, we present background concepts. In Sect. 3, we present the proposed framework. In Sect. 4, we present the results of the evaluation conducted in the context of this paper. In Sect. 5, we discuss related work. Section 6, concludes this work.

2   Background

In this section, we conceptually present *event logs* and *hierarchical clustering*.

Event Logs. Event logs are the primary source of data for almost any process mining analysis. An event log captures at which point in time an activity was executed in the context of an instance of a process. Consider Table 1a, in which we present a simplified example of an event log. The first column depicts the *Case identifier*, i.e., identifying a specific instance of a process, e.g., a concert ticket. The *Activity* column registers what activity was performed for the corresponding case. The *Timestamp* column registers at what time the activity occurred. Observe that other data attributes, related to the activities, are available in an event log as well, e.g., the *Resource* and *Cost* columns in Table 1a.

Table 1. Example event log (a), adopted from [1], describing a compensation request process for concert tickets and (b) an exemplary mapping of case identifiers to data attributes, i.e., each case id has an associated ticket class.

(a)

Case id	Event id	Activity	Timestamp	Resource	⋯
⋮	⋮	⋮	⋮	⋮	
⋮	⋮	⋮	⋮	⋮	⋯
1273	4632	register request	12-11-2019 11.02	Barbara	⋯
1274	4633	register request	12-11-2019 11.32	Jan	⋯
1273	4634	check ticket	12-11-2019 12.12	Stefanie	⋯
1274	4635	examine casually	12-11-2019 14.16	Jorge	⋯
1275	4636	register request	12-11-2019 14.32	Josep	⋯
1275	4637	examine thoroughly	12-11-2019 15.42	Marlon	⋯
1273	4638	examine thoroughly	13-11-2019 11.18	Barbara	⋯
1273	4639	decide	13-11-2019 15.34	Wil	⋯
1273	4640	reject request	13-11-2019 16.50	Arthur	⋯
⋮	⋮	⋮	⋮	⋮	
⋮	⋮	⋮	⋮	⋮	⋯

(b)

Case id	Ticket type
⋮	⋮
⋮	⋮
1273	VIP
1274	Basic
1275	Basic
1276	Plus
1277	Royal
⋮	⋮
⋮	⋮

For each process instance, multiple activities are logged over time, e.g., for the process instance with case-id 1273 we observe the sequence ⟨register request, check ticket, ..., reject request⟩. The execution of an activity in the context of a process instance is referred to as an *event* (identified by the *Event id* column). Observe that events carry auxiliary data payload, e.g., the *Resource* column. Typically, process instances (identified by the case identifier) also carry data attributes. Consider Table 1b, in which we depict an example of such a *case attribute*, corresponding to the event log in Table 1a, i.e., where each compensation request (Case id) is related to a ticket class. In general, different case attributes may exist, e.g., in what country/factory a product is produced, the ticket price, etc. Typically, a business owner is interested in assessing or comparing the execution of the process along the lines of such data attributes, e.g., in which branch of the company is the process executed most efficiently?

In the context of this paper, we assume an event log $L{\subseteq}\mathcal{C}$ to be a collection of cases ($\mathcal{C}$ denotes the universe of cases). Furthermore, given some $c{\in}\mathcal{C}$ data attribute of interest $d{\in}\mathcal{D}$ ($\mathcal{D}$ denotes the universe of data attributes), we assume that we are able to retrieve the value of attribute d by means of $\pi_d(c)$. In particular, for $\texttt{trace}{\in}\mathcal{D}$, we assume that $\pi_{\texttt{trace}}(c){\in}\mathcal{A}^*$ (where $\mathcal{A}^*$ denote the set of all sequences over the activity universe $\mathcal{A}$). Finally, we let $\hat{\mathcal{D}} = \mathcal{D}\backslash\{\texttt{trace}\}$ and we overload the notation for attribute projection (π) to sets of cases, i.e., given $L{\subseteq}\mathcal{C}$, we have $\pi_d(L){=}\{\pi_d(c) \mid c{\in}L\}$.

Hierarchical Clustering. The framework proposed in this paper builds a hierarchical clustering of cases. A detailed description of hierarchical clustering is out of scope, i.e., we refer to [27] for an elaborate overview. Informally, a hierarchical clustering of a set of (sets of) elements defines a hierarchy of clusters based on these aforementioned (sets of) elements. Mathematically, a hierarchical clustering is a binary tree of some height h. The bottom layer of the hierarchy, i.e., the nodes at height 0, is defined by the individual elements of the set (or an initial clustering). The top layer of the hierarchy, i.e., the root at height h, is the complete set of elements. A cluster C_i at height $0{<}i{<}h$ connects to, i.e., is a super-set of, at least one cluster C_{i-1} at height $i-1$ and a cluster C_j at height $0{\leq}j{<}i$. Given a distance measure on the clusters, a cluster combining two other clusters implies that the distance among these two clusters is minimal.

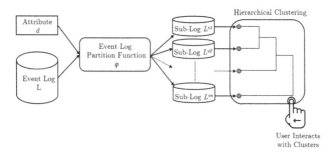

Fig. 1. Overview of the proposed framework. The event log is partitioned based on a user-defined data attribute $d{\in}\mathcal{D}$, i.e., by means of partition function φ. The corresponding resulting event logs, i.e., $L^{v_1}, L^{v_2}, ..., L^{v_n}$, are combined into a hierarchical clustering based on the behavior they describe.

3   Attribute-Driven Hierarchical Trace Clustering

Consider Fig. 1, in which we depict a schematic overview of the proposed framework for data-driven hierarchical clustering. Given *attribute* $d{\in}\mathcal{D}$ and event log $L{\subseteq}\mathcal{C}$, we partition the cases in the event log (using partition function φ). The partitioning groups cases together that have the same value for data attribute $d{\in}\mathcal{D}$. Hence, given that $|\pi_d(L)|{=}n$, the partitioning function φ yields n sub-logs

($L^{v_1}, ..., L^{v_n}$ in Fig. 1) that form a partition of the original event log. By definition, $|\pi_d(L')|=1$, $\forall L' \in \varphi(L, d)$, i.e., each sub-log defined by the partition is uniquely associated to one value $v \in \pi_d(L)$.

Given the initial partitioning as defined by $\varphi(L, d)$, we construct a *hierarchical clustering*, using the initial partitioning as a primary input, i.e., the hierarchy's nodes at height 0. To construct a hierarchy of clusters, we require a distance function, i.e., a *linkage criterion*, that allows us to compute the distance between the hierarchy clusters, i.e., the sub-logs. Formally, we require a linkage criterion of the form $\Delta \colon \mathcal{P}(\mathcal{C}) \times \mathcal{P}(\mathcal{C}) \to \mathbb{R}_{\geq 0}$. However, we are primarily interested in the linkage criterion in terms of the *control-flow behavior* of the cases, i.e., in terms of $\pi_{\texttt{trace}}(c) \in \mathcal{A}^*$. Using $\pi_{\texttt{trace}}(c) \in \mathcal{A}^*$, $\forall c \in \mathcal{C}$, we consider two forms of linkage criterion computation, i.e., *trace-based linkage computation* and *abstraction-based linkage computation*. Consider Fig. 2, in which we schematically depict the two different linkage computation strategies.

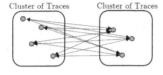

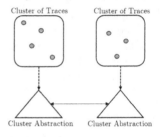

(a) *Trace-based linkage criterion computation*; Cluster distances are aggregations of the inter-trace distance of the elements of the different clusters.

(b) *Abstraction-based linkage criterion computation*; Inter-cluster distance is computed in terms of a distance based on abstractions of the cluster.

Fig. 2. Overview of the two different ways to compute linkage criteria considered.

In *trace-based linkage computation*, the linkage criterion Δ is computed by means of computing some aggregate over a trace-based distance function $\delta \colon \mathcal{A}^* \times \mathcal{A}^* \to \mathbb{R}_{\geq 0}$. Observe that a wide variety of such aggregates exists, e.g., *maximal linkage clustering* with, given $X, Y \subseteq \mathcal{C}$, $\Delta(X, Y) = \max\{\Delta(\pi_{\texttt{trace}}(c), \pi_{\texttt{trace}}(c')) \mid c \in X, c' \in Y\}$, etc. In *abstraction-based linkage computation*, we compute an abstraction of the behavior described by the collection of cases, which we subsequently use to compute the distance measure, e.g., by first translating the traces of each cluster into a prefix-tree and computing a distance function over the corresponding prefix trees.

4 Evaluation

Here, we evaluate the application of the framework. In Sect. 4.1, we present the implementation. In Sect. 4.2, we present the results of the evaluation.

4.1   Implementation

The proposed framework is implemented on top of the PM4Py [4] process mining library and is publicly available.[1] In the prototype (Fig. 3), the user can select which attribute to use as a main driver for the initial partitioning. In the left-hand side of the application, the hierarchical clustering is shown. The user can select different nodes of the clustering, to inspect the corresponding discovered process model, i.e., based on the event data belonging to the cluster. Different distance/linkage functions can be selected from the right-hand-side menu.

We used the implementation to conduct our experiments. In the implementation, several instantiations of the required distance/linkage functions, i.e., inter-trace distance function δ, linkage criteria and abstraction-based distances are available. We briefly describe these instantiations here.

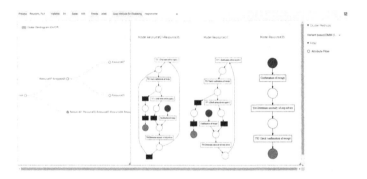

Fig. 3. Screenshot of an interactive prototype of the clustering framework.

Inter-Case Distance Metrics. Two different inter-case measures, i.e., instantiations of the δ-function, are available, i.e., *Levenshtein distance* and *Behavioral-Trace Distance*:

- *Levenshtein distance* [24](δ_L); Expresses the amount of edits (insertion, removal or replacements) we need, in order to transform a given sequence σ into another sequence σ'. The Levenshtein distance between $\langle a, b, c \rangle$ and itself is 0, whereas the Levenshtein distance between $\langle a, b, c \rangle$ and $\langle a, c \rangle$ is 1.
- *Behavioral-Trace Distance* (δ_B^α); Transforms two given traces into behavioral abstraction vectors, on which we compute a vector based distance. We consider two different behavioral vectors:
  - *Activity occurrence abstraction*, i.e., given $\sigma \in \mathcal{A}^*$, we define a vector $\vec{\sigma}_a \in \mathbb{N}^{\mathcal{A}}$, where $\vec{\sigma}_a(a) = |\{i \in \{1, ..., |\sigma|\} \mid \sigma(i) = a\}|$.

---

[1] https://github.com/caoyukun0430/pm4py-source/tree/yukun_paper and https://github.com/caoyukun0430/pm4py-ws/tree/dev-yukun.

- *Subsequent relation occurrence abstraction*, i.e., given $\sigma \in \mathcal{A}^*$, we define a vector $\vec{\sigma}_s \in \mathbb{N}^{\mathcal{A} \times \mathcal{A}}$, where $\vec{\sigma}_s(a, a') = |\{i \in \{1, ..., |\sigma| - 1\} \mid \sigma(i) = a \wedge \sigma(i+1) = a'\}|$.

Given the two aforementioned abstractions, and, parameter $\alpha \in [0, 1] \subseteq \mathbb{R}$, we define the behaviorally driven inter-trace distance function as:

$$\delta_B^\alpha(\sigma, \sigma') = \alpha \left(1 - \frac{\vec{\sigma}_a \cdot \vec{\sigma}_a'}{||\vec{\sigma}_a|| \; ||\vec{\sigma}_a'||} \right) + (1 - \alpha) \left(1 - \frac{\vec{\sigma}_s \cdot \vec{\sigma}_s'}{||\vec{\sigma}_s|| \; ||\vec{\sigma}_s'||} \right) \tag{1}$$

Trace-Based Linkage Criteria. Next to the two trace-based distance metrics, we implemented two trace-based linkage criteria, i.e., *unweighted average linkage clustering (UPGMA)* and *dual minimal match linkage criterion (DMM)*:

- *Unweighted Average Linkage Clustering (UPGMA)*; Given two clusters of cases, i.e., $L, L' \subseteq \mathcal{C}$, we compute:

$$\frac{1}{|L| \; |L'|} \sum_{c \in L} \sum_{c' \in L'} \delta \left(\pi_{\texttt{trace}}(c), \pi_{\texttt{trace}}(c') \right) \tag{2}$$

- *Dual Minimal Match (DMM)*; Given two clusters of cases, i.e., $L, L' \subseteq \mathcal{C}$, we compute:

$$\frac{1}{|L| + |L'|} \left(\sum_{c \in L} \min \left\{ \delta \left(\pi_{\texttt{trace}}(c), \pi_{\texttt{trace}}(c') \right) \mid c' \in L' \right\} + \sum_{c' \in L'} \min \left\{ \delta \left(\pi_{\texttt{trace}}(c), \pi_{\texttt{trace}}(c') \right) \mid c \in L \right\} \right) \tag{3}$$

Abstraction-Based Linkage Criteria. In the implementation, we consider the same abstraction as used in the inter-trace behavioral distance metric, i.e., behavioral abstraction vectors capturing activity occurrence and subsequent relations. However, we first aggregate the behavioral relations, i.e., based on the members of a cluster, prior to computing the behavioral vectors. or convenience, we formalize the two abstractions on the cluster-level.[2]

- *Activity occurrence abstraction*, i.e., given $C \in \mathcal{P}(\mathcal{A}^*)$, we define a vector $\vec{C}_a \in \mathbb{N}^{\mathcal{A}}$, where $\vec{C}_a(a) = \sum_{\sigma \in C} |\{i \in \{1, ..., |\sigma|\} \mid \sigma(i) = a\}|$.
- *Subsequent relation occurrence abstraction*, i.e., given $C \in \mathcal{P}(\mathcal{A}^*)$, we define a vector $\vec{C}_s \in \mathbb{N}^{\mathcal{A} \times \mathcal{A}}$, where $\vec{C}_s(a, a') = \sum_{\sigma \in C} |\{i \in \{1, ..., |\sigma| - 1\} \mid \sigma(i) = a \wedge \sigma(i+1) = a'\}|$.

Given the cluster vectors, we are able to compute the linkage criterion using the weighted cosine similarity-based distance metric, presented in Eq. 1.

In total, we consider 5 different instantiations of the framework, i.e., trace-based instantiations δ_L-UPGMA, δ_L-DMM, δ_B^α-UPGMA and δ_B^α-DMM and the vector-based abstraction-based linkage V-ABL.

---

[2] We formalize the abstractions on sets of sequences over $\mathcal{A}$, rather than elements $c \in \mathcal{C}$. Note that, given $c \in \mathcal{C}$, we are able to access the trace-view by means of $\pi_{\texttt{trace}}(c)$.

4.2   Results

Here, we present the results of the evaluation of applying attribute-driven hierarchical clustering. We briefly present qualitative results on real event data, after
which we quantitatively assess the approach on several real event logs.

Qualitative Evaluation on Real Event Data. In this section, we present the
qualitative clustering results of our framework applied to real event data. We
use the *Business Process Intelligence Challenge (BPIC)* 2017 Offer [14] event
data, which describes all the offers made for a *loan application processes.* We
select the *CreditScore* attribute containing 520 different values, ranging from 0
to 1145, since we are interested in discovering typical application behaviors on
customers under different credit scores.

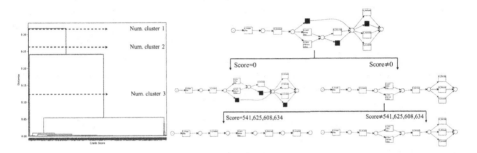

Fig. 4. Truncated dendrogram at the top three hierarchy levels with the corresponding
discovered process models ((BPIC) 2017 Offer [14] event data).

As depicted in Fig. 4, the algorithm separates the data into clusters with a
zero (0) and a *non-zero (≠0) Score* value. Corresponding event log sizes are
27735 and 15260 respectively. The underlying discovered models are relatively
similar (depicted on the right-hand side of Fig. 4), i.e., for Score values equal to
0, more skipping of activities is allowed. When we dive deeper into the data, we
observe that the *O_Cancelled* and the *O_Refused* activity dominate the sub-log
under the zero value (total occurrence 80.35%). At the same time, the *O_Accepted*
activity dominates the traces under non-zero values (94.38%). This difference
leads to the vast behavioral inter-cluster distances and explains the clustering
found. *In particular, this matches the expectation that people without a credit
score get rejected more often when applying for loans in reality.* As we traverse
the hierarchy further, we observe one outlier cluster containing only four cases
with credit scores "541, 608, 625, 634", in which only the single behavior *O_Sent
(online only)* is allowed. Compared to this behavioral cluster, the remaining
cluster excluding these four values shows a large variety of behavior. Therefore,
the clustering dendrogram in Fig. 4 provides us a clear understanding of different
customer behaviors under different credit scores as well as extracting outlier
behavior from the complete model.

Quantitative Evaluation on Real Event Data. In this section, we show the results of applying our implementation on a variety of different publicly available event logs. We first assess the impact of the different distance metrics and linkage criteria based on the "Receipt phase of an environmental permit application process ('WABO'), CoSeLoG project" event log ("Receipt log") [7]. Additionally, we assess how the technique performs on several different publicly available event logs.

In Fig. 5, we present average $F1$-, replay-fitness- and precision scores, obtained for the models discovered based on the Receipt log, for the five different implemented distance metrics and linkage criteria. Replay-fitness (range $[0, 1] \subseteq \mathbb{R}$) indicates how well a model describes a given event log (value 1 implies that the model describes all behavior in the event log). Precision (range $[0, 1] \subseteq \mathbb{R}$) quantifies the amount of additional behavior described by the model (value 1 implies that all described behavior is part of the input event log). The F1 score is the harmonic mean of replay-fitness and precision. We used the `responsible` attribute as a basis for the hierarchy, and, for abstraction based linkage, we use $\alpha = \frac{1}{2}$. The x-axis of the charts describes the number of separate clusters in the hierarchy, i.e., starting in the root of the hierarchy and traversing down. We observe an increase in the F1 score, which gradually decreases, yet, remains above the F1 score obtained for the model discovered based on the whole event log (the hierarchy's root). When going deeper into the hierarchy, the increase in the F1 score is explained by a higher increase in precision when compared to the corresponding decrease in replay-fitness. However, when traversing the hierarchy further, surprisingly, the replay-fitness slightly increases whereas the precision slightly drops. Upon inspection of the results, we observe that the higher levels of the hierarchy contain outlier clusters that have a relatively high corresponding precision value and relatively low replay-fitness value. However, deeper in the hierarchy, more similar clusters are "split-up" into the lower levels of the hierarchy. In some cases, the newly added fitness values in a deeper level in the hierarchy are higher than the average fitness values of the current level of the hierarchy, leading to a higher average fitness value in the next layer.

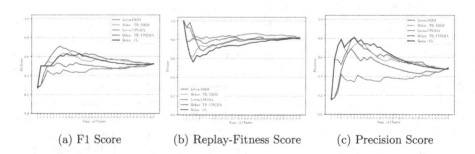

(a) F1 Score   (b) Replay-Fitness Score   (c) Precision Score

Fig. 5. Average F1, replay-fitness and precision scores obtained for the models discovered based on the Receipt log [7], for the different distance metrics and linkage criteria as described in this paper.

The same holds, symmetrically, for the corresponding precision values, explaining the results in Fig. 5.

In Fig. 6, we present the results F1, replay-fitness and precision scores obtained for the models discovered based on different publicly available event logs, using V-ABL. Observe that, in the plots provided here, we only plot the highest 23 levels of the hierarchy (in most cases the hierarchies tend to be higher). The event logs considered are BPIC 2012 [13], three different sub-logs (Control, Geo and Payment) of BPIC 2018 [15] and the Receipt log (also used in the previous section). Selection of the trace attributes is based on manual inspection of the results. We observe similar results on the different logs, i.e., slight increases of the F1 scores due to increases in precision and smaller decrease (and stabilizing) fitness values. Again, stabilization of the fitness values is due to the initial identification of outlier clusters.

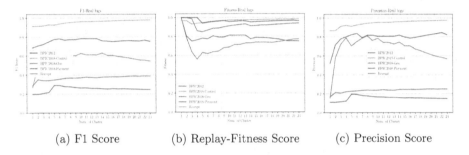

(a) F1 Score   (b) Replay-Fitness Score   (c) Precision Score

Fig. 6. Average F1, replay-fitness and precision scores obtained for the models discovered based on different publicly available event logs, based on abstraction-based linkage criterion. In all cases, we observe a slight increase in the F1 score, primarily driven by increases in precision.

5   Related Work

An overview of the field of process mining is out of the scope of this paper, i.e., we refer to [1] for an introduction to the field. In the remainder, we explicitly focus on related work in the area of *trace clustering*. In particular, we briefly provide a generalized overview of different existing clustering techniques, i.e., in general, the majority of the different clustering techniques proposed can be integrated into the framework presented in this paper.

Several authors focus on trace clustering based on a *feature vector based distance* [3, 6, 11, 12, 20, 22, 23, 25, 26, 28, 29]. These techniques consider different ways to extract specific features from the activity traces, which are subsequently translated into a feature vector. Examples of such features are event frequency, direct succession frequency, etc. For a given feature vector, different linkage criteria and clustering algorithms can be used to compute the final clustering. Alternatively the *syntactic distance* between traces, e.g., Levensthein, is used [5].

Another commonly used type of clustering focuses directly on *models*, rather than individual traces [9,11,19,21,25,30,31]. The goal of these methods here is to obtain a better model (i.e., higher precision, fitness) by merging traces into clusters. Typically traces are added into a cluster if they improve the quality of the cluster-based process model.

6   Conclusion

Processes executed in companies, hospitals, etc. are often performed in the context of an *execution context*, e.g., a customer type. Often, slight variations exist in the execution of the process for the different execution contexts. Manual comparison of the differences in execution, based on event data stored in event logs, is no longer feasible for more significant numbers of execution contexts. Hence, in this paper, we presented a hierarchical trace clustering framework that allows us to perform behavioral trace clustering over groups of traces, that share common data attributes. We evaluated the approach using several real data sets, based on several different behavioral comparison techniques. Our evaluation shows that the models described by the different clusters of the hierarchy are of better quality compared to the models discovered on the event data as a whole.

References

1. Aalst, W.: Data science in action. Process Mining, pp. 3–23. Springer, Heidelberg (2016). https://doi.org/10.1007/978-3-662-49851-4_1
2. Augusto, A., et al.: Automated discovery of process models from event logs: review and benchmark. IEEE Trans. Knowl. Data Eng. **31**(4), 686–705 (2019)
3. Bae, J., Caverlee, J., Liu, L., Yan, H.: Process mining by measuring process block similarity. In: Eder, J., Dustdar, S. (eds.) BPM 2006. LNCS, vol. 4103, pp. 141–152. Springer, Heidelberg (2006). https://doi.org/10.1007/11837862_15
4. Berti, A., van Zelst, S.J., van der Aalst, W.M.P.: Process mining for python (PM4Py): bridging the gap between process-and data science. In: Proceedings of the ICPM Demo Track 2019, co-located with 1st International Conference on Process Mining (ICPM 2019), Aachen, Germany, June 24–26, 2019, pp. 13–16 (2019)
5. Bose, R.J.C., Van der Aalst, W.M.: Context aware trace clustering: towards improving process mining results. In: Proceedings of the 2009 SIAM International Conference on Data Mining, pp. 401–412. SIAM (2009)
6. Bose, R.P.J.C., van der Aalst, W.M.P.: Trace clustering based on conserved patterns: towards achieving better process models. In: Rinderle-Ma, S., Sadiq, S., Leymann, F. (eds.) BPM 2009. LNBIP, vol. 43, pp. 170–181. Springer, Heidelberg (2010). https://doi.org/10.1007/978-3-642-12186-9_16
7. Buijs, J.: Receipt phase of an environmental permit application process ('wabo'), coselog project (2014). https://doi.org/10.4121/UUID:A07386A5-7BE3-4367-9535-70BC9E77DBE6
8. Buijs, J.: Environmental permit application process ('WABO'), CoSeLoG project. Eindhoven Univ. Technol. Dataset (2014). https://doi.org/10.4121/uuid:26aba40d-8b2d-435b-b5af-6d4bfbd7a270

9. Cadez, I., Heckerman, D., Meek, C., Smyth, P., White, S.: Model-based clustering and visualization of navigation patterns on a web site. Data Mining Knowl. Disc. **7**(4), 399–424 (2003)

10. Conforti, R., La Rosa, M., ter Hofstede, A.H.M.: Filtering out infrequent behavior from business process event logs. IEEE Trans. Knowl. Data Eng. **29**(2), 300–314 (2017)

11. De Koninck, P., Nelissen, K., Baesens, B., vanden Broucke, S., Snoeck, M., De Weerdt, J.: An approach for incorporating expert knowledge in trace clustering. In: Dubois, E., Pohl, K. (eds.) CAiSE 2017. LNCS, vol. 10253, pp. 561–576. Springer, Cham (2017). https://doi.org/10.1007/978-3-319-59536-8_35

12. de Medeiros, A.K.A., Guzzo, A., Greco, G., van der Aalst, W.M.P., Weijters, A.J.M.M., van Dongen, B.F., Saccà, D.: Process mining based on clustering: a quest for precision. In: ter Hofstede, A., Benatallah, B., Paik, H.-Y. (eds.) BPM 2007. LNCS, vol. 4928, pp. 17–29. Springer, Heidelberg (2008). https://doi.org/10.1007/978-3-540-78238-4_4

13. van Dongen, B.F.: Bpi challenge **2012** (2012). https://doi.org/10.4121/UUID:3926DB30-F712-4394-AEBC-75976070E91F

14. van Dongen, B.F.: Bpi challenge **2017** (2017). https://doi.org/10.4121/UUID:5F3067DF-F10B-45DA-B98B-86AE4C7A310B

15. van Dongen, B.F., Borchert, F.: Bpi challenge **2018** (2018). https://doi.org/10.4121/uuid:3301445f-95e8-4ff0-98a4-901f1f204972

16. Sani, M.F., van Zelst, S.J., van der Aalst, W.M.P.: Improving process discovery results by filtering outliers using conditional behavioural probabilities. In: Teniente, E., Weidlich, M. (eds.) BPM 2017. LNBIP, vol. 308, pp. 216–229. Springer, Cham (2018). https://doi.org/10.1007/978-3-319-74030-0_16

17. Fani Sani, M., van Zelst, S.J., van der Aalst, W.M.P.: Applying sequence mining for outlier detection in process mining. In: CoopIS, C&TC, and ODBASE 2018, Valletta, Malta, October 22–26, 2018, Proceedings, Part II, pp. 98–116 (2018)

18. Fani Sani, M., van Zelst, S.J., van der Aalst, W.M.P.: Repairing outlier behaviour in event logs. In: Abramowicz, W., Paschke, A. (eds.) BIS 2018. LNBIP, vol. 320, pp. 115–131. Springer, Cham (2018). https://doi.org/10.1007/978-3-319-93931-5_9

19. Ferreira, D., Zacarias, M., Malheiros, M., Ferreira, P.: Approaching process mining with sequence clustering: experiments and findings. In: Alonso, G., Dadam, P., Rosemann, M. (eds.) BPM 2007. LNCS, vol. 4714, pp. 360–374. Springer, Heidelberg (2007). https://doi.org/10.1007/978-3-540-75183-0_26

20. Greco, G., Guzzo, A., Pontieri, L., Sacca, D.: Discovering expressive process models by clustering log traces. IEEE Trans. Knowl. Data Eng. **18**(8), 1010–1027 (2006)

21. Hompes, B., Buijs, J., Van der Aalst, W., Dixit, P., Buurman, J.: Discovering deviating cases and process variants using trace clustering. In: 27th Benelux Conference on Artificial Intelligence (BNAIC), November, pp. 5–6 (2015)

22. Jung, J.-Y., Bae, J.: Workflow clustering method based on process similarity. In: Gavrilova, M.L., Gervasi, O., Kumar, V., Tan, C.J.K., Taniar, D., Laganá, A., Mun, Y., Choo, H. (eds.) ICCSA 2006. LNCS, vol. 3981, pp. 379–389. Springer, Heidelberg (2006). https://doi.org/10.1007/11751588_40

23. Jung, J.Y., Bae, J., Liu, L.: Hierarchical clustering of business process models. Int. J. Innovative Comput. Inf. Control **5**(12), 1349–4198 (2009)

24. Levenshtein, V.I.: Binary codes capable of correcting deletions, insertions, and reversals. Soviet Phys. Doklady. **10**, 707–710 (1966)

25. Lu, X., Tabatabaei, S.A., Hoogendoorn, M., Reijers, H.A.: Trace clustering on very large event data in healthcare using frequent sequence patterns. In: Hildebrandt, T., van Dongen, B.F., Röglinger, M., Mendling, J. (eds.) BPM 2019. LNCS, vol. 11675, pp. 198–215. Springer, Cham (2019). https://doi.org/10.1007/978-3-030-26619-6_14

26. Luengo, D., Sepúlveda, M.: Applying clustering in process mining to find different versions of a business process that changes over time. In: Daniel, F., Barkaoui, K., Dustdar, S. (eds.) BPM 2011. LNBIP, vol. 99, pp. 153–158. Springer, Heidelberg (2012). https://doi.org/10.1007/978-3-642-28108-2_15

27. Murtagh, F., Contreras, P.: Algorithms for hierarchical clustering: an overview, II. Wiley Interdiscip. Rev. Data Min. Knowl. Discov. **7**(6) (2017)

28. Song, M., Günther, C.W., van der Aalst, W.M.P.: Trace clustering in process mining. In: Ardagna, D., Mecella, M., Yang, J. (eds.) BPM 2008. LNBIP, vol. 17, pp. 109–120. Springer, Heidelberg (2009). https://doi.org/10.1007/978-3-642-00328-8_11

29. Song, M., Yang, H., Siadat, S.H., Pechenizkiy, M.: A comparative study of dimensionality reduction techniques to enhance trace clustering performances. Expert Syst. Appl. **40**(9), 3722–3737 (2013)

30. Sun, Y., Bauer, B.: A novel top-down approach for clustering traces. In: Zdravkovic, J., Kirikova, M., Johannesson, P. (eds.) CAiSE 2015. LNCS, vol. 9097, pp. 331–345. Springer, Cham (2015). https://doi.org/10.1007/978-3-319-19069-3_21

31. De Weerdt, J., vanden Broucke, S.K.L.M., Vanthienen, J., Baesens, B., : Active trace clustering for improved process discovery. IEEE Trans. Knowl. Data Eng. **25**(12), 2708–2720 (2013)

Process Outcome Prediction: CNN vs. LSTM (with Attention)

Hans Weytjens[(⊠)] and Jochen De Weerdt

Research Centre for Information Systems Engineering (LIRIS), KU Leuven,
Leuven, Belgium
{hans.weytjens,jochen.deweerdt}@kuleuven.be

Abstract. The early outcome prediction of ongoing or completed processes confers competitive advantage to organizations. The performance of classic machine learning and, more recently, deep learning techniques such as Long Short-Term Memory (LSTM) on this type of classification problem has been thorougly investigated. Recently, much research focused on applying Convolutional Neural Networks (CNN) to time series problems including classification, however not yet to outcome prediction. The purpose of this paper is to close this gap and compare CNNs to LSTMs. Attention is another technique that, in combination with LSTMs, has found application in time series classification and was included in our research. Our findings show that all these neural networks achieve satisfactory to high predictive power provided sufficiently large datasets. CNNs perform on par with LSTMs; the Attention mechanism adds no value to the latter. Since CNNs run one order of magnitude faster than both types of LSTM, their use is preferable. All models are robust with respect to their hyperparameters and achieve their maximal predictive power early on in the cases, usually after only a few events, making them highly suitable for runtime predictions. We argue that CNNs' speed, early predictive power and robustness should pave the way for their application in process outcome prediction.

Keywords: Process mining · Outcome prediction · Neural networks · LSTM · Attention · Convolutional Neural Networks

1 Introduction

Every organization will gain considerable advantage from the early outcome prediction of its ongoing processes: they can attempt to influence undesired outcomes for the better, make well-informed decisions further down the value chain, or realize efficiency gains. Process mining is the exercise of extracting information from event logs stored in computer systems with the aim of discovering or improving them. Within this realm, predictive process monitoring focuses on making predictions and has already applied a plethora of machine learning approaches, achieving varying degrees of success. Teinemaa et al. [10] benchmark several classical approaches such as random forests, gradient boosted trees (both based on decision trees), logistic regression and support vector machines using

© Springer Nature Switzerland AG 2020
A. Del Río Ortega et al. (Eds.): BPM 2020 Workshops, LNBIP 397, pp. 321–333, 2020.
https://doi.org/10.1007/978-3-030-66498-5_24

various datasets. These classical machine techniques sometimes rely heavily on manual feature engineering to represent the data, which is far from a trivial task. Deep learning techniques have enjoyed remarkable successes automatically representing data as a hierarchy of useful features. This led to a growing body of predictive process monitoring applications. Both Evermann et al. [4] and Tax et al. [9] use Long Short-Term Memory neural networks (LSTMs) to predict next events and time stamps in business processes. Camargo et al. [3] further refined this approach. Hinkka et al. [6] were amongst the first to apply them in process outcome prediction. Kratsch et al. [7] included LSTM in their comprehensive comparison of deep learning and classical approaches for outcome prediction. LSTMs sometimes suffer from their limited memory capacity, an issue that Bahdanau et al. [1] address with the Attention mechanism. Wang et al. [11] used LSTMs with Attention in their outcome prediction benchmarking study that also included bidirectional LSTMs and classic approaches. Convolutional Neural Networks (CNNs) work with fixed-size, spatially-organized data and are often associated with computer vision. Nevertheless, one-dimensional CNNs are also utilized for time series classification or sequence modeling [2]. Fawaz et al.'s [5] large-scale empirical study of deep learning methods for time series classification includes CNNs but not LSTMs. Pasqualdibisceglie et al. [8], however, opted for an original two-dimensional CNN approach to predict next events in processes.

Table 1. Related research in process mining and positioning of this paper.

Paper	Classic machine learning	LSTM	LSTM attention	CNN	Outcome prediction	Early prediction
Teinemaa [10]	x	–	–	–	x	x
Evermann [4]	–	x	–	–	–	–
Tax [9]	–	x	–	–	–	x
Camargo [3]	–	x	–	–	–	–
Hinkka [6]	–	x	–	–	x	x
Kratsch [7]	x	x	–	–	x	x
Wang [11]	x	x	x	–	x	x
Pasqual. [8]	–	x	–	2D	–	–
This paper	–	x	x	x	x	x

To the best of our knowledge, there is no published research investigating CNNs in process outcome prediction. Table 1 relates our paper to the research described above. It concentrates on comparing CNNs to LSTMs with and without Attention. We not only investigate classifying completed processes, but also look at predicting the outcomes of ongoing processes. We find that neural networks can make accurate and early predictions, provided the datasets are large enough. CNNs are much faster than LSTMs, but deliver very similar results. Attention does not improve the plain-vanilla LSTMs. All three models prove to be robust, to be relatively insensitive to hyperparameter changes. Intriguingly, the time-related features and even the ordering of the events seem to play a minor role at best for the quality of the models.

2   Solving the Learning Problem

Our objective, to benchmark different neural networks against each other, guided the methodological choices described in this section. The choices we made will not necessarily be optimal for any given learner on any given dataset. All datasets are event logs describing processes, often called cases or traces. These processes consist of events. A number of attributes, also called features or variables, describe the events. Every case is associated with a binary outcome, also called class or target, e.g. 'approved' vs. 'non-approved' in the case of a loan application process. In this paper, we use the words 'cases', 'events', 'features' and 'targets'. The word 'prefix' refers to ongoing, incomplete cases.

The learning problem is essentially to train a learner using a training dataset containing events, described by their features and organized in cases that are labeled with targets, with the goal of predicting the targets of unseen cases (complete or ongoing).

2.1   Models

Recurrent Neural Networks (RNN) are neural networks specifically designed to handle sequences of variable length. Since process mining datasets are organized in cases containing a variable number of events that can be chronologically sorted using their respective time stamps, RNNs are intuitively the first choice when applying neural networks to process prediction problems as testified by their extensive use in the form of LSTMs (the most prominent RNNs, specifically designed to treat longer-term dependencies) in [3,4,9].

An LSTM processes every sequence of events it is presented one time step at a time. At any given time step, it will pass a vector (aka 'state of the memory cell') containing information about the current and previous time steps to the next time step, until reaching the last one (as depicted by the dotted horizontal arrows in Fig. 1(a)) whose output is propagated to the next layer. The vector's fixed size, however, will inevitably limit its informational content. Especially for longer sequences, the information from the earlier inputs (events) risks dilution or loss. The Attention Mechanism shown in Fig. 1(b) was proposed to overcome this problem by retaining the outputs of all nodes in the hidden layer, scoring them, and then calculating a weighted average of the outputs using these scores to compute the final outcome of the model (lines from the nodes in the hidden layer to the output).

In contrast to LSTMs, Convolutional Neural Networks work with fixed-sized, spatially-organized data. A series of alternating convolution layers applying weight-sharing filters and dimension-reducing pooling layers enables the models to automatically recognize patterns and extract features from the input data. These features are then passed to a series of dense layers for classification (or regression). Two-dimensional CNNs are very commonly used in computer vision applications. Interpreting time as a spatial dimension, one-dimensional CNNs can be applied to sequence processing as well. Figure 1(c) shows such a 1-D CNN with the filters striding along the temporal axis.

2.2   Preprocessing

The data we used and the targets (outcomes that are based on certain events in the cases) we defined are described in Sect. 3. We labeled the cases by adding a target column to the original datasets and then clipped every case just before the event indicating its target value. To improve comparability, possibly at the detriment of the final result, we decided against incorporating any human domain knowledge. We made an exception for the following synthetic features, the same for all datasets, which were calculated as shown in Table 2.

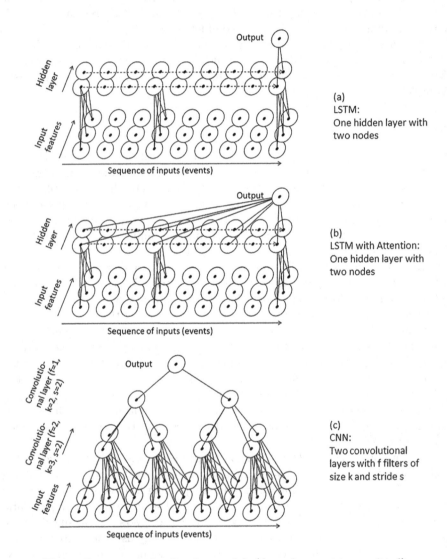

Fig. 1. Conceptual visualization models (dense layers at top omitted).

Since our datasets contain both range (e.g. 'nr_open') and categorical features (e.g. 'month'), we proceeded to map the labels of the categorical features to integers. All our models include an embedding layer, which maps these integer values into vectors that should ideally be similar for values with comparable properties. The length of these vectors was set to be one-fifth of the respective feature's vocabulary size (nearly always far below 10, rarely above), thereby realizing a substantial dimensionality reduction compared to one-hot encoding. The embeddings themselves are learned by the model. Finally, we also standardized (mean of zero and standard deviation of one) the range features.

Table 2. The synthetic features.

Name	Type	Explanation
nr_open	Range	nr. of open cases at time of every event's time stamp (=load)
Elapsed	Range	Time elapsed since start of case, marked by its first event
evTime	Range	Time since last event (0 for first event in the case)
SinceMidnight	Range	Time elapsed since midnight of previous day
Month	categ	Month of year
Day	categ	Day of month
Hour	categ	Hour of day
evNR	Range	Order nr of event in case

2.3 Feeding the Models

Since both types of LSTM learners require time sequences as inputs, we combined the vectors containing the features for all events in a given sequence into matrices with shape (sequence length x number of features). This was done by sliding a sequence-length-wide window over the events in a case. The first window only covered the first event in the case, the second one covered two, etc. The last window covered the case's sequence-length last number of events. Figure 2 visualizes this step, showing the matrix formed by the vectors associated with events 1–10.

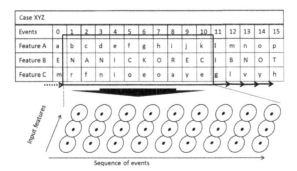

Fig. 2. A window with (sequence) length nine sliding over a sixteen-event case forms the input matrices for the model.

Padding was used to fill in the missing values for short cases and for the first (sequence-length - 1) windows of a case. Despite their one-dimensional character suggesting otherwise, our CNNs work with the same two-dimensional input data. The one-dimensionality refers to the filters covering the whole width (number of features) and striding in one direction only, along the longitudinal axis of the matrix (sequence length). In this fashion, every case produced a number of matrices in the processed dataset equal to its number of events.

3   Experimental Evaluation

In this section, we first present the datasets we chose for our experiments. We then describe the implementation of the experiments, including splitting the datasets into training, validation and test sets, hyperparameter tuning, measures against overfitting and physical infrastructure. Before finally describing our results, we devote a short paragraph to the metrics used.

3.1   Datasets

We selected a number of publicly available and widely-used datasets[1] containing both range and categorical features. To discover the suitability of deep learning models in different environments, we created variety by including small and large datasets, with short and long traces, varying degrees of class balance and of lower and higher quality:

- **BPIC_2012:** This dataset describes a loan application process at a Dutch bank. Every case has three possible targets: 'approved', 'declined' or 'canceled'. This multi-classification problem was transformed into three different binary classification problems for our purposes.
- **BPIC_2017:** This dataset is a higher-quality version of BPIC_2012 with both more examples and features that should facilitate better predictions. Note that the sum of the three targets exceeds 100% as sometimes 'canceled' cases are restarted and 'approved' or 'declined' cases become 'canceled' later.
- **Traffic fines:** Is an event log of a system managing traffic road fines. Fines are either paid in full or sent for credit collection. The latter is our target ('deviant'). Cases in this dataset are very short (avg. 3.3 events).
- **Sepsis cases:** Describes the pathways of patients through a hospital. We define three targets:
  - 28_days_EM: is the patient admitted to the emergency rooms within twenty eight days of his/her release from the hospital?
  - IC: does the patient enter the intensive care unit?
  - no_A_release: is the patient eventually released from the hospital for another reason than the most frequent 'A'?

---

[1] All datasets can be found at https://data.4tu.nl/repository/collection:event_logs_real (4TU Centre for Research Data).

Table 3. Statistics of the used datasets.

Dataset	Target (binary)	#Events	#Cases	Min events	Max events	Mean events	Med events	Pos events	Pos cases	#categ features	#range features
BPIC_	Approved	219,858	12,688	2	172	17.3	8	.40	.18	3	2
2012	Declined	219,858	12,688	2	172	17.3	8	.26	.60	3	2
	Canceled	219,858	12,688	2	172	17.3	8	.33	.22	3	2
BPIC_	Approved	1,071,054	31,417	7	175	34.1	30	.65	.55	11	6
2017	Declined	1,071,054	31,417	7	175	34.1	30	.12	.12	11	6
	Canceled	1,071,054	31,417	7	175	34.1	30	.44	.50	11	6
Traffic	Deviant	496,067	149,958	1	20	3.3	4	.48	.39	8	4
Sepsis	28_days	13,095	781	5	185	16.8	14	.16	.14	25	4
Sepsis	IC	10,841	781	3	60	13.9	13	.09	.14	25	4
Sepsis	no_A	13,182	781	5	183	16.9	14	.16	.14	25	4

Table 3 provides an overview of the used datasets, some of which contain highly imbalanced classes. In line with our decision to apply an identical methodology to all datasets, we decided against sampling techniques to restore balance. This negatively impacted results, especially for BPIC_2017 (declined) and Sepsis.

3.2 Implementation

We first preprocessed the data and reshaped it to fit our models as described in Sect. 2.2 and 2.3. A test set comprising the chronologically last 20% cases of each dataset was set apart for the final evaluation of the models' predictions. We ran every model/dataset combination 50 times with different values for three hyperparameters: sequence length, batch size and model size. The latter was accomplished by multiplying every layer's number of nodes by a multiplication factor. As the CNN models had one additional hyperparameter, kernel size, they were run 100 times. Prior to every run, the input data was reshuffled before separating a training set of 80% from a validation set of 20% of the examples. To avoid overfitting, we used an automatic stopping mechanism to halt training after five epochs without improvement of the metric on the validation set. Each experiment was run on the Google Cloud. Polyaxon running above Kubernetes allowed for parallel execution on multiple two-core, 13 GB, 2.0 GHz Intel Xeon Scalable Processors (Skylake) and either NVIDIA Tesla K80 or P100 (for BPIC_2017) GPUs.

3.3 Metrics

Given the class imbalance in some of our datasets and the binary nature of the targets, we opted for the area under the curve ROC (AUC_ROC) metric to evaluate our models. Since AUC_ROC yields a non-differentiable loss curve, we had to resort to accuracy to train our models. At the end of each epoch, we computed the AUC_ROC on the validation set for the early-stopping mechanism. After training, the model with the highest AUC_ROC score on the validation set was withheld to make predictions on the test set, on which we report below in this paper. We also calculated other metrics: F1-score, accuracy and AUC_PR (area under the precision-recall curve) and recorded computation times.

3.4   Results

The evaluation of our experiments' results allows us to formulate answers to the following questions: Which models perform better on what kind of datasets? What are the speed differences between the models? How robust are they with respect to their hyperparameters? Are early runtime predictions possible? We also make an observation about the relevance of timestamps.

CNNs Perform Comparably to LSTMs with and without Attention. Table 4 shows that neural networks can deliver useful results for large enough datasets (BPIC_2017, Traffic and BPIC_2012). However, none of our models could cope with the combined challenge of short, imbalanced datasets and many (sparse and possibly correlated) variables found in Sepsis. Whilst the BPIC_2012 predictions still beckon a great deal of caution (the F1-scores are extremely low for 'canceled'), the BPIC_2017 results are vastly superior and clearly demonstrate the benefits of improved data collection by the user. The lower F1- and AUC_PR-scores for 'declined' versus 'approved' and 'canceled' in all likelihood result from the dataset's class imbalance. Despite the extreme shortness of the cases in the Traffic dataset, the results are significantly better than random guesses. The three different models deployed performed equally well, suggesting they all manage to extract the same information from the data. CNNs proved to be a match for the state-of-the-art LSTM, confirming the findings of [2,5]. In our setting, no benefit was derived by adding an attention layer to the LSTM models.

CNNs Train Much Faster than LSTMs. Training times matter, especially since concept drift will require frequent training of models in production. CNN models remarkably outrun the other models during training—sometimes by over one order of magnitude—making them the model of choice for practitioners. Prediction times are negligible. Training times depend heavily on model architecture decisions, and hardware infrastructure, and are stochastic by nature. In our implementation as described in Subsect. 3.2, the run times of the fastest models were 23–60, 380–500, 270, and 17–40 s for BPIC_2012, BPIC_2017, Traffic and Sepsis respectively (BPIC_2017 was run on faster GPUs). Hyperparameter tuning multiplies total training times by the number of trainings performed.

All Models are Robust with Respect to their Hyperparameters. As for the hyperparameters, there is no consensus on the batch sizes: all available sizes (128, 256, 512, 1024) are used by the best models. One could expect longer sequence lengths to support better predictions for datasets with longer cases. Indeed, the sequence length is consistently about 1.5 times the median number events per case for the BPIC_2017 and Traffic datasets. This is less clear for BPIC_2012. We used a multiplication factor (between one and sixteen) to

Table 4. Results from best models in hyperparameter space, aggregated over all prefix lengths. 'Rel. Time' is the relative run time compared to the fastest run time for the respective experiment which is set at 100%.

Dataset	Target	Model	AUC_ROC	F1-score	Accu-racy	AUC_PR	Rel. Time	Batch Size	Seq. Length	Kernel Size	Model Size
BPIC_2012	approved	LSTM	0.79	0.67	0.72	0.74	668%	512	45		16
		Att	0.79	0.66	0.72	0.74	552%	512	35		16
		CNN	**0.80**	0.69	0.72	0.75	100%	256	15	8	16
	declined	LSTM	**0.76**	0.59	0.76	0.63	443%	1024	35		16
		Att	0.75	0.59	0.75	0.63	160%	512	5		8
		CNN	**0.76**	0.59	0.75	0.62	100%	512	35	4	8
	canceled	LSTM	**0.75**	0.35	0.79	0.50	1158%	128	5		2
		Att	**0.75**	0.35	0.79	0.50	1352%	128	5		4
		CNN	0.74	0.26	0.79	0.48	100%	1024	5	2	16
BPIC_2017	approved	LSTM	**0.93**	0.88	0.83	0.96	909%	128	47		4
		Att	**0.93**	0.88	0.84	0.97	946%	128	47		8
		CNN	**0.93**	0.88	0.83	0.97	100%	128	47	2	4
	declined	LSTM	**0.91**	0.66	0.92	0.69	200%	1024	47		16
		Att	**0.91**	0.66	0.92	0.67	240%	1024	47		2
		CNN	**0.91**	0.67	0.92	0.69	100%	1024	47	2	4
	canceled	LSTM	**0.92**	0.79	0.82	0.91	294%	1024	47		1
		Att	**0.92**	0.77	0.82	0.91	356%	512	47		4
		CNN	**0.92**	0.80	0.82	0.91	100%	256	47	8	4
Traffic	deviant	LSTM	0.75	0.66	0.66	0.63	170%	512	8		2
		Att	**0.77**	0.67	0.68	0.66	100%	512	6		2
		CNN	0.76	0.68	0.69	0.64	132%	128	6	3	2
Sepsis	28_days_EM	LSTM	0.51	0.03	0.84	0.10	845%	256	45		1
		Att	0.50	0.07	0.82	0.10	424%	128	5		2
		CNN	0.47		0.89	0.11	100%	512	25	4	4
	IC	LSTM	0.65	0.29	0.89	0.31	191%	256	45		4
		Att	0.64	0.34	0.89	0.34	125%	256	15		16
		CNN	**0.69**		0.90	0.16	100%	512	45	2	8
	no_A_release	LSTM	0.55	0.34	0.79	0.32	447%	128	45		4
		Att	0.56	0.27	0.78	0.32	100%	256	5		4
		CNN	0.57	0.30	0.79	0.34	103%	128	5	2	2

generate models of different widths. A preference for a certain size cannot be deduced from our experiments. The same holds true for the CNNs' kernel size. Fortunately, these hyperparameter differences do not oblige the practitioner to engage in very extensive hyperparameter tuning. The models are robust with respect to them as visualized in Fig. 3, where the AUC_ROC values hardly move with changes in the individual hyperparameters.

Early Runtime Predictions are Possible. Figure 4 provides insight into the earliness of learning based on the example of the BPIC_2012 dataset. The test set was sorted according to the prefix length of the (incomplete) cases. When basing predictions on up to the ten very first events of cases as in the left column of the figure, a pattern emerges: the outcomes predictions for all models are best for prefix lengths of three and four! Thus, early runtime predictions are possible as most of the information about case outcomes resides in the first few events.

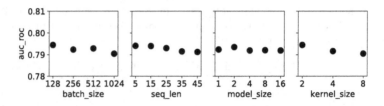

Fig. 3. Hyperparameter tuning (BPIC_2012 approved, CNN, average values for all relevant samples in 100 experiments).

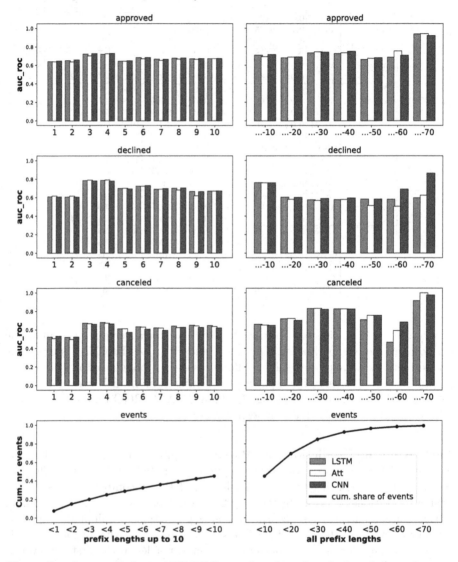

Fig. 4. Runtime predictions: AUC_ROC as a function of prefix length (nr. of events since start of case considered) for 3 BPIC_2012 datasets (max. prefix length of 70).

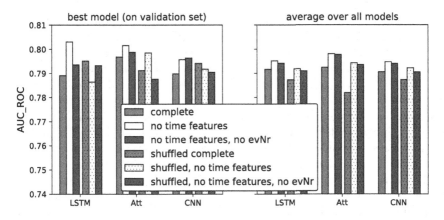

Fig. 5. Time-related features and events order and BPIC_2012 (approved) predictions. (Color figure online)

Waiting for more events to materialize does not seem to pay off for BPIC_2012, as shown in the second column of Fig. 4 where up to 70 events are considered[2].

Do Timestamps Matter? The aforementioned observations, combined with the CNNs' strong performance and the robustness with regards to sequence length as described earlier, lead to the conclusion that the characteristic timing feature of process mining problems does not play an important role in outcome prediction, at least not for the BPIC_2012 dataset (approved) shown in Fig. 5. The left-hand graph shows the results for the models that scored best on the validation sets as before. For more general conclusions, we enlarged the sample to all models trained in the hyperparameter space on the right side. The results without timestamp-related features always mildly outperform the base case (white better than grey). The effects of dropping evNr (the feature indicating the event's order in the case, red bars) and of shuffling (dotted bars) are not statistically significant. Apparently, the models are not learning from timestamp-related data, all information within the datasets is stored in the other features, an observation also made by [10].

4   Conclusion and Future Work

We found neural networks to be a useful tool for process outcome prediction, given sufficiently large datasets. CNNs deliver the same results as the state-of-the-art LSTMs at a fraction of the time and can therefore be recommended as first choice for practitioners. We found no benefit in the use of the Attention mechanism in the LSTM models. The comparison of BPIC_2017 and BPIC_2012

---

[2] Notice that the (weighted) average of the AUC_ROC of subsets will not necessarily match the AUC_ROC of the total set, as observed here.

clearly demonstrates how organizations can benefit from improved data collection. Based on further exploration of the results on the BPIC_2012 dataset, neural networks turned out to be robust with regards to their hyperparameters as well. The models often nearly reached full predictability after observing only the first few events of a case, suggesting that events critical to determining the case outcome often appear early. These factors support the usability of the models in practice, both for completed and (young) ongoing cases. The aim of this paper was to compare methods; we did not seek optimal predictions. Therefore, the published results would probably improve from balancing classes in the datasets and applying some domain-specific knowledge among other things.

Our conclusions on the models' performance and speed were based on experiments on several datasets with consistent results, and hence generalizable. However, the BPIC_2012 dataset is not necessarily representative of all process mining datasets. Therefore, future research considering a wider range of datasets could solidify our conclusions about hyperparameters, runtime predictions and timestamps. We expect most of our conclusions to be valid for next-event and duration prediction problems, as the approach and models used would be identical, but only further experiments could confirm that. The observed irrelevance of the timestamp-related features warrants further inquiry. Our preliminary research suggests that under the right circumstances, neural networks outperform the classic methods in [10] to the order of a few percentage points on average. Further investigation is required to strengthen these findings and derive recommendations for the deployment of the networks. Offering insights into how the neural networks' results were reached would contribute to the practitioner's confidence in them. Finally, one could gain deeper insight into the uncertainty of the model predictions (e.g., by using Bayesian techniques), to account for the probabilistic nature of the input data and model weights.

References

1. Bahdanau, D., Cho, K.H., Bengio, Y.: Neural machine translation by jointly learning to align and translate. In: Bengio, Y., LeCunn, Y. (eds) 3rd International Conference on Learning Representations ICLR 2015. Conference Track Proceedings (2015). https://dblp.org/rec/bib/journals/corr/BahdanauCB14
2. Bai, S.J., Kolter, J.Z., Koltun, V.: An empirical evaluation of generic convolutional and recurrent networks for sequence modeling. arXiv:1803.01271v2 (2018)
3. Camargo, M., Dumas, M., González-Rojas, O.: Learning accurate LSTM models of business processes. In: Hildebrandt, T., van Dongen, B.F., Röglinger, M., Mendling, J. (eds.) BPM 2019. LNCS, vol. 11675, pp. 286–302. Springer, Cham (2019). https://doi.org/10.1007/978-3-030-26619-6_19
4. Evermann, J., Rehse, J.-R., Fettke, P.: Predicting process behaviour using deep learning. Decis. Support Syst. 100, 129–140 (2017)
5. Ismail Fawaz, H., Forestier, G., Weber, J., Idoumghar, L., Muller, P.-A.: Deep learning for time series classification: a review. Data Min. Knowl. Disc. 33(4), 917–963 (2019). https://doi.org/10.1007/s10618-019-00619-1

6. Hinkka, M., Lehto, T., Heljanko, K., Jung, A.: Classifying process instances using recurrent neural networks. In: Daniel, F., Sheng, Q.Z., Motahari, H. (eds.) BPM 2018. LNBIP, vol. 342, pp. 313–324. Springer, Cham (2019). https://doi.org/10.1007/978-3-030-11641-5_25

7. Kratsch, W., Manderscheid, J., Roeglinger, M., Seyfried J.: Machine learning in business process monitoring: a comparison of deep learning and classical approaches used for outcome prediction. Bus. Inf. Syst. Eng. (2020). https://doi.org/10.1007/s12599-020-00645-0. https://link.springer.com/article/10.1007%2Fs12599-020-00645-0

8. Pasquadibisceglie, V., Appice, A., Castellano, G., Malerba, D.: Using convolutional neural networks for predictive process analytics. In: 2019 International Conference on Process Mining (ICPM) (2019). https://doi.org/10.1109/ICPM.2019.00028

9. Tax, N., Verenich, I., La Rosa, M., Dumas, M.: Predictive business process monitoring with LSTM neural networks. In: Dubois, E., Pohl, K. (eds.) CAiSE 2017. LNCS, vol. 10253, pp. 477–492. Springer, Cham (2017). https://doi.org/10.1007/978-3-319-59536-8_30

10. Teinemaa, I., Dumas, M., La Rosa, M., Maggi, F.M.: Outcome-oriented predictive process monitoring: review and benchmark. ACM Trans. Knowl. Discov. Data (TKDD) 13(2), 1–57 (2019). Article No. 17

11. Wang, J., Yu, D., Liu, C., Sun, X.: Outcome-oriented predictive process monitoring with attention-based bidirectional LSTM neural networks. In: 2019 IEEE International Conference on Web Services (ICWS)(2019). https://doi.org/10.1109/ICWS.2019.00065

Improving the State-Space Traversal of the eST-Miner by Exploiting Underlying Log Structures

Lisa L. Mannel[1]([✉]), Yannick Epstein[2], and Wil M. P. van der Aalst[1]

[1] PADS Group, RWTH Aachen University, Aachen, Germany
{mannel,wvdaalst}@pads.rwth-aachen.de
[2] RWTH Aachen University, Aachen, Germany
yannick.epstein@rwth-aachen.de

Abstract. In process discovery, the goal is to find, for a given event log, the model describing the underlying process. While process models can be represented in a variety of ways, Petri nets form a theoretically well-explored description language. In this paper, we present an extension of the process discovery algorithm eST-Miner. This approach computes the maximal set of non-redundant places, that are considered to be fitting with respect to a user-definable fraction of the behavior described by the given event log, by evaluating all possible candidate places using token-based replay. The number of candidate places is exponential in the number of activities, and thus evaluating all of them by replay is very time-consuming. To increase efficiency, the eST-miner organizes these candidates in a special search structure, that allows to skip large chunks of the search space, while still returning all the fitting places. While this greatly increases its efficiency compared to the brute force approach evaluating all the candidates, the miner is still very slow compared to other approaches. In this paper, we explore two approaches to increase the fraction of skipped candidates and thus the efficiency of the eST-Miner. The impact of the presented concepts is evaluated by various experiments using both real and artificial event logs.

Keywords: Process discovery · Petri nets · eST-Miner

1 Introduction and Related Work

Most corporations and organizations support their processes using information systems, while recording behavior that can be extracted in the form of *event logs*. Each event in such a log has a name identifying the executed activity (activity name), an identification mapping the event to some execution instance (case id), a time stamp showing when the event was observed, and often extended meta-data of the activity or process instance. In the field of *process discovery*, we utilize the event log to identify relations between the activities (e.g. preconditions, choices, concurrency), which are then expressed within a process

© Springer Nature Switzerland AG 2020
A. Del Río Ortega et al. (Eds.): BPM 2020 Workshops, LNBIP 397, pp. 334–347, 2020.
https://doi.org/10.1007/978-3-030-66498-5_25

model. This is non-trivial for various reasons. We cannot assume that the given event log is complete, as some possible behavior might be yet unobserved. Also, real life event logs often contain noise, which we would like to filter out. Correctly classifying behavior as noise can be hard to impossible. An ideal process model can reproduce all behavior contained in an event log, while not allowing for unobserved behavior. It should represent all dependencies between events and at the same time be simple enough to be understandable by a human interpreter. Computation should be fast and robust to noise. Usually, it is impossible to fulfill all these requirements at the same time. Thus, different algorithms focus on different quality criteria, while neglecting others. As a result, the models returned for a given event log can differ significantly.

Many existing discovery algorithms abstract from the full information given in a log and/or generate places heuristically, in order to decrease computation time and complexity of the returned process models. While this is convenient in many applied settings, the resulting models often are underfitting and sometimes even unsound. Examples are the Alpha Miner variants [1], the Inductive Mining family [2], Split-Miner [3], genetic algorithms or Heuristic Miner. In contrast to these approaches, which are not able to (reliably) discover complex model structures, algorithms based on region theory [4–7] discover models whose behavior is the minimal behavior representing the log. On the downside, these approaches are known to be rather time-consuming, cannot handle low-frequent behavior, and tend to produce complex, overfitting models which can be hard to interpret.

In [8] we introduce the discovery algorithm eST-Miner. This approach aims to combine the capability of finding complex control-flow structures like longterm-dependencies with an inherent ability to handle low-frequent behavior while exploiting the token-game to increase efficiency. The basic idea is to evaluate all possible places to discover a set of fitting ones. Efficiency is significantly increased by skipping uninteresting sections of the search space. This may decrease computation time immensely compared to the brute-force approach evaluating every single candidate place, while still providing guarantees with regard to fitness and precision.

While traditional region-theory uses a global perspective to find a set of feasible places, the eST-Miner evaluates each place separately, that is from a local perspective. This allows us to effectively filter infrequent behavior place-wise. Additionally, we are able to easily enforce all kinds of constraints definable on the place level, e.g., constraints on the number or type of arcs, token throughput or similar.

The most severe limitation of the eST-Miner is its high computation time, even on small event logs. This is due to the extensive search of whole candidate space, as well as the even more time-consuming removal of the many so-called *implicit* places during a post-processing. These implicit places are fitting with respect to the log, but do not restrict behavior of the Petri net, and thus they unnecessarily clutter the model. To tackle these performance problems, we present two approaches, one of which does not change the result, while the other one does. We aim to decrease computation time by further reducing the searched fraction of the candidate space, and already discarding a large number of fitting but uninteresting places during the search, thus speeding up both the search and the post-processing phase.

In Sect. 2 we provide basic notation and definitions. Afterwards, we briefly review the basics of the standard eST-Miner (Sect. 3). Our new concepts are introduced in Sects. 4 and 5, and their experimental evaluation is presented in Sect. 6. Finally, Sect. 7 concludes this work by summarizing our work and findings and suggesting possibilities for future work.

2 Basic Notations, Event Logs, and Process Models

A set, e.g. $\{a, b, c\}$, does not contain any element more than once, while a multiset, e.g. $[a, a, b, a] = [a^3, b]$, may contain multiples of the same element. By $\mathbb{P}(X)$ we refer to the power set of the set X, and $\mathbb{M}(X)$ is the set of all multisets over this set. In contrast to sets and multisets, where the order of elements is irrelevant, in sequences the elements are given in a certain order, e.g., $\langle a, b, a, b \rangle \neq \langle a, a, b, b \rangle$. We refer to the i-th element of a sequence σ by $\sigma(i)$. The size of a set, multiset or sequence X, that is $|X|$, is defined to be the number of elements in X. We define activities, traces, and logs as usual, except that we require each trace to begin with a designated start activity ($\blacktriangleright$) and end with a designated end activity ($\blacksquare$). Note that this is a reasonable assumption in the context of processes, and that any log can easily be transformed accordingly.

Definition 1 (Activity, Trace, Log). *Let $\mathcal{A}$ be the universe of all possible activities (e.g., actions or operations), let $\blacktriangleright \in \mathcal{A}$ be a designated start activity and let $\blacksquare \in \mathcal{A}$ be a designated end activity. A trace is a sequence containing $\blacktriangleright$ as the first element, $\blacksquare$ as the last element and in-between elements of $\mathcal{A} \setminus \{\blacktriangleright, \blacksquare\}$. Let $\mathcal{T}$ be the set of all such traces. A log $L \subseteq \mathbb{M}(\mathcal{T})$ is a multiset of traces.*

In this paper, we use an alternative definition for Petri nets. We only allow for places connecting activities that are initially empty (without tokens), because we allow only for traces starting with $\blacktriangleright$ and ending with $\blacksquare$. These places are uniquely identified by the set of input activities I and output activities O. Each activity corresponds to exactly one activity, therefore, this paper refers to transitions as activities.

Definition 2 (Petri nets). *A Petri net is a pair $N = (A, \mathcal{P})$, where $A \subseteq \mathcal{A}$ is the set of activities including start and end ($\{\blacktriangleright, \blacksquare\} \subseteq A$) and $\mathcal{P} \subseteq \{(I|O) \mid I \subseteq A \land I \neq \emptyset \land O \subseteq A \land O \neq \emptyset\}$ is the set of places. We call I the set of ingoing activities of a place and O the set of outgoing activities.*

Given an activity $a \in A$, $\bullet a = \{(I|O) \in \mathcal{P} \mid a \in O\}$ and $a\bullet = \{(I|O) \in \mathcal{P} \mid a \in I\}$ denote the sets of input and output places. Given a place $p = (I|O) \in \mathcal{P}$, $\bullet p = I$ and $p\bullet = O$ denote the sets of input and output activities.

Definition 3 (Overfed/Underfed/Fitting Places, see [9]). *Let $N = (A, \mathcal{P})$ be a Petri net, let $p = (I|O) \in \mathcal{P}$ be a place, and let σ be a trace. With respect to the given trace σ, p is called*

- underfed, *denoted by* $\nabla_\sigma(p)$, *if and only if* $\exists k \in \{1, 2, ..., |\sigma|\}$ *such that* $|\{i \mid i \in \{1, 2, ...k - 1\} \wedge \sigma(i) \in I\}| < |\{i \mid i \in \{1, 2, ...k\} \wedge \sigma(i) \in O\}|$,
- overfed, *denoted by* $\triangle_\sigma(p)$, *if and only if*
  $|\{i \mid i \in \{1, 2, ...|\sigma|\} \wedge \sigma(i) \in I\}| > |\{i \mid i \in \{1, 2, ...|\sigma|\} \wedge \sigma(i) \in O\}|$,
- fitting, *denoted by* $\square_\sigma(p)$, *if and only if not* $\nabla_\sigma(p)$ *and not* $\triangle_\sigma(p)$.

We extend these notions to the log whole log using the noise parameter: with respect to a log L and parameter $\tau \in [0, 1]$, p is called

- underfed, *denoted by* $\nabla^\tau_L(p)$, *if and only if* $|\{\sigma \in L \mid \nabla_\sigma(p)\}|\backslash|L| > 1 - \tau$,
- overfed, *denoted by* $\triangle^\tau_L(p)$, *if and only if* $|\{\sigma \in L \mid \triangle_\sigma(p)\}|\backslash|L| > 1 - \tau$,
- fitting, *denoted by* $\square^\tau_L(p)$, *if and only if* $|\{\sigma \in L \mid \square_\sigma(p)\}|\backslash|L| \geq \tau$.

Definition 4 (Behavior of a Petri net). *We define the* behavior *of the Petri net* $(A, \mathcal{P})$ *to be the set of all fitting traces, that is* $\{\sigma \in \mathcal{T} \mid \forall p \in \mathcal{P} : \square_\sigma(p)\}$.

Note that we only allow for behaviors of the form $\langle \blacktriangleright, a_1, a_2, \ldots a_n, \blacksquare \rangle$ (due to Definition 1) such that places are empty at the end of the trace and never have a negative number of tokens.

3 Introducing the eST-Miner

We briefly introduce the original eST-Miner first presented in [8]. As input, the algorithm takes a log L and a parameter $\tau \in [0, 1]$, and returns a Petri net as output. A place is considered *fitting*, if a fraction τ of traces in the event log is fitting. Inspired by language-based regions, the basic strategy of the approach is to begin with a Petri net whose transitions correspond exactly to the activities used in the given log. From the finite set of unmarked, intermediate places, the subset of all fitting places is computed and inserted. To facilitate further computations and human readability, all unneeded, i.e., implicit places are removed from this intermediate result in a post-processing step.

The algorithm uses token-based replay to discover all fitting places out of the set of possible candidate places. To avoid replaying the log on the exponential number of candidates (i.e. all pairs of subsets of activities, $(2^{|A|} - 1)^2$, it organizes the potential places as a set of trees, such that certain properties hold. When traversing the trees using a depth-first-strategy, these properties allow to cut off subtrees, and thus candidates, based on the replay result of their parent. This greatly increases efficiency, while still guaranteeing that all fitting places are found. An example of such a tree-structured candidate space is shown in Fig. 1. Note the incremental structure of the trees, i.e., the increase in distance from the roots corresponds to the increase of input (red edges) and output (blue edges) activities. However, the organization of candidates within the same depth and their connections to other candidates is not fixed, but defined by the order of ingoing activities ($>_i$) and outgoing activities ($>_o$). Additionally, note that blue edges are always part of a purely blue subtree, while red edges may connect subtrees that contain blue edges as well.

Definition 5 (Complete Candidate Tree). *Let A be a set of activities and let $>_i, >_o$ be two total orderings on this set of activities. A* complete candidate tree *is a pair $CT = (N, F)$ with $N = \{(I|O) \mid I \subseteq A \backslash \{\blacksquare\} \wedge O \subseteq A \backslash \{\blacktriangleright\} \wedge I \neq \emptyset \wedge O \neq \emptyset\}$. We have that $F = F_{red} \cup F_{blue}$, with*

$$F_{red} = \{((I_1|O_1),(I_2|O_2)) \in N \times N \mid |O_2| = 1 \wedge O_1 = O_2$$
$$\wedge \exists a \in I_1 \colon (I_2 \cup \{a\} = I_1 \wedge \forall a' \in I_2 \colon a >_i a')\} \text{ (red edges)}$$
$$F_{blue} = \{((I_1|O_1),(I_2|O_2)) \in N \times N \mid I_1 = I_2$$
$$\wedge \exists a \in O_1 \colon (O_2 \cup \{a\} = O_1 \wedge \forall a' \in O_2 \colon a >_o a')\} \textbf{ (blue edges)}.$$

If $((I_1|O_1),(I_2|O_2)) \in F$, we call the candidate $(I_1|O_1)$ the child *of its* parent $(I_2|O_2)$.

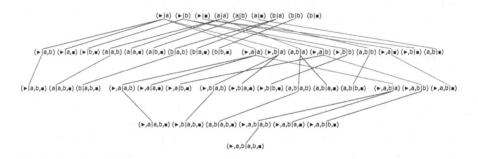

Fig. 1. Example of a tree-structured candidate space for the set of activities $\{\blacktriangleright, A, B, \blacksquare\}$, with orderings $\blacksquare >_i B >_i A >_i \blacktriangleright$ and $\blacksquare >_o B >_o A >_o \blacktriangleright$.

The runtime of the original eST-Miner strongly depends on the number of candidate places skipped during the search for fitting places. The approach uses results of evaluating a place p to skip subtrees of that place, that are known to be unfitting. For example, if 80% of the traces cannot be replayed because p is empty and does not enable the next activity in the trace, i.e., $\triangledown_L^{0.8}(p)$, then at least 80% will not allow for a place p' with even more output activities, i.e. we know that $\triangledown_L^{0.8}(p')$. With respect to the tree-structured candidate traversal, this indicates that all purely blue (outgoing activity is added) subtrees of p can be cut off. If p was overfed, we could cut off all purely red (ingoing activity added) subtrees, respectively.

While this results in a significant decrease in computation time compared to the brute force approach, the algorithm is still slow compared to most other discovery approaches. In this paper, we seek to maximize the number of skipped candidates and thus decrease runtime. We introduce two different heuristic strategies aiming to improve the discovery phase of the eST-Miner. The first strategy is based on organizing the candidates within the tree structure in such a way that the amount of skipped candidate places is maximized. By skipping more candidates, we need to evaluate fewer places and thus terminate faster, without compromising the result, i.e., the discovered set of fitting places remains the same. The second strategy adds additional cut-off criteria to the tree traversal. It heuristically determines certain subtrees to be uninteresting and thus skippable. This way we do not only speed up the search phase, but also significantly reduce the number of implicit places discovered, leading to less time needed for post-processing. However, the returned Petri nets may differ from the original variant.

4   Optimizing the Tree Structure

The positioning of candidate places within the tree-like search structure CT (Definition 5) is directly defined by the two orderings $>_i$ and $>_o$ on the set of all activities A. Consider a place $p = (I|O)$. A red child place of p is a place p_{red} with $p_{\mathrm{red}}\bullet = p\bullet$ and $\bullet p_{\mathrm{red}} = \bullet p \cup \{a\}$, such that $a \in A$ and $\forall b \in \bullet p\colon a >_i b$ (also, $|p \bullet| = 1$, but we focus on the order here). A blue child place of p is a place p_{blue} with $\bullet p_{\mathrm{blue}} = \bullet p$ and $p_{\mathrm{blue}}\bullet = p \bullet \cup\{a\}$, such that $a \in A$ and $\forall b \in p\bullet\colon a >_o b$. Note that the number of children of p is directly defined by the two orderings as well: if $a \in O$ (respectively $a \in I$) is the maximal activity with respect to $>_o$ (respectively $>_i$) of the outgoing (respectively ingoing) activities of p, then the number of blue (respectively red) children of p equals $|\{b \in A \mid b >_o a\}|$ (respectively $|\{b \in A \mid b >_o a\}|$).

We can skip all purely blue subtrees of places which are underfed, and thus we want to maximize the number of blue descendants of such places. Similarly, we want to maximize the number of skippable red descendants of overfed places. Experimental results [8] have shown that varying the orderings has a significant effect on the number of places that are cut off and thus the runtime, without changing the final result.

Computing an optimal traversal ordering is unfeasible, and thus we investigate easily computable approximations. In the following, we present heuristic strategies for choosing the orderings $>_i$ and $>_o$ aiming to maximize the number of cut off places. Consider the event log $L = \big[\langle \blacktriangleright, a, b, b, \blacksquare\rangle, \langle \blacktriangleright, c, b, b, b, b, \blacksquare\rangle^3\big]$ as a motivational example. We observe that the activity b occurs comparatively often in each trace. Thus, places which have b as an outgoing activity are likely to be underfed for each trace. To increase the number of places which are cut off, we want to maximize the number of blue children for such places. Similarly, places with b as incoming activity are likely to be overfed, and thus we would like to maximize their number of red children.

The intuitive idea illustrated by the example leads to a variety of metrics definable on the activities of the event log, aiming to quantify this intuition. The *Absolute Activity Frequency* counts the number of occurrences of an activity accumulated over all traces in the log. The *Absolute Trace Frequency* counts the number of traces in which an activity occurs. The *Average Trace Occurrence* is defined by the average number of occurrences of an activity in a trace of the log. Finally, by the *Average First Occurence Index* of an activity, we refer to the first index at which an activity occurs in a trace, averaged over the whole log.

Definition 6 (Metrics on Log Properties). *Let L be an event log and $A \subseteq \mathbb{A}$ be the set of activities, which occur in the log. We assign numerical values to these activities using the following functions:* [1]

- Absolute Activity Frequency:
  $absAF: A \to \mathbb{N}, absAF(a) = \sum_{\sigma \in L} |\{i \in \{1, ..., |\sigma|\} \mid \sigma(i) = a\}|$
- Absolute Trace Frequency: $absTF: A \to \mathbb{N}, absTF(a) = |[\sigma \in L \mid a \in \sigma]|$
- Average Trace Occurrence: $avgTO: A \to \mathbb{Q}, avgTO(a) = \frac{\sum_{\sigma \in L} |\{i \in \{1,...,|\sigma|\} | \sigma(i) = a\}| / |\sigma|}{|L|}$
- Average First Occurence Index:
  $avgFOI: A \to \mathbb{Q}, avgFOI(a) = \frac{\sum_{\sigma \in L} min\{i \in \{1,2,...,|\sigma|\} | \sigma(i) = a\}}{absTF(a)}$

If $absAF(a)$ is high, we expect many tokens to be produced (consumed) for places that have a as an ingoing (outgoing) activity during replay of the log, and thus such places are more likely to be underfed (overfed). The same holds for high $absTF(a)$ and $avgTO(a)$. If $avgFOI(a)$ is low, we can expect the activity a to generate or consume tokens early on during the replay of a trace. Places which have outgoing activities with low average first occurrence index are more likely to be underfed, as their output activities may require tokens early on during replay, where none might be available.

Definition 7 (Orderings Based on Metrics). *Let L be an event log and $A \subseteq \mathbb{A}$ be the set of activities, which occur in the log. Based on the metrics given in Definition 6 we propose the following orderings:*

- $absAF(a) > absAF(b) \Leftrightarrow a <_i^{absAF} b \Leftrightarrow a <_o^{absAF} b$ *(high frequencies first)*
- $absTF(a) > absTF(b) \Leftrightarrow a <_i^{absTF} b \Leftrightarrow a <_o^{absTF} b$ *(high frequencies first)*
- $avgTO(a) > avgTO(b) \Leftrightarrow a <_i^{avgTO} b \Leftrightarrow a <_o^{avgTO} b$ *(high occurrences first)*
- $avgFOI(a) > avgFOI(b) \Leftrightarrow a <_i^{avgFOI} b \Leftrightarrow a >_o^{avgFOI} b$
  (early activities first for ingoing, last for outgoing activities)

Experimental results investigating and comparing the impact of the presented orderings are presented in Sect. 6.

5   Pruning Uninteresting Subtrees

Our second strategy adds an additional, heuristic criterion to identify and skip uninteresting candidate subtrees. We notice, that fitting places returned by the eST-Miner often have no evidence, and sometimes even have counter-evidence, in the event log. For example, consider the event log $L = [\langle \blacktriangleright, a, c, d, e, \blacksquare \rangle,$

---

[1] Note that $\sum_{\sigma \in L} f(\sigma)$ and $[\sigma \in L | f(\sigma)]$ operate on multisets, i.e., if the same trace σ appears multiple times in L, this is taken into account.

$\langle \blacktriangleright, b, c, d, f, \blacksquare \rangle]$. The place $p_1 = (a, b|e, f)$ is perfectly fitting with respect to this log. However, it describes dependencies of f on a and e on b, which have no evidence in the event log. The places $p_2 = (a|e)$ and $p_3 = (b|f)$, which are fitting and thus discovered by eST-Miner as well, describe the dependencies much better and make the place p_1 superfluous. We aim to skip p_1 and its whole subtree, since all contained candidates describe the unsupported dependencies and could be replaced by better places contained in different subtrees.

In the following, we introduce a heuristic approach assigning an *interest score* to each place based on the *eventually-follows relations* between its ingoing and outgoing activities. This score is defined in such a way, that it can only decrease with increasing depth, i.e., every place is at most as interesting as its parent. This property allows us to skip whole subtrees based on the score assigned to the root of this subtree.

Definition 8 (Eventually-Follows Relation). *Let $a, b \in \mathcal{A}$ be two (possibly equal) activities. We say that b eventually follows a in a trace σ, if a a occurs in σ and later b occurs in σ. Formally, $a \leadsto_\sigma b := \exists i, j \in \{1, ..., |\sigma|\}: \big(i < j \wedge \sigma(i) = a \wedge \sigma(j) = b \big).$*

The interest score based on the eventually-follows relation is based on the intuition, that a place is interesting only if all pairs of ingoing and outgoing activities have more evidence than counter-evidence in the event log. The relation between evidence and acceptable counter-evidence is defined by the parameter λ.

Definition 9 (Interest Score). *Let L be an event log and $A \subseteq \mathcal{A}$ the set of activities which occur in L. For a pair of activities $(a, b) \in A \times A$ (possibly $a = b$), we define the interest score as*

$$\leadsto_L \big((a, b)\big) := \frac{|[\sigma \in L \mid a \leadsto_\sigma b]|}{max(1, |[\sigma \in L \mid a \in \sigma \wedge b \in \sigma]|)}.$$

We define the interest score of a place $p = (I|O)$ as

$$is_L(p) = min\big(\{\leadsto_L \big((a, b)\big) \mid a \in I \wedge b \in O\}\big).$$

The place p is called λ-interesting, if for a user-definable parameter $\lambda \in [0, 1]$ we have that $is_L(p) \geq \lambda$.

The interest score can be directly integrated into the eST-Miners search phase to skip additional subtrees. In particular, whenever we encounter a place p that is not λ-interesting, we conclude that none of its descendants is λ-interesting, and thus we can skip the whole subtree rooted in p (Prop. 1). This follows directly from the incremental construction of the tree structure (Definition 5) and the fact that the minimum is taken when computing $is(p)$ (Definition 9).

Proposition 1 (Interest Score Cannot Increase). *Let L be an event log and $\mathcal{A} \subseteq \mathbb{A}$ the set of activities, which occur in L. Let CT be the search structure of the eST-Miner. Let $p = (I|O)$ be a place with $I \subseteq \mathcal{A}$ and $O \subseteq \mathcal{A}$. Let $\lambda \in [0, 1]$ be a parameter. If p is not a λ-interesting place, then none of its descendants in CT is λ-interesting.*

Table 1. Overview of the event logs used in our experiments. SepsisA was obtained from the original log by removing the 9 least frequent activities. TeleclaimsT contains the 85% most frequent traces of the original log.

Log type	Log name	Abbreviation	Activities	Trace variants	Source
Real-Life	Sepsis-activity filtered	SepsisA	9	642	[10]
	Road Traffic Fines Management	RTFM	11	231	[11]
Artificial	Teleclaims-trace filtered	TeleclaimsT	10	12	[12]
	Repairexample	Repair	12	77	[12]

Table 2. Runtimes of the search phase with different heuristics applied, represented as percentage of the original eST-Miner's (lexicographical ordering for $>_i$ and $>_o$) search phase runtime, with minimal and maximal times given as absolute values. All values are averaged over values for $\tau \in \{1.0, 0.9, 0.8, 0.7, 0.6, 0.5\}$. We can see that the impact of the different activity orderings (Sect. 4) on the time performance is much lower than for the λ-interesting pruning strategy (Sect. 5).

	Repair			RTFM			SepsisA		
	Mean	min [s]	max [s]	Mean	min [s]	max [s]	Mean	min [s]	max [s]
absAF	57.84%	34	63	74.69%	4075	11196	77.13%	48	140
absTF	76.13%	50	77	72.95%	3994	10983	77.99%	56	139
avgTO	63.05%	35	74	75.15%	3972	11671	74.37%	54	131
avgFOI	57.67%	33	56	64.94%	3272	10384	69.29%	46	120
$\mathrm{is}_L(p) \geq 1.0$	4.94%	1.56	2.69	0.13%	12.91	17.06	0.55%	0.43	0.55

The impact of applying this pruning approach to eST-Miner is evaluated in Sect. 6.

6 Experimental Evaluation

We implemented the eST-Miner and our proposed extensions within the Python PM4Py framework for process mining (Python version 3.7.1). Our experiments are all executed on an Intel Core i5 ($2 \times 2.6\,\mathrm{GHz}$) with $8\,\mathrm{GB}$ of RAM, running MacOS Mojawave (10.14.5). We evaluated real and artificial event logs as presented in Table 1.

6.1 Evaluation of Optimizing Orderings

We investigate the impact of the different choices for $>_i$ and $>_o$ as presented in Defnition 7 on the time needed for the eST-Miner search phase. We compare the resulting times to a base case, given by the performance on lexicographic (i.e. random) orderings. An overview of our results for different logs is given in Table 2. The time needed for the search phase when applying the different orderings is averaged over different values of τ and presented as fractions of the runtime needed on the lexicographical ordering.

The table shows that when applying our proposed orderings, searching the candidate space requires at most 78% of the timed needed with the lexicographical ordering for all tested event logs. In many cases we achieve a runtime of less than two thirds of this base case. Based on the presented results, we can derive that the ordering based on *average first occurrence index* clearly leads to the shortest runtime for all tested logs. The other strategies, *absolute activity frequency, absolute trace frequency* and *average trace occurrence* all lead to a significant improvement over the lexicographical ordering, but none performs consistently better than the other on all logs.

We conclude that by choosing the orderings $>_i$ and $>_o$ in a sophisticated way, we are able to perform the search phase in about 60% to 80% of the original runtime, while returning the same set of fitting places.

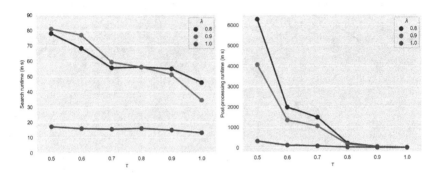

Fig. 2. Investigating runtimes for search (left) and post-processing phase (right) of the ip-eST variant on the RTFM log, for different values of τ and λ.

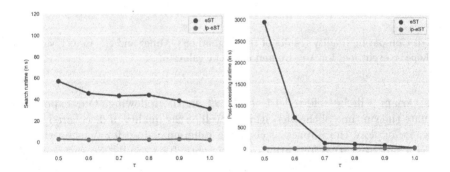

Fig. 3. Comparing runtimes of the original eST-Miner and the ip-eST variant for search (left) and post-processing phase (right) on the **Repair** event log, for $\lambda = 1.0$ and different values of τ.

6.2   Evaluation of Pruning Uninteresting Subtrees

We investigate the potential of the interesting places heuristic as presented in Sect. 5 to prune uninteresting subtrees and thus speed up the original eST-Miner.

Table 3. Time needed for the post-processing step when applying the ip-eST variant with $\lambda = 1.0$, represented as the percentage of time needed by the original eST-Miner, averaged over values for $\tau \in \{1.0, 0.9, 0.8, 0.7, 0.6, 0.5\}$.

	Repair	SepsisA	TeleclaimsT
Postprocessing	0.47%	0.17%	0.03%

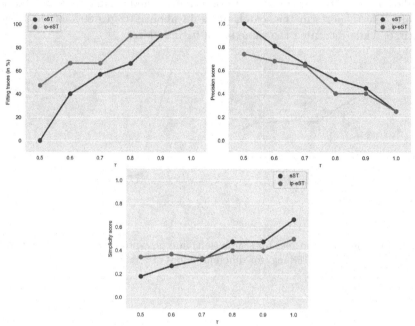

Fig. 4. Comparing quality results of the original eST-Miner and the ip-eST variant on the Repair event log, for $\lambda = 1.0$ and different values of τ.

This variant will be referred to as *ip-eST* in the following. Our experiments evaluate the runtime of the algorithm as well as the quality of discovered models. First, we explore the impact of choosing different values for λ using the RTFM log as a representative for various experiments. The results are summarized in Fig. 2. We conclude that the improved performance we achieve using a value of $\lambda = 1.0$ rapidly deteriorates for lower values of λ. Since our goal is improved time performance, we focus on high λ values in our other experiments, showing that even for $\lambda = 1.0$ model quality remains acceptable. Table 2 summarizes the drastically increased performance for $\lambda = 1.0$, averaged over different values for τ, for various logs as a percentage of the time needed by the original eST-Miner's search phase.

We choose the Repair log to represent our results on time performance. The huge impact of applying the ip-eST variant rather than the original eST-Miner is clearly visible in Fig. 3. For all τ, the search time of the ip-eST variant is only a very small fraction of the standard eST-Miner's search time. The difference becomes larger for smaller values for τ, since the time needed by the original miner increases while the ip-eST variant's search phase runtime remains low. For $\tau = 0.5$, the search phase of the ip-eST-Miner is more than 100 times faster. Since the ip-eST variant returns significantly fewer fitting places than the original eST-Miner, in particular for lower values of τ, the runtime of the post-processing step is greatly decreased. An overview is given in Table 3. The difference between those variants increases as τ decreases, peaking in the ip-eST variants post-processing being 4500 times faster than the standard eST-Miner's post-processing at $\tau = 0.5$.

Definition 10 (Simplicity). *We define the simplicity of a Petri net $N = (A, P)$ based on the fraction of nodes being activities:* $simp(N) = \frac{|A|}{|P|+|A|}$.

In Fig. 4, we investigate the fitness (token-based replay fitness [12]), precision [13], and simplicity (Definition 10) of the models returned by the standard eST-Miner and the ip-eST variant for different values of τ. To represent the results of our experiments on the various logs, we choose the `Repair` event log. For other logs, the approach produces similar results. Note, that for the models discovered by the original eST-Miner and the ip-eST variant, the performance is very similar for all quality metrics and all values of τ. The models returned by the standard eST-Miner do have a slightly higher precision and slightly lower fitness, which is to be expected since non-implicit fitting places may be skipped during the search phase. It is worth noticing that the ip-eST variant is still capable of discovering long-term dependencies, one of the main features of the eST-Miner.

Recall, that a high number of places results in a low simplicity score. Thus, we expect the ip-eST variant to return models that score at least as high as models discovered by the standard eST-Miner, since it discovers less fitting places. Our experiments confirm this expectation for small values of τ, where the eST-Miner discovers a lot more fitting places than the ip-eST variant using its high λ-value for aggressive pruning. For high τ values, we get reversed results. This can be explained by the few fitting places skipped by the ip-eST variant resulting in significantly less places being removed during post-processing. This phenomenon is illustrated in Fig. 5.

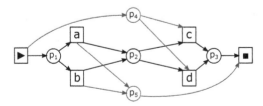

Fig. 5. For the log $[\langle \blacktriangleright, a, c, \blacksquare \rangle^5, \langle \blacktriangleright, b, c, \blacksquare \rangle^5, \langle \blacktriangleright, b, d, \blacksquare \rangle^5]$, the eST-Miner generates the places shown in black, in particular p_2. The ip-eST variant prunes p_2 due to the uninteresting dependency (a, d). Instead, it generates places like p_4 and p_5.

In summary, our experiments have shown a strong boost to the time performance compared to the standard eST-Miner, while we observe only small differences in the quality of discovered models. We conclude that the ip-eST variant seems to restrict the search space in an effective and adequate way and can reliably discover models of similar quality as returned by the eST-Miner in a small fraction of the time.

7   Conclusion

In this paper, we introduced two approaches to improve the time performance of the eST-Miner. The first strategy is based on arranging the candidate places in the tree-like search structure in such a way, that the amount of skipped unfitting candidates is approximately maximized. This decreases the time needed for the search phase to 60–80% of the time needed with random ordering, while returning exactly the same set of fitting places. The second strategy is based on heuristically classifying subtrees as uninteresting and thus skippable based on easily computable log properties. This thus not only greatly decreases runtime of the search phase, but also of the post-processing step, since significantly less fitting places are discovered. Our experiments show that most of the skipped fitting places seem to be implicit, since the quality of the models remains comparable. The computation time of the search phase is pushed below 5% of the time needed by the original algorithm. The post-processing step takes less than 0.5%. Moreover, both approaches can be combined.

Compared with many other discovery algorithms, the eST-Miner returns models with high scores in fitness and particular precision. On the downside, computation times are high and the models often complex. The introduced strategies offer a significant decrease in computation time and make the eST-Miner a competitive discovery approach on event logs with a small number of activities.

The presented approaches are clearly very promising with respect to discovering high-quality models faster. We see potential to further improve time performance as well as model quality by investigating additional variants of both our heuristics. Also, discovering reasonable dependencies between the parameters λ and τ would be interesting. A combination of these heuristics with other variants of eST-Miner, e.g. the uniwired variant [14], is clearly possible. In addition to improved computation speed, we may be able to use our approaches to simplify the returned models without loosing important structures.

Acknowledgments. We thank the Alexander von Humboldt (AvH) Stiftung for supporting our research.

References

1. Wen, L., van der Aalst, W.M.P., Wang, J., Sun, J.: Mining process models with non-free-choice constructs. Data Mining Knowl. Disc. **15**(2), 145–180 (2007)

2. Leemans, S., Fahland, D., van der Aalst, W.M.P.: Discovering block-structured process models from event logs - a constructive approach. In: Application and Theory of Petri Nets and Concurrency (2013)
3. Augusto, A., Conforti, R., Dumas, M., La Rosa, M., Polyvyanyy, A.: Split miner: automated discovery of accurate and simple business process models from event logs. Knowl. Inf. Syst. **59**(2), 251–284 (2018). https://doi.org/10.1007/s10115-018-1214-x
4. Badouel, E., Bernardinello, L., Darondeau, P.: Petri Net Synthesis. TTCSAES. Springer, Heidelberg (2015). https://doi.org/10.1007/978-3-662-47967-4
5. Lorenz, R., Mauser, S., Juhás, G.: How to synthesize nets from languages: a survey. In: Proceedings of the 39th Conference on Winter Simulation: 40 Years! The Best is Yet to Come, WSC 2007. IEEE Press (2007)
6. van der Werf, J.M.E.M., van Dongen, B.F., Hurkens, C.A.J., Serebrenik, A.: Process discovery using integer linear programming. In: van Hee, K.M., Valk, R. (eds.) PETRI NETS 2008. LNCS, vol. 5062, pp. 368–387. Springer, Heidelberg (2008). https://doi.org/10.1007/978-3-540-68746-7_24
7. Carmona, J., Cortadella, J., Kishinevsky, M.: A region-based algorithm for discovering petri nets from event logs. In: Business Process Management BPM (2008)
8. Mannel, L.L., van der Aalst, W.M.P.: Finding complex process-structures by exploiting the token-game. In: Donatelli, S., Haar, S. (eds.) PETRI NETS 2019. LNCS, vol. 11522, pp. 258–278. Springer, Cham (2019). https://doi.org/10.1007/978-3-030-21571-2_15
9. van der Aalst, W.M.P.: Discovering the "glue" connecting activities - exploiting monotonicity to learn places faster. In: It's All About Coordination - Essays to Celebrate the Lifelong Scientific Achievements of Farhad Arbab (2018)
10. Mannhardt, F.: Sepsis cases - event log (2016)
11. De Leoni, M., Mannhardt, F.: Road traffic fine management process (2015)
12. van der Aalst, W.M.P.: Process Mining: Data Science in Action. Springer, Heidelberg (2016). https://doi.org/10.1007/978-3-662-49851-4
13. Munoz-Gama, J., Carmona, J.: A fresh look at precision in process conformance. In: BPM (2010)
14. Mannel, L.L., van der Aalst, W.M.P.: Finding uniwired Petri Nets using eST-miner. In: Di Francescomarino, C., Dijkman, R., Zdun, U. (eds.) BPM 2019. LNBIP, vol. 362, pp. 224–237. Springer, Cham (2019). https://doi.org/10.1007/978-3-030-37453-2_19

8th International Workshop on Declarative, Decision and Hybrid Approaches to processes (DEC2H 2020)

8th International Workshop on Declarative, Decision and Hybrid Approaches to Processes (DEC2H 2020)

The rules, premises and outcomes that drive the decisions in the conduction of processes are often implicit in process flows, process activities or even tacit knowledge in the head of employees. However, the analysis and automation of decision processes should be made explicit in order to let computers analyse or execute business process models. The declarative modelling paradigm aims to directly express the business rules or constraints underlying a given process. The paradigm has gained momentum, and in recent years several declarative notations have emerged, including, a.o., DECLARE, Dynamic Condition Response (DCR) Graphs, Decision Modelling and Notation (DMN), and fragment-based Case Management (fCM). Recently, there has been a surge of interest in *hybrid* approaches, combining the strengths of declarative and procedural modelling paradigms.

In the workshop on Declarative, Decision and Hybrid approaches to processes (DEC2H), we are interested in the application and challenges of decision- and rule-based modelling in all phases of the Business Process Management lifecycle: identification, discovery, analysis, redesign, implementation and monitoring.

DEC2H 2020 is the eighth edition of the workshop – formerly known as DeHMiMoP (Workshop on Declarative/Decision/Hybrid Mining and Modelling for Business Processes) till its sixth edition. DEC2H 2020 attracted six high-quality international submissions. Each paper was reviewed by at least three members of the Program Committee. Out of the submitted manuscripts, the top four were accepted for presentation. In the proceedings, three of those are published as full research papers and one as a short paper.

The workshop began with the invited talk of Andrea Burattin. The talk focussed on two key challenges and solutions related to the automated execution of processes through autonomous systems: observing the current status of the process with live data analysis, and understanding the models used for training autonomous systems. Varvoutas and Gounaris propose new heuristics for the optimisation of data-centric, information-intensive processes. Focussing on the Product Data Model (PDM) approach, they evaluate the improvements brought by their implemented heuristics and compare them against existing solutions. Etikala et al. introduce Text2Dec, a technique to automatically extract Decision Requirements Diagrams (DRD) for DMN from text-based documents. To that end, their solution resorts to text-mining and natural language processing tools and techniques. Haarmann and Weske present an approach to extend fCM models with data object cardinalities that play the role of safety and reachability constraints, that is, global rules to be never violated and goals to be eventually satisfied, respectively. Galrinho et al. describe their new declarative data-centric process language named REDA (Reactive Data), with an analysis and discussion of its syntax and semantics. REDA expresses data and computation elements in a graph-based representation.

We thank the authors for their noteworthy contributions and the members of the Program Committee for their invaluable help in the reviewing and discussion of the manuscripts. We hope that the reader will benefit from the reading of these papers to know more about the latest advances in research about declarative, decision, and hybrid approaches to business process management.

September 2020   Søren Debois
                                                                 Claudio Di Ciccio
                                                                       Tijs Slaats
                                                                   Jan Vanthienen

Organization

Workshop Organisers

Søren Debois	IT University of Copenhagen, Denmark
Claudio Di Ciccio	Sapienza University of Rome, Italy
Tijs Slaats	University of Copenhagen, Denmark
Jan Vanthienen	KU Leuven, Belgium

Program Committee

Rafael Accorsi	PwC, Switzerland
Bart Baesens	KU Leuven, Belgium
Andrea Burattin	Technical University of Denmark, Denmark
Josep Carmona	Universitat Politècnica de Catalunya, Spain
João Costa Seco	Universidade Nova de Lisboa, Portugal
Massimiliano de Leoni	University of Padua, Italy
Johannes De Smedt	University of Edinburgh, UK
Jochen De Weerdt	KU Leuven, Belgium
Chiara Di Francescomarino	Fondazione Bruno Kessler, Italy
Rik Eshuis	Eindhoven University of Technology, Netherlands
Robert Golan	DBmind Technologies Inc., United States
María Teresa Gómez-López	Universidad de Sevilla, Spain
Xunhua Guo	Tsinghua University, China
Thomas Hildebrandt	University of Copenhagen, Denmark
Amin Jalali	Stockholm University, Sweden
Krzysztof Kluza	AGH University of Science and Technology, Poland
Fabrizio M. Maggi	University of Tartu, Estonia
Andrea Marrella	Sapienza University of Rome, Italy
Marco Montali	Free University of Bozen-Bolzano, Italy
Artem Polyvyanyy	University of Melbourne, Australia
Hajo A. Reijers	Utrecht University, The Netherlands
Flavia M. Santoro	Universidade do Estado do Rio de Janeiro, Brazil
Stefan Schönig	Universität Regensburg, Germany
Lucinéia H. Thom	Universidade Federal do Rio Grande do Sul, Brazil
Han van der Aa	Humboldt University of Berlin, Germany

Wil M. P. van der Aalst   RWTH Aachen University, Germany
Barbara Weber   University of St. Gallen, Switzerland
Mathias Weske   Hasso Plattner Institute, University of Potsdam,
                                 Germany

A Robot May Not Injure a Human Being: Improving (Declarative) Processes

Andrea Burattin [ID]

Software and Process Engineering, Technical University of Denmark,
Lyngby, Denmark
andbur@dtu.dk

Abstract. One of the benefits of managing business processes is the ability to improve them, for example, by applying the BPM lifecycle. An increasing effort is put in devising autonomous systems capable of providing such improvements, for example by analyzing historic data or by applying optimization heuristics. This presentation will focus on two key problems related to these technologies: observing the *current* status of the process and understanding the models used for training autonomous systems. The former suggests analyzing live data (as opposed to historic executions) to ensure that suggested improvements are timely and relevant. The latter relates to the algorithms underlying autonomous systems that require some type of training (that directly affects the way processes are improved): if we do not understand in all details the models use for training, we will end up with unexpected results. Examples of solutions for the two problems will be presented as well.

Evaluation of Heuristics for Product Data Models

Konstantinos Varvoutas and Anastasios Gounaris$^{(\boxtimes)}$

Department of Informatics, Aristotle University of Thessaloniki, Thessaloniki, Greece
{kmvarvou,gounaria}@csd.auth.gr

Abstract. Product Data Model (PDM) is an example of a data-centric approach to modelling information-intensive business processes, which offers flexibility and facilitates process optimization. It is declarative, and as such, there may be multiple workflow designs that can produce the end product. To this end, several heuristics have been proposed. The contributions of this work are twofold: (i) we propose new heuristics that capitalize on established techniques for optimizing data-intensive workflows; and (ii) we extensively evaluate the existing solutions. Our results shed light on the merits of each heuristic and show that our proposal can yield significant benefits in certain cases. We provide our implementation as an open-source product.

Keywords: Data-centric processes · Process optimization · PDM

1   Introduction

Data-centric approaches have been emerging in the last two decades as an alternative to the more mainstream activity-oriented modelling approaches for business processes [7,10,14]. We quote from [14] that *"the central idea behind data-centric approaches is that data objects/elements/artifacts can be used to enhance a process-oriented design or even to serve as the fundament for such a design. This has certain advantages, varying from increasing flexibility in process execution and improving reusability to actually being able to capture processes where data play a relevant role."*

In this work, we focus on a particular data-centric modelling approach, namely a *Product Data Model (PDM)*-oriented one, for which the main driver, as also reported in [14], is process optimization apart from flexibility; this approach is tailored to information-intensive processes and it is declarative. As such, it focuses on describing what is needed in order to deliver an information product rather than the exact way to achieve this goal. To fulfill the latter aspect, the declarative model is accompanied by a method to generate workflow designs, which is referred to as *Product Based Workflow Support (PBWS)* [18]. PBWS presents a set of heuristics for PDMs with a view to enhancing the performance on a case-by-case manner. PBWS improves upon a previous method, called *Product Based Workflow Design (PBWD)* [13], where the burden of defining

© Springer Nature Switzerland AG 2020
A. Del Río Ortega et al. (Eds.): BPM 2020 Workshops, LNBIP 397, pp. 355–366, 2020.
https://doi.org/10.1007/978-3-030-66498-5_26

the sequence of actions rests with the workflow designer, while PBWD merely assists this task through presenting the alternatives.

Heuristic solutions are intuitive in their rationale, easy to implement and are of low computational complexity. However, the existing solutions fail to benefit from established techniques in the areas of optimization for workflows for data analytics and database query execution plans, which adopt principled cost-based approaches [8]. Inspired by such techniques, in our heuristics, we suggest to consider both the time/cost of each operation in a PDM model and the probability of this operation to lead to an early termination of the process, thus saving time and resources. More specifically, we make a twofold contribution:

1. We propose a new heuristic for choosing the next operation to be performed in a PDM for a specific case to optimize time duration and/or cost. Our proposal comes in three flavors and is based on established query processing and data-centric workflow technology, and is of low computational complexity.
2. We perform an extensive experimental evaluation of the available heuristics and we show that our proposal yields benefits in terms of time, cost or a combination of both compared to previous heuristics, on the average case. However, there is no globally dominant solution, in the sense that in specific cases, existing heuristics may behave better. We provide the open source of all the heuristics and the experiments, so that interested third parties can repeat our work and extend the set of heuristics and/or test cases[1].

The remainder of this paper is structured as follows. In Sect. 2, we present the PDM underpinnings of the techniques evaluated. Section 3 discusses existing solutions from PBWS and introduces our proposal along with implementation details. We evaluate the candidate techniques in Sect. 4. The next section deals with the related work and we conclude in Sect. 6, where we also briefly discuss limitations and future work.

2   Background: The Product Data Model

A PDM is used to represent the structure of a workflow product in a rooted graph-like manner, similar to a Bill of Material [11,18]. PDMs describe the required elements for yielding the desired product in the root, where example (informational) products include decision on whether to grant an approval to a specific admission request, approval of a mortgage application, and so on. More specifically, the vertices (or equivalently nodes) in this structure correspond to data elements, that is the information that is processed in the workflow. Each node has a value assigned to it, which typically differs between process instances (cases). In Fig. 1, we present the PDM for a classical mortgage example, which will also be used in the comparison section. The final product of the process is to determine the value of the root (or top or end product) node. Values are determined from the bottom to the root as specified by the arcs (graph edges),

---

[1] https://github.com/kmvarvou/pdm_heuristics.

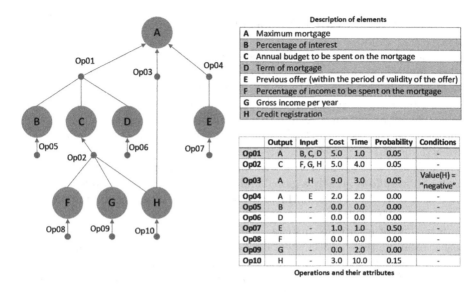

Fig. 1. The PDM mortgage example from [18]

which are called operations. These arcs represent actions that are applied on the valued data elements to produce values for nodes downstream. Each operation can have zero or more input data elements while producing the value for exactly one output data element. An operation is represented by a tuple, which consists of the output element and a set of input elements, e.g. $(A, \{B, C, D\})$ for $Op01$ in the figure, which means that $Op01$ can be applied only if B, C and D have been produced and *may* lead to the generation of the value of A. A data element may be determined through multiple operations, e.g., in the figure, A can be determined in three manners and, for a specific case, the process terminates as soon as one of them manages to complete. In the case where an element has zero input elements, it is called a leaf element, and commonly, it is provided as input to the process; in the figure, elements such as B, F and E are leaf elements.

In summary, the PDM describes the operations that can be performed to produce the top element, along with their inter-dependencies. Not all operations need to be executed. To allow for cost-based decisions, each operation has the following metadata (an example is in the table at the bottom-right of Fig. 1):

- *Cost*, which represents the cost associated with executing the operation.
- *Time*, which represents the time required for the complete execution of the operation.
- *Probability*, represents the probability that an operation is executed unsuccessfully, therefore not producing its output element.
- *Conditions*, which represent requirements regarding the value of the input elements. These requirements must be met, if the process is to be executed, meaning that the existence of all input elements of an operation is not sufficient.

3 Deriving Workflow Designs

PDM does not specify per se how the end information product is created but allows multiple workflow designs to produce the desired information product. As mentioned above, there are cases where multiple workflow designs lead to the production of an output element. Usually, in these cases, the alternative paths have different execution costs and time durations. This gives rise to the following optimization problem: *which paths of operations to choose for a specific case in order to optimize given quantitative objectives of cost and time?*

There are two high-level strategies for the calculation of an optimal execution path of a workflow, namely a global and a local one. A global strategy considers the effect of each decision on future steps. It takes into account the complete set of alternative paths that produce the end product to optimize the execution performance of each case. Instead, a local strategy adopts a step-by-step approach, meaning that, at each step, it examines the set of operations available for execution and chooses the best one, according to a particular metric, e.g. cost of execution. As explained in [18], a global strategy does not scale. For this reason, in this work, we exclusively deal with low-polynomial local strategy heuristics.

More specifically, the local strategies used by the PBWS method in [18] comprise the following heuristics (the last one is not explicitly mentioned in [18] but it is trivial to include it):

1. *Random*: the operation is randomly selected from the set of executable operations.
2. *Lowest Cost*: the operation with the lowest cost is selected.
3. *Shortest Time*: the operation with the shortest time is selected.
4. *Lowest Failure Probability*: the operation with the lowest probability of not being executed successfully is selected.
5. *Shortest Distance to Root Element*: the operation with the shortest distance to the root element (measured in the total number of operations) is selected.
6. *Shortest Remaining Process Time*: the operation with the shortest remaining processing time (measured as the sum of the processing times of the operations on the path to the root element) is selected.
7. *Shortest Remaining Cost*: the operation with the shortest remaining cost (measured as the sum of the costs of the operations on the path to the root element) is selected.

We discuss implementation details at the end of this section.

3.1 Rank-Based Heuristics

Our approach relies on treating productions rules in a manner that resembles knock-out activities, and their optimal ordering, which also bears similarities to the way data analytics operators and database joins are ordered [1,8,9]. A knock-out activity is an activity whose execution leads directly to the completion of the process. For example, the execution of $Op03$ in Fig. 1, which produces the root element A is a knock-out activity/production rule. Then, the optimal ordering needs to take into account the probability of an operation to produce the end element, either directly or indirectly as a sequence of operations starting with that operation, and the corresponding cost or execution duration.

More specifically, at each step, we consider all the operations ready for execution (i.e., those with all inputs present) and we choose the one with the highest *rank* value. The rank value of an operation Op is defined as follows:

$$rank(Op) = \frac{\prod_{Op' \in \pi(Op)} 1 - Probability(Op')}{\sum_{Op' \in \pi(Op)} Cost(Op')}$$

where $\pi(Op)$ is the path from Op (including) to the root.

Example: in the example in Fig. 1, assume that in the current state all leaf elements have been produced already except element E, for which $Op07$ was not executed successfully. Thus, in the next step, there are two available production rules for execution, namely $Op02$ and $Op03$. Based on their attributes they have the following ranking value: $rank(Op02) = 0.9025/10 = 0.09025$. While $Op02$ may not lead to a process termination directly, we consider $Op02$ as part of a path that leads indirectly to the root, that is the path: $Op02 \rightarrow Op01 \rightarrow A(end\ state)$. Therefore, we use as probability of this knock-out path, the probability of *success* of the operations in the whole path, which is $(1 - Probability(Op02)) * (1 - Probability(Op01)) = 0.95 * 0.95 = 0.9025$ and as cost, the aggregate cost of the whole path, which is $Cost(Op01) + Cost(Op02) = 5 + 5 = 10$. On the other hand, $rank(Op03) = 0.95/9 = 0.105556$. The probability 0.95, in the nominator, is the probability of the *successful* execution of $Op03$ because it is the successful execution of $Op03$ that produces the root element A and therefore, completes the workflow execution. Based on these values, $Op03$ is selected for execution. □

This heuristic comes in three flavors. The first one uses the formula above. The other two modify the denominator in the rank formula and employ (i) the sum of the operation duration times and (ii) the sum of the product of the cost and times, respectively. As such, they focus more on time duration and a combination of both metrics, respectively.

3.2   Implementation Issues

All the heuristics conform to a generic template, shown in Algorithm 1, which produces, for each case, the sequence of steps (operations) chosen; this sequence is captured in the variable WF. The operations metadata are mapped to a HashMap variable, where the key is the operation id and the value is nested and consists of all attributes in the table of the example in Fig. 1. Based on such a structure, the *executableList* can be found through a simple traversal of the hashmap, taking into account the contents of the *availableList*. This occurs once at the beginning and then, *executableList* keeps being updated. Then the *availableList* is scanned to choose the *nextOp* operation according to the chosen local strategy (line 4). The time complexity of this algorithm, for the first 4 heuristics, is $O(n(n+V))$, where n is the number of operations and V the number

---

Algorithm 1. PDM heuristics template

Require: (1) Operations metadata as depicted in the table in Fig. 1. (2) A set of the already produced data elements named *"availableList"*; if none this set is initialized to ∅.

```
 1:  WF ← {}
 2:  executableList ← operations that can be executed
 3:  while executableList ≠ ∅ and !availableList.contains(root_element) do
 4:      nextOp ← select operation for execution from executableList
 5:      WF ← {WF, nextOp}
 6:      execute nextOp
 7:      if execution is successful then
 8:          add output elements of executed operation to availableList
 9:          update executableList based on new available elements
10:      end if
11:  end while
12:  return WF
```

---

of nodes in the graph. This is because, in each step at most n operations are examined, and there are n steps at most. Also, for an operation to be inserted in the *executableList*, up to V elements checks need to be performed. A fast implementation employs a priority queue for supporting the choice of *nextOp* in each iteration, but it is beyond the scope of this work to discuss such details.

However, a more important point is that the three last existing heuristics along with our proposal need to process the path from a given operation to the root. For this, we need to employ another auxiliary structure, where the PDM model is seen as a typical graph with as many vertices as the data elements and directed edges for each data element in the input of another data element pointing to that element. Since finding the shortest path from a root element is at most $O(n log V)$ using an algorithm such as Dijkstra, the complexity of the relevant techniques is the previous complexity multiplied by this factor, i.e., $O(n^2(n + V) log V)$. In addition to the PDM's data elements, this graph also contains an artificial starting vertex. This vertex represents the initial state of execution where no elements have been produced. It covers operations, such as $Op08$, $Op09$ and $Op10$ in the example, which, otherwise, cannot be represented as edges connecting vertices.

However, in the rank-based solutions more problems arise due to the product in the fraction nominator. In addition, we have to deal with the cases, where multiple paths from a data element to the root element exist, as is the norm. We distinguish between two cases. First, if there exists an operation that directly leads to the root element, we consider this edge as the complete path. Second, if such an operation does not exist, we cannot rely on shortest paths with edges either non weighted or weighted according to the cost or time as we do for calculating the denominator, which is calculated using Dijkstra's algorithm in a straightforward manner. Our procedure is as follows with regards to the nominator of the rank fraction. We use the probability of failure of data element production, exactly as provided in the metadata table to assign weights to the graph edges. Then, as the representative path of the operator Op, we choose the path for which the sum of these probabilities is the smallest one, i.e., we have again reduced the problem to a shortest path one. For this path, we compute the product of the success probability values; the success probability of each operation (or graph edge) is 1 minus the failure probability.

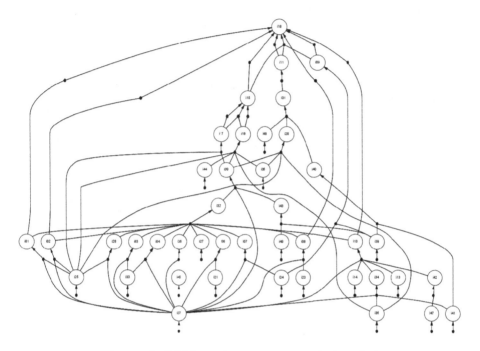

Fig. 2. The PDM social insurance example from [13]

4   Evaluation

In this section, we evaluate our proposed approach using two PDMs. The first one is the mortage example that was presented in Sect. 2, while the second one, shown in Fig. 2, represents a larger process from a social insurance company [13]. To evaluate the 7 existing heuristics and our 3 rank-based flavors, we created 100K random cases for each of the two PDMs. The cost and time attributes were assigned (integer) values in the [1,10] range, i.e., these attributes may differ up to an order of magnitude. The probability of failure was assigned values in the [0.0, 0.999] range, i.e., we consider the complete range from guaranteed success to almost certain failure. All distributions are uniform. We do not explicitly consider the probability of meeting conditions, since this probability can be seen as covered by the failure probability.

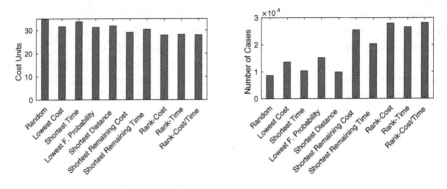

Fig. 3. The average cost per case (left) and number of cases where each heuristic achieved the best result (right) regarding the mortgage PDM.

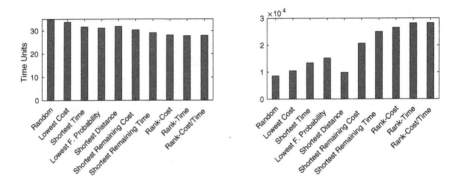

Fig. 4. The average execution time per case (left) and number of cases where each heuristic achieved the best result (right) regarding the mortgage PDM.

The results are summarized in Figs. 3, 4, 5 and 6. Each figure corresponds to one of the four combinations of the two objectives of cost and execution time of the process and the two PDMs. We start our discussion with the first two result figures that refer to the smaller PDM of Fig. 1. In all figures, the left barchart depicts average behavior, while the right one shows the number of cases where a heuristic exhibited the best performance for that specific case. The main observations are as follows:

1. The rank-based heuristics proposed in this work are the best performing ones both when cost and when time is the optimization objective. Their relative difference is small and does not exceed 1.1%, which means that the rank function effectively covers both objectives in all three flavors.
2. Choosing randomly the next operation incurs approximately 25% higher cost and 25% higher time compared to our solutions. The best performing heuristics from the existing ones are *Shortest Remaining Cost* for the cost objective

and *Shortest Remaining Time* for the time objective. These heuristics are on average only 4.2% and 4.6% worse than the rank-based ones, respectively.
3. There is no globally dominant solution. As the two right barcharts show, even the random heuristic yields the best performance in some cases. On average, for each case there are 1.86 heuristics that yield the best performance regardless of the exact objective (it can be observed in the figures that the barcharts do not sum to 100K). The most common winners are the rank-based heuristics, but each individual flavor is the best in no more than 28.5% of the cases.

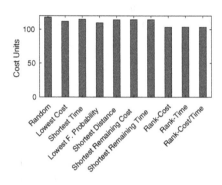
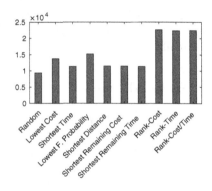

Fig. 5. The average cost per case (left) and number of cases where each heuristic achieved the best result (right) regarding the social insurance PDM.

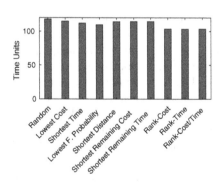
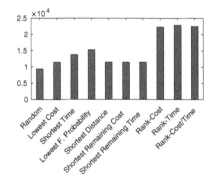

Fig. 6. The average execution time per case (left) and number of cases where each heuristic achieved the best result (right) regarding the social insurance PDM.

Next, we move our attention to Figs. 5 and 6, which refer to the largest PDM in Fig. 2. In this PDM, in summary, the rank-based heuristics are still the best ones in the average case, with even smaller relative differences between the three flavors (less than 0.4%). The random heuristic is only 14% worse than the best heuristic on the average case. But the best performing heuristic from the ones in [18] has now become the one that chooses the operation with the lowest failure probability; this heuristic is 6.3% worse that the rank-based solutions. Finally, in each case, 1.52 heuristics achieve the top performance on average.

Similar and even better results are observed when the optimization objective is the product of cost and time (no details are provided due to space limitations). Two additional significant points are: (i) Our proposals are superior to the best performing existing heuristic by up to more than 20 times. (ii) The overhead time to run the heuristics is extremely low: on a Ryzen 5 3600x CPU with 16 GB RAM, our rank-based heuristics take less than 1.34 milliseconds for each case of the social insurance PDM; the other heuristics are even faster and the times are negligible for the small PDM.

5 Related Work

As stated in the introduction, an increasing amount of data-centric approaches have been developed as part of a general trend in the area of Business Process Management (BPM). Despite this recent interest, Business Process Improvement or Redesign, one of the key areas of BPM remains relatively undeveloped in terms of automated algorithmic solutions. In a recent survey that aims to evaluate several data-centric process approaches, this lack of focus on process optimization or redesign is highlighted [14]. Out of the 14 methods examined, only 2 of them identify the objective of business process optimization as a motive for their development. These two methods are Product Based Workflow Design (PBWD) [13] and its extension, Product Based Workflow Support (PBWS) [18], upon which we build our work.

A significant part of recent research in BPs targets variability between process models aiming at the same high-level objectives [15]. For example, the work in [15] is motivated by the fact that the same goal in different municipalities is performed using different equivalent processes and, to manage such variability, it introduces the configurable process trees. This methodology allows a specific set of process models to be selected according to several criteria. This bears some similarity to the way PBWS exploits the existence of alternative paths in order to optimize each case's performance. The main difference lies in the fact that these alternatives are different paths of the same, already existing PDM model, while in [15], there is an attempt to create a model that contains alternative paths to cover all rationales. In such a context, proposals like [4] deal with the problem of extracting alternative models, whereas the issue of assessing the quality of different process model configurations [3] has also been explored.

Additionally, there are proposals considering variant optimization objectives, such as the techniques in [1], where a set of heuristics is introduced for optimizing the metrics of resource utilization, maximal throughput and execution (cycle) time. These heuristics consider changing the relative ordering of activities, enforcing parallel execution and activity merging, but they cannot be applied to PDMs (at least in a straightforward manner). Finally, our work relates to declarative process models [5,12]; e.g., our workflow design solution can be seen as a promising means to derive executable model structures out of such declarative models although providing a complete methodology to achieve this remains an open issue.

Regarding data-centric workflows, a lot of effort has been put towards finding the best sequential order of flow tasks for objectives such as minimization of the sum of the costs of these tasks or the bottleneck cost, or the maximization of the utilization of each execution processor, and so on [2,6,8]. All these proposals aim to optimize a single criterion, but there are also proposals that target multi-objective data flow optimization, such as the algorithms in [16,17]. Despite some initial efforts in [9], transferring the results of data analytics workflow optimization to business process workflows is still a topic in its infancy.

6 Conclusions and Future Work

This work focuses on processes modelled according to the declarative PDM paradigm and aims to evaluate both existing and novel heuristics for yielding workflow designs on a case-by-case basis. Inspired by data analytics, we use the notion of rank, which combines the probability to produce the root element and the cost to achieve this in a single metric. In our experiments, we show that rank-based heuristics exhibit the best performance on average, but in specific cases, each of the 10 heuristics examined in this work may be the dominant one.

Our work suffers from the same limitations as the heuristics in [18]: we optimize on a case-by-case basis without seeing the process as a whole, e.g., in terms of resource utilization and without considering parallel task execution. Apart from addressing these limitations, we aim to follow three directions as future work: (i) to better handle the information that a data element may need input by multiple elements, when computing path costs (which is now implicitly ignored); (ii) to devise hybrid methodologies that switch between heuristics in a specific case, motivated by our key observation that there is no globally dominant solution; and (iii) to transfer similar techniques to other declarative modelling approaches, such as [12].

Acknowledgment. The research work was supported by the Hellenic Foundation for Research and Innovation (H.F.R.I.) under the "First Call for H.F.R.I. Research Projects to support Faculty members and Researchers and the procurement of high-cost research equipment grant" (Project Number:1052, Project Name: DataflowOpt). We would like also to thank Dr. Georgia Kougka for her comments and help.

References

1. van der Aalst, W.M.P.: Re-engineering knock-out processes. Decis. Support Syst. **30**(4), 451–468 (2001). https://doi.org/10.1016/S0167-9236(00)00136-6

2. Agrawal, K., Benoit, A., Dufossé, F., Robert, Y.: Mapping filtering streaming applications. Algorithmica **62**(1–2), 258–308 (2012). https://doi.org/10.1007/s00453-010-9453-6
3. Buijs, J.C.A.M., van Dongen, B.F., van der Aalst, W.M.P.: Discovering and navigating a collection of process models using multiple quality dimensions. In: Lohmann, N., Song, M., Wohed, P. (eds.) BPM 2013. LNBIP, vol. 171, pp. 3–14. Springer, Cham (2014). https://doi.org/10.1007/978-3-319-06257-0_1
4. Buijs, J.C.A.M., van Dongen, B.F., van der Aalst, W.M.P.: Mining configurable process models from collections of event logs. In: Daniel, F., Wang, J., Weber, B. (eds.) BPM 2013. LNCS, vol. 8094, pp. 33–48. Springer, Heidelberg (2013). https://doi.org/10.1007/978-3-642-40176-3_5
5. Chawla, N., King, I., Sperduti, A.: User-guided discovery of declarative process models (2011)
6. Deshpande, A., Hellerstein, L.: Parallel pipelined filter ordering with precedence constraints. ACM Trans. Algorithms **8**(4), 1–38 (2012)
7. Henriques, R., Rito Silva, A.: Object-centered process modeling: principles to model data-intensive systems. In: zur Muehlen, M., Su, J. (eds.) BPM 2010. LNBIP, vol. 66, pp. 683–694. Springer, Heidelberg (2011). https://doi.org/10.1007/978-3-642-20511-8_62
8. Kougka, G., Gounaris, A., Simitsis, A.: The many faces of data-centric workflow optimization: a survey. Int. J. Data Sci. Anal. **6**(2), 81–107 (2018). https://doi.org/10.1007/s41060-018-0107-0
9. Kougka, G., Varvoutas, K., Gounaris, A., Tsakalidis, G., Vergidis, K.: On knowledge transfer from cost-based optimization of data-centric workflows to business process redesign. In: Hameurlain, A., Tjoa, A.M. (eds.) Transactions on Large-Scale Data- and Knowledge-Centered Systems XLIII. LNCS, vol. 12130, pp. 62–85. Springer, Heidelberg (2020). https://doi.org/10.1007/978-3-662-62199-8_3
10. Künzle, V., Reichert, M.: Philharmonicflows: towards a framework for object-aware process management. J. Softw. Maintain. **23**(4), 205–244 (2011)
11. Orlicky, J.A., Plossl, G.W., Wight, O.W.: Structuring the bill of material for MRP. In: Lewis, M., Slack, N. (eds.) Operations Management: Critical Perspectives on Business and Management, vol. 58. Taylor & Francis, New York (2003)
12. Pesic, M., Schonenberg, H., van der Aalst, W.M.P.: Declare: full support for loosely-structured processes. In: 11th IEEE International Enterprise Distributed Object Computing Conference (EDOC 2007), pp. 287–300 (2007)
13. Reijers, H.A., Limam, S., van der Aalst, W.M.P.: Product-based workflow design. J. Manag. Inf. Syst. **20**(1), 229–262 (2003)
14. Reijers, H.A., et al.: Evaluating data-centric process approaches: does the human factor factor in? Softw. Syst. Model. **16**(3), 649–662 (2016). https://doi.org/10.1007/s10270-015-0491-z
15. Schunselaar, D.: Configurable process trees : elicitation, analysis, and enactment. Ph.D. thesis, Department of Mathematics and Computer Science, October 2016. Proefschrift
16. Simitsis, A., Wilkinson, K., Dayal, U., Castellanos, M.: Optimizing ETL workflows for fault-tolerance. In: 2010 IEEE 26th International Conference on Data Engineering (ICDE 2010), pp. 385–396 (2010)
17. Simitsis, A., Wilkinson, K., Castellanos, M., Dayal, U.: Optimizing analytic data flows for multiple execution engines. In: Proceedings of the 2012 ACM SIGMOD International Conference on Management of Data, pp. 829–840 (2012)
18. Vanderfeesten, I.T.P., Reijers, H.A., van der Aalst, W.M.P.: Product-based workflow support. Inf. Syst. **36**(2), 517–535 (2011)

Text2Dec: Extracting Decision Dependencies from Natural Language Text for Automated DMN Decision Modelling

Vedavyas Etikala[(⊠)] [iD], Ziboud VanVeldhoven [iD], and Jan Vanthienen [iD]

Leuven Institute for Research on Information Systems (LIRIS),
KU Leuven, Leuven, Belgium
vedavyas.etikala@kuleuven.be

Abstract. Decisions are of significant value to organisations. Business decisions are often written down in textual documents, and modelling them is a tedious and time-consuming task. Although decision modelling has seen a surge of interest since the introduction of the Decision Model and Notation (DMN) standard, limited research has been conducted regarding automatically extracting decision models from the text. In this paper, we propose a text mining technique to automatically extract the decisions and their dependencies from natural language text to build the decision requirements diagram. A case-based evaluation is shown for the proposed mining approach with promising results. This approach can serve as a groundwork for further research in the field of decision automation.

Keywords: Decision Model and Notation (DMN) · Business decision management · Natural language processing (NLP)

1 Introduction

Efficient decision modelling adds significant value to organisations in managing their recurrent yet essential business decisions such as granting a loan, determining credit card eligibility, or diagnosing a patient. Representing business knowledge as decision models not only increases the interpretability of otherwise complex decision processes but also paves the road towards automation of business decision management (BDM).

BDM, which concerns the entire process of modelling, managing, and enacting decisions present in the organisation, has gained increased interest in recent years. Since the introduction of Decision Model and Notation (DMN) as a decision modelling standard by the Object Management Group (OMG) in 2015, modelling decision knowledge at a higher level of abstraction has been made possible and reliable [11,22]. Successful decision modelling requires understanding the business knowledge and learning the modelling technique [24]. Both of these

© Springer Nature Switzerland AG 2020
A. Del Río Ortega et al. (Eds.): BPM 2020 Workshops, LNBIP 397, pp. 367–379, 2020.
https://doi.org/10.1007/978-3-030-66498-5_27

tasks are time-consuming. Despite DMNs recent popularity [5,8,11,17,18,23], little research has been conducted to improve these areas. Furthermore, most of the business decision knowledge is still stored and shared in textual documents [13,20].

Hence, we propose a novel Text2Dec framework that uses state-of-the-art text mining and natural language processing (NLP) to extract decision dependencies as Decision Requirements Diagrams (DRD) from text-based documents. As far as we know, this is the first attempt to address the transformation of text into DMN decision requirements. We follow a three-step methodology inspired by similar works in other modelling domains [1,10,21]: (i) understanding the textual descriptions of decisions, (ii) coding an ensemble of tailored NLP techniques for detecting patterns and identifying the concepts of decision models, and (iii) generating the DRD. The method is evaluated with a real-world case of "Employee Health Assessment".

This paper is structured as follows. Section 2 introduces related work on decision modelling and DMN. Next, we explain the scope of the proposed approach in Sect. 3. Section 4 presents the methodology of the Text2Dec framework for automated decision dependency extraction. In Sect. 5, we present a case study to evaluate the proposed approach. Section 6 contains the discussion, challenges, and future work. Finally, Sect. 7 concludes the research.

2   Motivation and Related Work

2.1   Decision Modelling and DMN

Decision models, represented in DMN, are knowledge structures that capture not only business decisions but also their dependencies and logic. The models are intended to be both executable and understandable by all stakeholders. A DMN decision model consists of two levels. First, the decision requirements level in the form of a DRD (or graph) is used to model the requirements of the decisions and the dependencies between the different constructs in the decision model. Secondly, the decision logic level is used to specify the detailed logic for each decision, usually in the form of decision tables. The DRD consists of a small set of constructs: rectangles to depict decisions, corner-cut rectangles for business knowledge models, and ovals to represent data input as shown in Fig. 1. Requirements are depicted with arrows, e.g.. an information requirement indicates that a decision requires the information from input data or the result of another decision.

Definition 1. An *entity* is the business object paying a pivot role in the given decision scenario. e.g.: person, weather or discount.

Definition 2. A *concept* is an attribute of an entity, e.g.: loyalty of a customer, height of a patient, loan qualification (an information item in DMN).

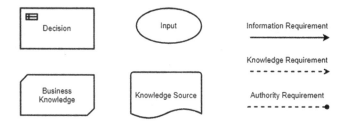

Fig. 1. Components and requirements of DMN

Definition 3. A *dependency* is either an information requirement or an authority requirement or a knowledge requirement in DMN, which is represented as a link between components of the DRD. A dependency link is mathematically represented as a tuple dep $= (\mathcal{A}, \mathcal{B}, \mathcal{D})$, where $\mathcal{A}$ is an action verb, $\mathcal{B}$ is a base concept, and $\mathcal{D}$ is a derived concept.

Definition 4. A *base concept* is an input information item in DMN. It is not the result of a decision.

Definition 5. A *derived concept* is a concept derived from one or more other concepts through a decision. The derived concept is dependent on these other concepts. E.g, *"Body Mass Index value"* is the derived concept from the base concepts *"height"* and *"weight"* of the patient.

Definition 6. A *DMN DRD* is a tuple in the form of DRD $= (\mathcal{D}, \mathcal{I}, \mathcal{R})$ where $\mathcal{D}$ is a set of decisions, $\mathcal{I}$ is a set of input information items identified as concepts and $\mathcal{R}$ is set of requirements identified as dependencies.

2.2   DMN Modelling

Decision models are usually constructed manually by domain modellers or business analysts based on the provided documentation and/or the knowledge acquired from the interaction with domain experts. Several modelling guidelines have already been proposed [2,16,24] to obtain complete, consistent and implementable decision models. However, the process remains a manual effort, even with tool support, and is therefore difficult and time-consuming.

Recently, some knowledge discovery approaches have been proposed for extracting parts of the decision model from structured sources, such as process event logs, historical data or process model flows:

- Decision rules can be discovered from historical cases as part of the large area of business analytics. In the context of process discovery, mining the decision logic at decision points in a process is called decision mining. The structure of the decisions can be derived too [4,6].
- When a business process model is available, a decision model can be extracted from the process by identifying the decision points and data dependencies in

the process flow. The result is a decision model (including the DRD) and a more simple process model [2,3].

- Methodologies to mine decision models together with process models from extensive decision-process logs are also proposed, producing separate but integrated decision and process models [4,9].

Despite these works on automating modelling from structured sources, limited research has been conducted on extracting decision models from textual sources. Decisions, however, are usually described in text. Hence, (semi)-automatic extraction of decision models from text would be a highly beneficial endeavor. In related business modelling domains, extraction from text is not a new field:

- A number of techniques and tools have been suggested to extract process models from text [1,12,19].
- Rule extraction from legal and business texts has been researched in e.g. [7,10]. Extracting the attributes and their values for correct logical statements is the point of focus in this domain.

Existing model extraction approaches detect sentence patterns to identify tasks and control structures such as "approve the claim" or "mail the client" [15]. The decision logic and requirements extractors, however, need to identify the relevant information that exists both explicitly in declarative statements and also implicitly in conditional statements. Conceptually, decision information is not just limited to a single sentence. Therefore, adequate paragraph-level mining is needed to analyse semantic and syntactic clues for language.

3   Scope of the Proposed Approach

Because deriving the entire detailed decision logic from a complex text would be an immense task and would produce a long list of logical rules without an appropriate structure, we follow the DMN guidelines and separate the decision requirements level from the decision logic level. Decision logic, because it has to deal with specific values for input and outcome information items, has to be much more precise than dependencies between concepts. Once the decisions and their requirements are identified, it will be easier to isolate and construct the decision logic in a (semi)-automatic way. The exact value of the outcome of a decision is not the first concern but rather the observation that a decision depends on, e.g. one specific input information item is the result of another decision. In this stage of the research, only information requirements are considered. Consider the following running example to understand the practical benefit of being able to automatically structure textual knowledge as a DRD:

Example: *"The health risk level of a patient should be assessed from the obesity level, waist circumference and the sex of the patient. Furthermore, the degree of obesity should be determined from the BMI value and sex of the patient. Patient's height and weight are considered to calculate his BMI value. If the weight of the*

patient given in kgs and height of patient given in meters, then the BMI value is weight/(height*height)."[1]
The concepts need to identified, extracted, and categorised into base concepts (height, weight, sex, waist circumference) and derived concepts (BMI value, obesity level, health risk level) based on the semantic role using syntactic patterns of nouns around the action verbs (assess, consider, determine, is). Each pair of base and derived concept can then be represented as dependency tuples (action, base concept, derived concept). Finally, using the domain-independent heuristics, the tuples can be converted into a DRD as shown in Fig. 2.

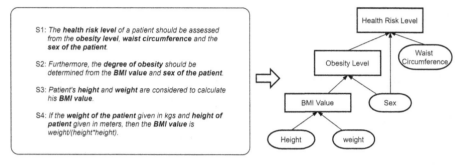

Fig. 2. The DRD of the health risk level description

4 Text2Dec Framework

To identify, extract and represent the decision requirement knowledge into a DRD from natural language documents, we have developed a novel Text2Dec framework. This approach follows three stages and is shown in Fig. 3. We use the above given *'health risk level'* example to illustrate the proposed methodology.

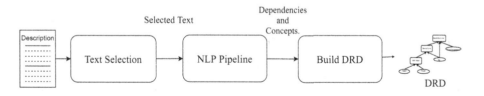

Fig. 3. Text2Dec framework

---

[1] Taken from https://www.nhlbi.nih.gov/files/docs/guidelines/prctgd_c.pdf.

Table 1. Sentence patterns considered to extract dependencies.

Pattern	Example	Base concepts	Derived concepts
Passive A<= B	Patient's BMI value is determined from his height	Height	BMI value
Active A => B	Patient's height determines his BMI value	Height	BMI value
Conditional A => B	Unless the season is summer, do not plan a barbeque	Season	Plan a barbeque
Conditional A<= B	A customer is loyal, if his annual sales are high	Annual sales	Customer

4.1 Stage I: Requirement Text Selection

In the first stage, the software automatically identifies the sentences containing decision dependency patterns as shown in Table 1 from the given input text using regular expressions and a list of predefined verbs that is used to identify requirement statements. Only sentences in these formats are considered for further processing. Where A and B are concepts, and an arrow <= indicates dependency.

4.2 Stage II: NLP Pipeline

In the second stage, using the open-source tool kits standford's coreNLP[2], NLTK[3], neuralcoref[4], and SpaCy libraries[5], an NLP pipeline has been built in python code and applied on the running example. This pipeline consists of six steps to extract the dependencies from the text as shown in Fig. 4.

Step a: Preprocessing. The algorithm reads the selected text and preprocesses it by removing determinants such as *the*, *a*, and *an*. For example, the sentence *"The Risk Level is assessed from the BMI Level and waist circumference"* is converted to *"Risk level is assessed from BMI level and waist circumference."*

Step b: Coreference Resolution. In this step, coreferences are identified and mapped based on semantic equivalence following the coreference resolution technique. Coreferences are the synonymous terms and phrases that occur in the text. They are resolved by replacing all coreferences with their first occurrence's term. For the running example, *degree of obesity* is replaced with *obesity level.*

Step c: Anaphora Resolution. Anaphoras such as cross-referred pronouns are detected and fixed by replacing pronouns with their referring owners. The

---

[2] https://nlp.stanford.edu/software/.
[3] https://www.nltk.org/.
[4] https://github.com/huggingface/neuralcoref.
[5] https://spacy.io/usage/.

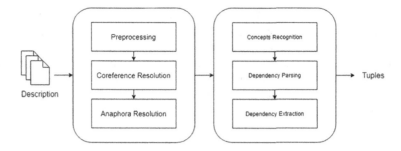

Fig. 4. NLP pipeline

ownership of the noun terms is also propagated over the conjunctions. This step is important because both the entity or the attribute of an entity could be the concept of interest. For the running example, the pronoun *his* is mapped with *patient's*. Also, the ownerships are simplified, e.g.. *"of the patient"* is replaced with *"patient's"*. These both resolutions also help to correctly identify the intermediate decisions.

The result of these steps is: ***patient's*** health risk level should be assessed from ***patient's obesity level***, ***patient's*** waist circumference and ***patient's*** sex. Furthermore, ***patient's obesity level*** should be determined from ***patient's*** BMI value and ***patient's*** sex ...

Step d: Concept Recognition. At sentence level, the syntactic details are exploited by breaking each sentence into tokens. We process each sentence to recognise different concepts based on noun phrases, e.g. *health risk level*. To convert the noun phrases into concepts, a chunking technique has been used to merge noun tokens based on the parts-of-speech (POS)[6] tags such as NOUN for a noun, and PRON for a pronoun and PROPN for a proper noun. The result is that *[health:NOUN][risk:NOUN][level:NOUN]* is converted into *[health risk level:NOUN]*. In our risk level example, this step automatically generates a concept list c1: health risk level, c2: obesity level, c3: waist circumference, c4: sex from the statement "health risk level should be assessed from obesity level, waist circumference and sex".

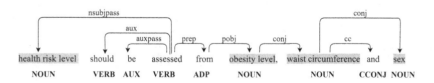

Fig. 5. Dependency parse tree for the example statement

---

[6] https://universaldependencies.org/u/pos/.

Step e: Dependency Parsing. In this step, a verb-based dependency parse tree is built. To confirm dependency between concepts, we match the detected action verb with a predefined list of action verbs such as *require, decide, select determine, asses, calculate, consider* .... For the conditional statements when the main action verb is not detected, auxiliary verbs such as *"is"* are considered as action verbs. For example, a parse tree formed for the example statement is shown in Fig. 5. Even though *should* is a *verb* along with the root verb *assessed*, it is not considered as an action verb due to the lack of dependent concepts. Afterwards, a compact dependency parse tree is formed, highlighting only the concepts and action verbs for each sentence as shown in Fig. 6. Hence, the output of this step is a compact dependency parse tree for each sentence.

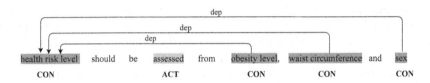

Fig. 6. Compacted parse tree for the example statement

Step f: Dependency Extraction. In this step, the concepts are linked with the identified dependencies from the parse tree. Concepts are classified into base concepts and derived concepts depending on the role they are playing in the statement. In the example statement, the concept *health risk level* is playing a passive subject role (*nsubjpass*) with the action verb *assessed*. Therefore, it will be labelled as a derived concept. The concepts from step d and the dependencies from step e are joined, and a set of dependency tuples in the format *(relation, base concept, derived concept)* is generated based on the labelled concepts. The result of this example is a set of tuples consisting of { *(assess, obesity level, health risk level), (assess, waist circumference, health risk level), (assess, sex, health risk level),...*}. By using set data structures, duplication of tuples is avoided.

Once tuple sets for all sentences from the text are formed, a unified set of tuples is generated. This process depends on the order of the statements in the given description. Here, base concepts are filtered by removing the concepts that are identified as derived concepts in the later sentences and labelled as initial inputs to the DRD. A decision set is constructed where each decision contains a set of inputs and outputs. The final result is a DRG. In the running example, the DRG formed is { *(decisions, inputs, requirements)*} with decisions = { *([obesity level, waist circumference, sex],[health risk level]),([BMI value, sex],[obesity level]),([weight, height],[BMI value])*}, inputs = { *waist circumference, sex, weight, height*}, requirements = { *(assess, obesity level, health risk level), (assess, waist circumference, health risk level), (assess, sex, health risk level),....*}.

4.3   Stage III: DRD Construction

In the final step of our framework, an XML file is generated based on the extracted concepts and their dependency relationships mentioned in the DRG. Only the connected graph component of the obtained DRG of the target decision is converted into a DRD. The XML file can be read by popular DMN tools such as Camunda[7]. For the running example generated DRD is shown in Fig. 2. Other DMN constructs such as knowledge sources, business knowledge models, knowledge or authority requirements are not included yet.

5   Evaluation

We evaluated the Text2Dec framework on a simple "Prepayment" example and a larger "Employee Health Evaluation" case inspired from the example DRD given in [14]. The decision descriptions are designed to be natural and to contain various patterns of implicit and explicit dependencies and pose multiple challenges. The first description reads *"Prepayment is not required for loyal customers when the OrderAmount is small. A loyal customer is defined as such if his AnnualSales is high and his Customeryears is more than 5."*. The Text2Dec approach generates the dependency tuples and corresponding DRD as shown in Fig. 7.

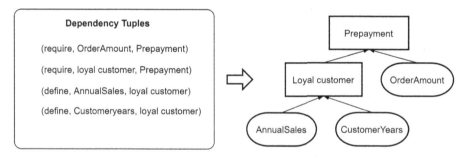

Fig. 7. The Dependencies and DRD of the Customer prepayment

The second description reads: *"Health evaluation criteria consider both physical health score and mental health score of a patient. The physical health score is determined from the physical fitness score, BMI based health risk level and healthiness of senses. Physical fitness score is calculated from the sex of a patient and results of various tests such as strength test, coordination test, agility test, stamina test and speed test. The patient's risk level should be assessed by determining the level of obesity based on BMI value, also on the waist circumference. The obesity level or degree of overweight should be assessed by determining the*

---
[7] https://camunda.com/dmn/.

BMI value. If the weight of the patient given in kgs and length of patient given in meters, then the BMI value is weight/(length*length). The healthiness of the senses is calculated from the results of eye and hearing tests. Health evaluation also depends on the score of Mental health, which is determined from the EQ test result and the IQ test score. An IQ of a patient is assessed from testing his verbal, math and abstract levels."

Table 2. Extracted dependencies from the health evaluation case

SNO	Base concept	Action verb	Derived concept
1	Physical health score	Consider	Health evaluation criteria
2	Mental health score	Consider	Health evaluation criteria
3	Physical fitness score	Determined	Physical health score
4	BMI based health risk level	Determined	Physical health score
5	Healthiness of senses	Determined	Physical health score
6	Sex	Calculated	Physical fitness score
...	...	...	...
22	Math	Assessed	IQ test score
23	Abstract	Assessed	IQ test score

By applying the Text2Dec framework on the second description, the following dependencies are extracted, shown in Table 2. Next, these dependencies are automatically transformed into an XML file which can be read by Camunda as shown in Fig. 8.

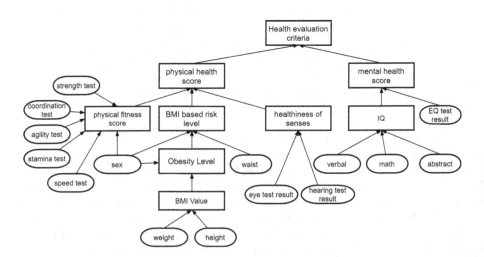

Fig. 8. The DRD of the health evaluation example

To assess the quality of the generated DRDs, we compared them with manually modelled DRDs. The generated DRDs have promising results: the structure of the DRD, and the number of nodes, information items, and information requirements stay the same. However, the obtained decision labels are slightly different. For example, the "math test result" is detected as "math" in our approach resulting in a small loss of semantic meaning.

6 Discussion

In this paper, we presented the Text2Dec framework for extracting DRDs from textual descriptions. The obtained DRDs correspond with those manually modelled, given the assumptions stated below, showcasing the promising applicability of our work. Automatically extracting DRDs can be useful for rapid prototyping. These prototype models can then be manually completed into full models with the information needed for the execution (e.g. decision logic). The presented framework can serve as a cornerstone for automatic decision model extraction from text.

This methodology faces several challenges. First, the ambiguity in language is hard to grasp. Therefore, we assume that the input decision descriptions consist of non-ambiguous and grammatically sound full English sentences. Nevertheless, our methodology allows for expanding the vocabulary and detection of patterns. Secondly, the main decision must be correctly identified for the DRD construction which can be hard in long texts. Thus, the Text2Dec framework relies on the assumption that the description is sequential, contains no irrelevant information or redundancies, and contains only one main decision. Otherwise, this could lead to unconnected components in the intermediate decisions in the DRD. The applied coreference resolution is a crucial step in this regard to link the cross-referred concepts in the text. It is used to identify the links between different implicit dependencies between the components of the DRD. Thirdly, not all derived inputs are inputs in the real models. Some of these are knowledge sources or business knowledge models. This distinction is not yet implemented and is needed to make models more clean and consistent.

For future work, we will extend the Text2Dec framework to automatically extract the decision logic found in textual documents in the form of decision tables. This way, we can automatically derive a fully functional decision model from the text. Moreover, we would like to extend our work towards full end-to-end automation for decision knowledge and decision support applications. Secondly, we plan to evaluate the performance of the proposed methodology quantitatively. To analyse the complexity of the generated models in realistic decision support systems, we will apply the Text2Dec framework on a series of real-world cases which also includes the challenges of ambiguity and redundancy. Opportunities exist for fellow researchers to investigate automatic conformance checking with the textual guidelines, conversational decision support agents and explainability in decision making.

7 Conclusion

In this study, we contribute to the BDM and DMN research by presenting a novel Text2Dec framework for automating the extraction of decision dependencies from business decisions' descriptions. The evaluation shows that the generated DRDs correspond with the manually developed models with in terms of structure and semantics. This work paves the way towards automated decision modelling and logic extraction.

References

1. van der Aa, H., Di Ciccio, C., Leopold, H., Reijers, H.A.: Extracting declarative process models from natural language. In: Giorgini, P., Weber, B. (eds.) CAiSE 2019. LNCS, vol. 11483, pp. 365–382. Springer, Cham (2019). https://doi.org/10.1007/978-3-030-21290-2_23
2. van der Aa, H., Leopold, H., Batoulis, K., Weske, M., Reijers, H.A.: Integrated process and decision modeling for data-driven processes. In: Reichert, M., Reijers, H.A. (eds.) BPM 2015. LNBIP, vol. 256, pp. 405–417. Springer, Cham (2016). https://doi.org/10.1007/978-3-319-42887-1_33
3. Batoulis, K., Meyer, A., Bazhenova, E., Decker, G., Weske, M.: Extracting decision logic from process models. In: Zdravkovic, J., Kirikova, M., Johannesson, P. (eds.) CAiSE 2015. LNCS, vol. 9097, pp. 349–366. Springer, Cham (2015). https://doi.org/10.1007/978-3-319-19069-3_22
4. Bazhenova, E., Buelow, S., Weske, M.: Discovering decision models from event logs. In: Abramowicz, W., Alt, R., Franczyk, B. (eds.) BIS 2016. LNBIP, vol. 255, pp. 237–251. Springer, Cham (2016). https://doi.org/10.1007/978-3-319-39426-8_19
5. Calvanese, D., Dumas, M., Laurson, Ü., Maggi, F.M., Montali, M., Teinemaa, I.: Semantics and analysis of DMN decision tables. In: La Rosa, M., Loos, P., Pastor, O. (eds.) BPM 2016. LNCS, vol. 9850, pp. 217–233. Springer, Cham (2016). https://doi.org/10.1007/978-3-319-45348-4_13
6. Campos, J., Richetti, P., Baião, F.A., Santoro, F.M.: Discovering business rules in knowledge-intensive processes through decision mining: an experimental study. In: Teniente, E., Weidlich, M. (eds.) BPM 2017. LNBIP, vol. 308, pp. 556–567. Springer, Cham (2018). https://doi.org/10.1007/978-3-319-74030-0_44
7. Danenas, P., Skersys, T., Butleris, R.: Natural language processing-enhanced extraction of SBVR business vocabularies and business rules from UML use case diagrams. Data Knowl. Eng. 19 (2020)
8. Dasseville, I., Janssens, L., Janssens, G., Vanthienen, J., Denecker, M.: Combining DMN and the knowledge base paradigm for flexible decision enactment. Supplementary Proceedings of the RuleML 2016 Challenge 1620 (2016)
9. De Smedt, J., Hasić, F., vanden Broucke, S.K., Vanthienen, J.: Holistic discovery of decision models from process execution data. Knowl.-Based Syst. **183**, 15 (2019)
10. Dragoni, M., Villata, S., Rizzi, W., Governatori, G.: Combining NLP approaches for rule extraction from legal documents. In: MIREL (2016)
11. Figl, K., Mendling, J., Tokdemir, G., Vanthienen, J.: What we know and what we do not know about DMN. Enterp. Model. Inf. Syst. Archit. Int. J. Concept. Model. **13**(2), 1–16 (2018)

12. Friedrich, F., Mendling, J., Puhlmann, F.: Process model generation from natural language text. In: Mouratidis, H., Rolland, C. (eds.) CAiSE 2011. LNCS, vol. 6741, pp. 482–496. Springer, Heidelberg (2011). https://doi.org/10.1007/978-3-642-21640-4_36

13. Froelich, J., Ananyan, S.: Decision support via text mining. In: Handbook on Decision Support Systems (2008)

14. Hasic, F., Vanthienen, J.: Complexity metrics for DMN decision models. Comput. Stand. Interfaces **65**, 15–37 (2019)

15. Honkisz, K., Kluza, K., Wisniewski, P.: A concept for generating business process models from natural language description. In: KSEM (2018)

16. Janssens, L., Bazhenova, E., Smedt, J.D., Vanthienen, J., Denecker, M.: Consistent integration of decision (DMN) and process (BPMN) models. In: CAiSE Forum. CEUR Workshop Proceedings, vol. 1612, pp. 121–128. CEUR-WS.org (2016)

17. Janssens, L., De Smedt, J., Vanthienen, J.: Modeling and enacting enterprise decisions. In: Krogstie, J., Mouratidis, H., Su, J. (eds.) CAiSE 2016. LNBIP, vol. 249, pp. 169–180. Springer, Cham (2016). https://doi.org/10.1007/978-3-319-39564-7_17

18. Kluza, K., Honkisz, K.: From SBVR to BPMN and DMN models. proposal of translation from rules to process and decision models. In: Rutkowski, L., Korytkowski, M., Scherer, R., Tadeusiewicz, R., Zadeh, L.A., Zurada, J.M. (eds.) ICAISC 2016. LNCS (LNAI), vol. 9693, pp. 453–462. Springer, Cham (2016). https://doi.org/10.1007/978-3-319-39384-1_39

19. Sànchez-Ferreres, J., Burattin, A., Carmona, J., Montali, M., Padró, L.: Formal reasoning on natural language descriptions of processes. In: BPM (2019)

20. Silver, B.: DMN Method and Style, 2nd Edition: A Business Practitioner's Guide to Decision Modeling. Cody-Cassidy Press (2018)

21. Sintoris, K., Vergidis, K.: Extracting business process models using natural language processing (NLP) techniques. In: 2017 IEEE 19th Conference on Business Informatics (CBI), vol. 1, pp. 135–139 (2017)

22. Taylor, J., Fish, A., Vanthienen, J., Vincent, P.: Emerging standards in decision modeling. In: Intelligent BPM Systems: Impact and Opportunity, pp. 133–146. BPM and Workflow Handbook series, iBPMS Expo (2013)

23. Valencia-Parra, Á., Parody, L., Varela-Vaca, Á.J., Caballero, I., Gómez-López, M.T.: DMN for data quality measurement and assessment. In: Di Francesco-marino, C., Dijkman, R., Zdun, U. (eds.) BPM 2019. LNBIP, vol. 362, pp. 362–374. Springer, Cham (2019). https://doi.org/10.1007/978-3-030-37453-2_30

24. Vanthienen, J., Dries, E.: Illustration of a decision table tool for specifying and implementing knowledge based systems. Int. J. Artif. Intell. Tools **3**(2), 267–288 (1994)

Data Object Cardinalities in Flexible Business Processes

Stephan Haarmann$^{(\boxtimes)}$ and Mathias Weske

Hasso Plattner Institute, University of Potsdam, 14482 Potsdam, Germany
{stephan.haarmann,mathias.weske}@hpi.de

Abstract. Business process models are an important tool for business process management, used to specify, enact, and analyze organizations' processes. Traditional process modeling languages are particularly well suited for highly structured, predictable processes but lead to complex models when faced with flexible data-centric processes such as knowledge-intensive ones. Different approaches address this gap. However, while data is central in many modeling languages for knowledge-intensive processes, the impact of a data model (containing classes, associations, and cardinalities) on the process behavior has not been investigated holistically. We extend the fragment-based Case Management (fCM) approach with data cardinalities and discuss the impact of cardinalities on process execution and analysis, as well as flexibility.

Keywords: Knowledge-intensive processes · Process modeling · Data modeling · Cardinalities

1 Introduction

Business process management (BPM) helps practitioners to elicit, design, enact, and analyze processes. Most BPM activities employ process models to specify processes [20]. Traditional BPM methods focus on highly structured, repetitive, and fully-defined processes. At the same time, knowledge work plays an important role in companies. Corresponding knowledge-intensive processes are highly flexible and driven by decisions. The processes depend on data and may require adaptation during execution. Such processes cannot be modeled concisely using traditional modeling languages, such as BPMN. New languages address the requirements of knowledge-intensive processes. In general, these languages strive for providing guidance to the knowledge workers while maintaining a beneficial degree of flexibility. Since knowledge-intensive processes are driven by data-based decisions, it is necessary to link the domain model with the process definition. Domain models describe the data types available and possible relationships between corresponding objects. However, most approaches neglect certain aspects common in domain models, such as cardinalities.

In this paper, we investigate the impact of cardinality constraints on the behavior of processes. While the approach is applicable to BPMN, it is especially useful to flexible processes because they have few behavioral constraints.

© Springer Nature Switzerland AG 2020
A. Del Río Ortega et al. (Eds.): BPM 2020 Workshops, LNBIP 397, pp. 380–391, 2020.
https://doi.org/10.1007/978-3-030-66498-5_28

Therefore, we extend the fragment-based Case Management (fCM) [6] approach with cardinalities. We distinguish between global cardinality constraints that must never be violated and goal cardinalities that must be satisfied eventually. These constraints impact the processes' behavior directly. Furthermore, we discuss consequences for BPM-related tasks such as process modeling, enactment, and analysis as well as the impact on the flexibility. Adding constraints may limit the knowledge workers' scope of actions drastically.

This paper is structured as follows: we provide an overview of the fCM approach as well as cardinalities in data models in Sect. 2. In Sect. 3, we informally refine fCM's execution semantics using cardinalities. We discuss the implication on BPM and knowledge workers flexibility in Sect. 4. Related work is presented in Sect. 5. Finally, Sect. 6 concludes our work.

2 Preliminaries and Motivating Example

Knowledge-intensive processes evolve around data and have many variants. The decisions of knowledge-workers scope a case by discarding some variants. Capturing such processes in a concise but expressive model requires a tight integration of data and behavior. We use the hybrid modeling approach fCM, which combines activity-centric process *fragments* with data flow and object behavior.

Fragments contain relatively few activities, which are connected through control flow. Conditional flow can be realized using exclusive gateways, and a fragment may start with an event that triggers a new case. Fragments share data: all objects that belong to the same case can be accessed by all fragments of that case. Fragments can be executed repeatedly, and instances of the same and different fragments can run concurrently. However, data requirements of one fragment instance may be satisfied by progressing another. This leads to synchronization. Usually, a set of activities is enabled, meaning that it has been reached by control flow and all data requirements are satisfied. In this situation, it is the knowledge workers' responsibility to choose the next activity. Furthermore, the knowledge worker can terminate the case once the termination condition is satisfied.

An activity can read, write, and update a set of data objects. Just as activities in BPMN, activities have input and outputs sets partitioning the inputs and outputs in potentially overlapping groups. Whenever an activity is executed, all objects of one input set are read, and all objects of one output set are written. The combinations of inputs with outputs are further limited by *object life cycles* (OLC). An activity may only perform transitions included in the OLCs.

A class is a type for a set of objects. In fCM, a domain model contains all classes for one case and binary associations among them [5]. A dedicated case class describes case objects: there is exactly one such object for each case. When an activity creates two objects whose classes are associated, the two objects are associated as well. Associations are also created if one object is read and another one is created. When objects whose classes are associated are read, the objects must be associated. Thus far, cardinalities play no role in the domain model.

For illustration, consider the process of submitting a paper to a conference till the notification of the authors. The fragments of the case model are depicted in Fig. 1. Once the start event *conference scheduled* occurs (fragment **f1**), all other fragments (**f2–f8**) are control flow enabled. However, activities have data requirements, data objects that are read, and cannot be executed before these are satisfied. Thus, the call for papers (CfP) must be published first. Next either the CfP can be sent out (**f1**) or the conference can be opened for submissions (**f2**). Once the conference has been opened, papers can be submitted (**f3**), updated (**f4**), and the conference can be closed for submissions (**f2**). After closing the submission, reviewers are assigned (**f5**), reviews are created (**f6**), papers are accepted or rejected (**f7**), before the notifications are sent to the authors (**f8**). Once a fragment has been control-flow enabled, it is not disabled. Even if it is executed, for example, when a paper is submitted (**f3**), it can be executed again if the data requirements are met. The case can be closed, if the termination condition "*conference is closed for submissions*" is satisfied.

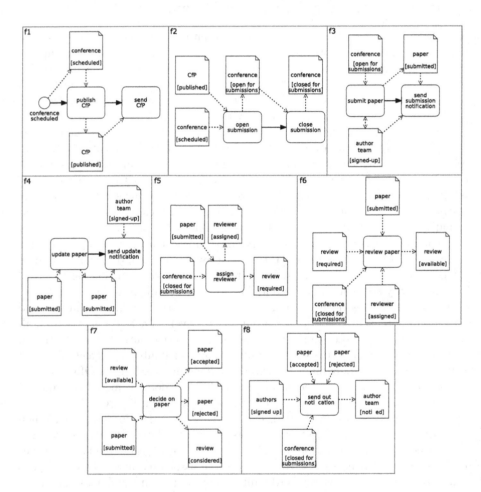

Fig. 1. Fragments of a paper submission and reviewing process for a conference

The example's domain model is depicted in Fig. 2. It contains the case class *Conference* and the classes *CallForPapers, Extension, Paper, Review*, and *Reviewer* as well as the depicted relationships. The domain model is represented as a data model. Cardinalities are commonly used in data models to constrain the number of instances of a class in relationship to other objects. While there exist different notations and semantics for cardinalities, we use and extend UML's multiplicities. The extension, a second pair of cardinalities with a leading ◇, is explained later. A cardinality constrain is composed of a lower bound and an upper bound. The lower bound is a natural number greater or equal to zero; an upper bound is any natural number greater than zero or infinite (unbounded). We exclude infinite from our discussion, because a case should never depend on an unbounded number of objects. An arbitrary but specific bound should always be set. In the example, every paper is in relationship with up to four reviews, but each review is associated with only one paper and one reviewer. A reviewer is responsible for at least one, but possibly ten reviews and papers.

The set of objects in a case is represented by an object model. As the case progresses new objects are created, and consequently the object model evolves. The cardinality constraints must never be violated in the domain model. However, usually the case, including the object model, must reach a certain configuration to terminate. A traditional domain model does not contain this information. Therefore, we extend the domain model with a second pair of cardinalities that must hold *eventually*. We call these *goal cardinalities* and mark them with a leading ◇. Such cardinalities exist in the object-centric behavioral constraints (OCBC) modeling language [1]. Goal cardinalities refine global ones by setting an equal or higher lower bound and a smaller or equal upper bound. In the example, one paper is always associated with 0..4 reviews. However, eventually there must be at least three. Setting the respective goal cardinality to 3..4 ensures this.

3   Interference Between Cardinalities and Process Logic

In case models, data model, object life cycles, and process logic form a symbiosis. The definition of one usually contains information about or relevant to the others. For example, fragments define the process logic based on activities creating, reading, and updating data objects, whose types are specified in the domain model and whose state transitions are defined by the OLCs. In the following, we showcase the effect of data cardinalities on the execution semantics of flexible processes specified as fCM models.

3.1   Data Cardinalities Apply to Fragments

Flexible processes often require repeated execution of the same activity or the same sequence of activities specified as a fragment. Consequently, some fragments may be executed arbitrarily often. In the example (Fig. 1), *assign reviewer* (**f5**) and *update paper* (**f4**) can be executed arbitrarily often before the case

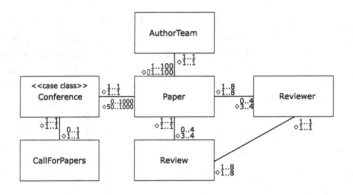

Fig. 2. Domain model for the paper submission example

terminates. Some activities in these fragments create data objects, e.g., *assign reviewer* creates a *review* and a *reviewer* objects. However, the domain model specifies cardinalities restricting the number of objects with respect to a *paper* object. Since there are at most four reviews for each paper, we know that the fragment *assign reviewer* must not be executed more than four times, respectively. Furthermore, the upper bound of both global and goal cardinalities should always be equal because objects cannot be deleted (in fCM); thus, once a case exceeds goal cardinalities, the violation cannot be resolved.

It is possible that multiple fragments contain activities creating objects of a certain type. In this case, the upper bound does not apply to each fragment individually but to the group of respective fragments. However, not all fragments have upper bounds. Some fragments do not create new data objects. If these fragments are unbounded, i.e., can be executed arbitrarily often, they remain unbounded in the presence of cardinalities. Other fragments may allow the knowledge worker to decide whether an object is created or not, i.e., through exclusive branches or different input and output sets. In this case, the branch or the output-set rather than the complete fragment is disabled.

Lower bounds can be used to specify a minimum number of executions. For now, we focus on lower bounds of goal cardinalities. If a domain model states how many objects of a certain type have to exist eventually (lower bound of the goal cardinality), then the fragments creating such objects have to be executed at least as often as necessary to satisfy the lower bound. In the example, there eventually must be three to four reviews for each paper. Since the activity *assign review* assigns a single review, it must be executed at least three times.

3.2   Cardinalities Reflect in Input- and Output-Sets

If two classes are associated by a *global* 1-to-1 relationship (both upper and lower bounds are 1), no respective object can exist without a corresponding object of the other class. Additionally, each object is linked to exactly one counterpart. Thus, the number of objects is the same for both classes. On a process level, this can only be accomplished by jointly creating both objects: any activity that creates an instance of one class must also create an object of the other. For the example in Fig. 1, assume the relationship between *review* and *reviewer* to be such a 1-to-1 association. Then, *assign reviewer* must create both objects.

Now, we adapt the relationship by setting the lower bound for one class to zero. This class, we call it *dependent*, requires objects of the other class, which we call *supporter*, but not the other way around. This constellation allows another possibility additionally to joint creation: Objects of the dependent class can be created if the activity also reads an object of the supporter. This means, in order to create a dependent object the supporting object must either be part of the input set (it is read) or the same output set (joint creation).

Thus far, we investigated global cardinalities with a lower bound of either 0 or 1. In theory, the lower bound can be any natural number. Whenever the lower bound is positive, this bound is the minimal number of supporter objects required to create one or multiple dependent objects. Just as before, supporters must be available during creation of dependents. Thus, it is necessary that activities read or write multiple objects of one type, the supporter type, at once.

We adapt the fragment that accepts or rejects a paper (Fig. 3, cf. Fig. 1). We store the decision in an additional data object, it is associated to a paper and the three to four corresponding reviews. The lower bound for the review class for the decision-to-review association is three. Meaning, for each decision object there always will be at least three review. This requires all three reviews to be present during the creation of a decision, namely when executing *decide on paper*. Consequently, the activity must read multiple review objects. Previously, such behavior could not be specified in fCM. We reuse BPMN's list data object to indicate that multiple objects of a certain class are read or written.

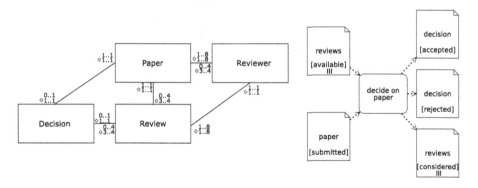

Fig. 3. The adapted domain model featuring a decision class with associations to a paper and a set of reviews. It requires the activity to read and write multiple review objects at once.

3.3   Process Logic Realizes Transitive Associations

The process may realize constraints that cannot be captured in a plain domain model. Consider the triangular associations between paper, reviewer, and review. The domain model does not capture that a review must be related to the paper that is associated to the review's reviewer. Such a property can, for example, be expressed using the object constrain language [15]:

```
context Review inv: self.paper = self.reviewer.paper
```

However, the process logic may manifest this property without explicit definition: review and reviewer are created in a single step during which a paper is read. This establishes the desired relationship among the process. Even if the assignment and review creation is split, a paper and a reviewer is necessary to create a review. fCMs semantics assert that these two are related and, thus, establishes the relationship nevertheless [5].

4   Discussion of Implications

Combining process and data models, including cardinalities, may have significant impact on the behavior. Furthermore, there are implications for various business process management activities as well as the flexibility of the processes. The latter is a crucial property for knowledge work. In this section, we discuss both types of implications.

4.1   Cardinalities During Design and Analysis

Data models impose structural properties on the objects written, read, and updated by the activities of the process. A proper integration of the data structure and the process logic is necessary. This leads to refined execution semantics (cf. Sect. 3). Also, cardinalities may impact other BPM activities.

Obviously, to refine the process behavior with data cardinalities, it is necessary to capture them. However, this can be done at different points in the case model. In the previous section, we modeled global and goal cardinalities within the domain model. At the same time, global lower boundaries greater than one are reflected in the process model: respective output and input objects are marked as list objects to show that multiple objects are read or written.

The refined execution semantics impact process model analysis. Cardinalities may set lower and upper bounds for the execution of certain fragments of the case model. This, in response, prunes the model's state space. State space analysis plays an important role in the verification of process models, e.g., to check soundness properties or compliance rules [8].

Usually, the knowledge workers have the flexibility to decide how often fragments need to be executed and how many objects are required. The state space of an fCM case model can be infinite because fragments can be executed repeatedly creating new objects every time. Cardinalities prevent the unbounded creation of data objects and, thereby, limit the state space.

In [8], the state space of fCM models was limited by manually setting upper bounds for how often each fragment may be executed (instantiated). Cardinalities in the domain model can imply such domain-specific limits for the case. However, some fragments may still be unbound and require the manual limit.

Besides the influence of cardinalities on the behavior, process logic and data model can be conflicting. Fragments or activities may not be enabled because they cannot be executed without violating the cardinality constraints. This may stop the process from progressing or reaching the termination condition. The following properties must be checked:

- mandatory relationships must be reflected by reading the supporter object or co-creating both the supporter and the dependent objects.
- lower bounds greater than one lead to batch reading/creation of objects
- in any case, the lower bound of the goal cardinalities must be realizable

Since cardinalities may have significant impact on the process, a data model with cardinality constraints may enhance the analysis of process data. In process discovery, considering a data model may improve the precision of the discovered model and/or reduce its fitness. It may explain decisions taken during process execution (e.g., looping behavior) or be used during event correlation to separate cases. OCBC models contain many of these aspects and have been successfully mined from object aware event-logs [12]. Furthermore, there are plans to investigate database schemata to derive a data model for data-centric process models [3].

4.2   Cardinalities and the Role of the Knowledge-Worker

The core idea of fCM is to support knowledge workers by providing guidance when possible and by allowing flexibility when necessary. In general, those two goals counteract each other; more guidance comes at the cost of less flexibility and vice versa. It is, therefore, important to balance flexibility and guidance.

Cardinality constraints are a tool to balance flexibility and guidance. Without cardinalities, it is up to the knowledge workers to decide how often a repeatable fragment is executed and how many objects are required. By applying cardinalities, we move this decision partly to design time providing general lower and upper bounds. The knowledge workers' decision must lay within these bounds. The more limiting these bounds are the less flexible is the process execution.

The lower bound of global cardinalities requires the knowledge workers to mandatory read or write an object while the global upper bounds must never be exceeded; thus it limits the number of executions. The goal lower bound dictates how many objects there have to be at least to terminate a case. It thereby requires a certain number of fragment instances. The goal cardinalities upper bound set an upper limit and should be the same as the global upper bound, because fCM does not allow data objects to be deleted.

Traditional fCM. If we set all lower bounds to zero and all upper bounds to a sufficiently large number, we effectively recreate the behavior of an fCM model without cardinality constraints. For a given class, each activity reads and writes an object. Instantiating an object by executing a respective activity or object is generally optional because cardinalities do not enforce the existence of any object. Furthermore, while there exists a theoretical upper bound, it can be set so high that it plays no role in practice. With such cardinalities, knowledge workers have full flexibility.

Mandatory and Optional Objects. By allowing the lower bound of the goal cardinality to be either zero or one, we introduce mandatory objects. Such objects are required for terminating a case. Given the distinction between optional and mandatory objects, knowledge workers receive guidance in the means of obligatory information and corresponding fragments, which allows knowledge workers to plan ahead. Besides this, the case may contain optional fragments and allow repetitive execution.

Bounded Objects. The repetitive execution can be limited through meaningful upper bounds, which dictate how many objects can exist at most. This provides knowledge workers with guidance because fragments or activities are automatically disabled if the specified boundary has been reached. Thereby, most fragments can be bound to a maximum number of instances, for example, to limit very expensive activities, such as an expert's review in an insurance claim. Upper bounds provide a safety net assuring that cases do not create indefinite amounts of objects.

Batch Processing. Allowing the global lower bound to be any value greater than one introduces batch processing since multiple data objects must be read or written at once. Thus far, fCM did not support batch processing. However, the knowledge worker could consider multiple objects sequentially before making a decision. Using cardinalities, we guide the knowledge worker by explicitly stating where batch processing is required. While this may limit the freedom of the knowledge worker, it can be empowering because batch processing can potentially improve the process' efficiency.

BPMN Semantics. Cardinality constraints are the most restricting if they are either 0..1 or 1..1. Objects are either optional or mandatory and there is never more than one object for a given class. This means that object creating fragments are executed at most once. Furthermore, in many cases, data requirements enforce a rather strict ordering of the fragments. Arguably, such a process does not require the flexibility of a case management approach and can easily be represented in a BPMN model[1]. However, due to low variance, the model provides strict guidance.

5   Related Work

While no approach to knowledge-intensive processes has been widely adopted in industry, many respective modeling languages are researched. DECLARE [16] and DCR-Graphs [7] scope a process through declarative constraints on the

---

[1] Assuming that run-time adaptability is not required.

execution-order of activities. Since these approaches do not support data, Montali et al. extend DECLARE with data conditions without cardinalities [13]. The guard stage milestone (GSM) approach, defines (data) pre-conditions for stages [9]. If satisfied, the stage can be entered and the contained activities can be executed. GSM does not support cardinalities.

Data-centric approaches emphasize the role of data in processes. [19] provides an overview on data-centric process management software. PHILharmonicFlows [11] is an approach defining object life cycles for classes and dependencies among states of different classes. A recent extension by Steinau et al. [18], introduces the relational process structure and support n-to-m interactions among objects. Therefore, cardinality constraints are used. However, temporal aspects, i.e., goal cardinalities, are not supported. Similarly, Proclets [2] can be used to model object behavior and synchronous m-to-n communication [4]. Like PHILharmonicFlows, goal cardinalities are not supported.

DB nets by Montali et al. [14], describe data-centric processes extending colored Petri nets [10]. DB nets distinguish between local data, in the scope of a single process instance, and persisted data. The latter is described by a data model and can be constraint, e.g., using cardinalities. While the process' local data may violate the constraints, persistent data must not. A process can access and update persistent data. The approach supports neither goal cardinalities for the persistent model, nor any type of cardinality constraints for local data. Nevertheless, DB nets can be a powerful tool to formalize data-centric process models and might be used to describe fCM's execution semantics formally.

Hybrid process modeling approaches combine multiple paradigms, e.g., DCR-KiPN [17] combines activity-centric and declarative process modeling. Object-Centric Behavioral Constraints (OCBC) by van der Aalst combine declarative and data-centric modeling [1]. In addition to declarative constraints among activities, data classes are connected by associations; and activities as well as constraints are linked to classes or associations. Furthermore, links and associations have cardinalities that must hold globally or eventually. However, OCBC rejects a traditional notion of a case and defines temporally global constraints.

6 Conclusion

Knowledge-intensive processes are driven by informed and interconnected decisions of knowledge workers. Since data is required for decisions making, many languages for knowledge-intensive processes integrate data models with process models. While the extend of the integration may differ, data cardinalities received little attention. In this work, we addressed this gap by exploring the role of data cardinalities in flexible processes. Therefore, we extended fCM case models with cardinality constraints that must always hold or that must hold upon termination. A cardinality constraint for a given class and relationship defines lower and upper bounds specifying how often a single object takes part in the relationship at least or at most. Since objects and their relationships are created by the process, cardinality constraints limit the behavior of the process. We showed that cardinalities can specify whether activities and fragments are optional or mandatory. Furthermore, we used cardinalities to model batch behavior—reading/writing multiple objects while executing an activity once.

During process modeling, cardinality constraints must be captured. Usually, cardinalities are part of the domain model but may be reflected in the process model. Furthermore, additional analysis tasks can be performed, such as the proper integration of process logic and domain model or the analysis of soundness with respect to cardinality constraints. At run-time, instances must adhere to the constraints. When analyzing execution logs, detecting and respecting cardinalities may provide additional insights and enhance existing methods.

Cardinality constraints refine the possible actions decreasing the knowledge workers' flexibility. In case management, it is important balance guidance and flexibility. We discussed how cardinalities can be used to fine tune this balance: cardinalities may impose no additional constraints or structure an otherwise unstructured process: cardinality constraints should neither be ignored nor applied thoughtlessly. The right upper and lower boundaries can guide knowledge workers during execution without sacrificing flexibility where it is crucial. Furthermore, cardinalities may similarly refine the multi-instance activities in BPMN.

So far, our discussion is based on a single example and informal semantics. In future work, we will define formal execution semantics, and it should be evaluated on real world scenarios. DB nets [14] and/or colored Petri nets [10] may provide the formal means to define refined execution semantics. Furthermore, formal correctness criteria, such as an adapted soundness notion, should be defined.

References

1. van der Aalst, W.M.P., Artale, A., Montali, M., Tritini, S.: Object-centric behavioral constraints: integrating data and declarative process modelling. In: Proceedings of the 30th International Workshop on Description Logics, Montpellier, France, 18–21 July 2017 (2017). http://ceur-ws.org/Vol-1879/paper51.pdf
2. van der Aalst, W.M.P., Barthelmess, P., Ellis, C.A., Wainer, J.: Proclets: a framework for lightweight interacting workflow processes. Int. J. Coop. Inf. Syst. **10**(4), 443–481 (2001). https://doi.org/10.1142/S0218843001000412
3. Breitmayer, M., Reichert, M.: Towards the discovery of object-aware processes. In: Proceedings of the 12th ZEUS Workshop on Services and their Composition, Potsdam, Germany, 20–21 February 2020, pp. 1–4 (2020). http://ceur-ws.org/Vol-2575/paper1.pdf
4. Fahland, D.: Describing behavior of processes with many-to-many interactions. In: Donatelli, S., Haar, S. (eds.) PETRI NETS 2019. LNCS, vol. 11522, pp. 3–24. Springer, Cham (2019). https://doi.org/10.1007/978-3-030-21571-2_1
5. Haarmann, S., Weske, M.: Correlating data objects in fragment-based case management. In: Abramowicz, W., Klein, G. (eds.) BIS 2020. LNBIP, vol. 389, pp. 197–209. Springer, Cham (2020). https://doi.org/10.1007/978-3-030-53337-3_15
6. Hewelt, M., Weske, M.: A hybrid approach for flexible case modeling and execution. In: La Rosa, M., Loos, P., Pastor, O. (eds.) BPM 2016. LNBIP, vol. 260, pp. 38–54. Springer, Cham (2016). https://doi.org/10.1007/978-3-319-45468-9_3

7. Hildebrandt, T.T., Mukkamala, R.R.: Declarative event-based workflow as distributed dynamic condition response graphs. In: Proceedings Third Workshop on Programming Language Approaches to Concurrency and communication-cEntric Software, PLACES 2010, Paphos, Cyprus, 21st March 2010, pp. 59–73 (2010). https://doi.org/10.4204/EPTCS.69.5
8. Holfter, A., Haarmann, S., Pufahl, L., Weske, M.: Checking compliance in data-driven case management. In: Di Francescomarino, C., Dijkman, R., Zdun, U. (eds.) BPM 2019. LNBIP, vol. 362, pp. 400–411. Springer, Cham (2019). https://doi.org/10.1007/978-3-030-37453-2_33
9. Hull, R., et al.: Introducing the guard-stage-milestone approach for specifying business entity lifecycles. In: Bravetti, M., Bultan, T. (eds.) WS-FM 2010. LNCS, vol. 6551, pp. 1–24. Springer, Heidelberg (2011). https://doi.org/10.1007/978-3-642-19589-1_1
10. Jensen, K., Kristensen, L.M.: Coloured Petri Nets. Springer, Heidelberg (2009). https://doi.org/10.1007/b95112
11. Künzle, V., Reichert, M.: PHILharmonicFlows: towards a framework for object-aware process management. J. Softw. Maint. 23(4), 205–244 (2011). https://doi.org/10.1002/smr.524
12. Li, G., de Carvalho, R.M., van der Aalst, W.M.P.: Automatic discovery of object-centric behavioral constraint models. In: Abramowicz, W. (ed.) BIS 2017. LNBIP, vol. 288, pp. 43–58. Springer, Cham (2017). https://doi.org/10.1007/978-3-319-59336-4_4
13. Montali, M., Chesani, F., Mello, P., Maggi, F.M.: Towards data-aware constraints in declare. In: Proceedings of the 28th Annual ACM Symposium on Applied Computing, SAC 2013, Coimbra, Portugal, 18–22 March 2013, pp. 1391–1396 (2013). https://doi.org/10.1145/2480362.2480624
14. Montali, M., Rivkin, A.: DB-Nets: on the marriage of colored petri nets and relational databases. Trans. Petri Nets Other Model. Concurr. 12, 91–118 (2017). https://doi.org/10.1007/978-3-662-55862-1_5
15. Object Management Group (OMG): Object constraint language (OCL) (2014). https://www.omg.org/spec/OCL/2.4/PDF
16. Pesic, M., Schonenberg, H., van der Aalst, W.M.P.: DECLARE: full support for loosely-structured processes. In: 11th IEEE International Enterprise Distributed Object Computing Conference (EDOC 2007), 15–19 October 2007, Annapolis, Maryland, USA, pp. 287–300 (2007). https://doi.org/10.1109/EDOC.2007.14
17. Santoro, F.M., Slaats, T., Hildebrandt, T.T., Baiao, F.: DCR-KiPN a hybrid modeling approach for knowledge-intensive processes. In: Laender, A.H.F., Pernici, B., Lim, E.-P., de Oliveira, J.P.M. (eds.) ER 2019. LNCS, vol. 11788, pp. 153–161. Springer, Cham (2019). https://doi.org/10.1007/978-3-030-33223-5_13
18. Steinau, S., Andrews, K., Reichert, M.: The relational process structure. In: Advanced Information Systems Engineering - 30th International Conference, CAiSE 2018, Tallinn, Estonia, 11–15 June 2018, Proceedings, pp. 53–67 (2018). https://doi.org/10.1007/978-3-319-91563-0_4
19. Steinau, S., Marrella, A., Andrews, K., Leotta, F., Mecella, M., Reichert, M.: DALEC: a framework for the systematic evaluation of data-centric approaches to process management software. Softw. Syst. Model. 18(4), 2679–2716 (2018). https://doi.org/10.1007/s10270-018-0695-0
20. Weske, M.: Business Process Management - Concepts, Languages, Architectures. Springer, Heidelberg (2019). https://doi.org/10.1007/978-3-662-59432-2

Author Index

Agarwal, Prerna 142, 219
Agarwal, Shubham 181
Alam, Syed Mehtab 263
Allard, Tony G. 232

Becker, Jörg 129
Bergmann, Ralph 95
Boltenhagen, Mathilde 281
Brockman, Sarah 232
Brunk, Jens 129

Cao, Yukun 308
Chakraborti, Tathagata 181
Chambers, Alexander J. 232
Compagnucci, Ivan 108
Corradini, Flavio 108

Dallasega, Patrick 263
De Weerdt, Jochen 321
Dechu, Sampath 142, 219
Depaire, Benoît 295

Epstein, Yannick 334
Etikala, Vedavyas 367

Faes, Christel 295
Fani Sani, Mohammadreza 281
Fornari, Fabrizio 108

Gounaris, Anastasios 355
Gupta, Monika 142, 219
Gzara, Lilia 60

Haarmann, Stephan 380
Hassani, Marwan 17
Hotel, Olivier 60

Indihar Štemberger, Mojca 194
Isahagian, Vatche 181

Jafari, Pooria 49
Janssenswillen, Gert 295
Johnston, Ian A. 232

Khazaeni, Yasaman 181
Kirss, Kristin Kamilla 251

Lüftenegger, Egon 83
Luo, Ben B. 232

Malburg, Lukas 95
Mannel, Lisa L. 334
Marengo, Elisa 263
Matzner, Martin 70, 129
Milani, Fredrik 251

Nutt, Werner 263

Park, Gyunam 206
Poels, Geert 5
Polini, Andrea 108
Prasannakumar, Sushruth 142

Qafari, Mahnaz Sadat 155

Rahman, Arif Ur 263
Re, Barbara 108
Revoredo, Kate 129
Rizk, Yara 181
Rizzi, Romeo 168

Seiger, Ronny 95
Serebrenik, Alexander 219
Shing, Leslie 232
Simm, Joosep 30
Softic, Selver 83
Steiner, Jamie 30
Stringfellow, Amy M. 232
Swarup, Daivik 142

Tang, Willi 70
Tater, Tarun 219
Tenschert, Johannes 70
Tiezzi, Francesco 108
Triaa, Wafa 60
Truu, Ahto 30

Underwood, Sophie J. 232

van der Aalst, Wil M. P.   155, 206, 281, 334
van Dongen, Boudewijn F.   17
Van Looy, Amy   49
Van Veldhoven, Ziboud   367
van Zelst, Sebastiaan J.   308
VanDam, Courtland   232
Vanthienen, Jan   367
Varvoutas, Konstantinos   355
Verdonck, Michaël   5
Verjus, Hervé   60
Villa, Tiziano   168

Weber, Barbara   95
Weinzierl, Sven   129
Weske, Mathias   380
Weytjens, Hans   321
Wollaber, Allan   232

Zaman, Rashid   17
Zavatteri, Matteo   168
Zebec, Aleš   194
Zilker, Sandra   129

Printed in the United States
By Bookmasters